Student Solutions Manual Volume 1

Sears and
Zemansky's

UNIVERSITY
PHYSICS with Modern Physics
11th Edition

Young & Freedman

A. LEWIS FORD
Texas A&M University

PEARSON
Addison
Wesley

San Francisco • Boston • New York
Capetown • Hong Kong • London • Madrid • Mexico City
Montreal • Munich • Paris • Singapore • Sydney • Tokyo • Toronto

ISBN 0-8053-8777-3

CONTENTS

PREFACE

This *Student Solutions Manual*, Volume 1, contains detailed solutions for approximately one third of the Exercises and Problems in Chapters 1 through 20 of the Eleventh Edition of *University Physics* by Hugh Young and Roger Freedman. The Exercises and Problems included in this manual are selected solely from the odd-numbered Exercises and Problems in the text (for which the answers are tabulated in the back of the textbook). The Exercises and Problems included were not selected at random but rather were carefully chosen to include at least one representative example of each problem type. The remaining Exercises and Problems, for which solutions are not given here, constitute an ample set of problems for you to tackle on your own. In addition, there are the Challenge Problems in the text for which no solutions are given here.

This manual greatly expands the set of worked-out examples that accompanies the presentation of physics laws and concepts in the text. This manual was written to provide you with models to follow in working physics problems. The problems are worked out in the manner and style in which you should carry out your own problem solutions.

The author will gratefully receive comments as to style, points of physics, errors, or anything else relating to this manual. The companion volume *Student Solutions Manual, Volumes 2 and 3* is also available from your college bookstore.

A. Lewis Ford
Physics Department
Texas A&M University
College Station, TX 77843
ford@physics.tamu.edu

CHAPTER 1
UNITS, PHYSICAL QUANTITIES, AND VECTORS

Exercises 1, 7, 9, 17, 23, 25, 35, 37, 39, 41, 43, 45, 47, 51, 55
Problems 59, 67, 69, 71, 75, 77, 79, 85, 89, 91, 93

Exercises

1.1 **IDENTIFY:** Unit conversion problem. Target variable is distance in km.

SET UP: 1.00 in. = 2.54 cm

EXECUTE: 1.00 mi = 1.00 mi(5280 ft/1 mi)(12 in./1 ft)(2.54 cm/1 in.)
(1 m/10^2 cm) (1 km/10^3 m) = 1.61 km

EVALUATE: Note that the units all cancel to give the desired units of the answer.

1.7 **IDENTIFY:** Unit conversion problem. Target variables are speed in km/h and m/s.

SET UP: 1 mi = 1.609km, 1 km = 10^3 m and 1 h = 3600 s.

EXECUTE:
a) 1450 mi/h = (1450 mi/h)(1.609 km/1 mi) = 2.33×10^3 km/h

b) 2.33×10^3 km/h = (2.33×10^3 km/h)(10^3 m/1 km)(1 h/3600 s) = 647 m/s

EVALUATE: 1 km is less than 1 mile, so the speed is larger in km/h than in mi/h.
1 m/s is greater than 1 km/h so the speed is smaller in m/s.

1.9 **IDENTIFY:** Unit conversion problem. Target variable is consumption in mi/gal.

SET UP: Use conversion factors from Appendix E.

EXECUTE: 15.0 km/L = (15.0 km/L)(0.6214 mi/1 km)(3.788 L/1 gal) = 35.3 mi/gal

EVALUATE: A km is smaller than a mi by a factor of about 0.5 and a gal is larger than a L by about a factor of 4, so the consumption in mi/gal should be larger by a factor of 2.

1.17 **IDENTIFY:** Calculate the average volume and diameter and the uncertainty in these quantities.

SET UP: Using the extreme values of the input data gives us the largest and smallest values of the target variables and from these we get the uncertainty.

EXECUTE:
a) The volume of a disk of diameter d and thickness t is $V = \pi(d/2)^2 t$.

The average volume is $V = \pi(8.50 \text{ cm}/2)^2(0.050 \text{ cm}) = 2.837 \text{ cm}^3$. But t is given to only two significant figures so the answer should be expressed to two significant figures: $V = 2.8$ cm^3.

We can find the uncertainty in the volume as follows. The volume could be as large as $V = \pi(8.52 \text{ cm}/2)^2(0.055 \text{ cm}) = 3.1$ cm^3, which is 0.3 cm^3 larger than the average value. The volume could be as small as $V = \pi(8.52 \text{ cm}/2)^2(0.045 \text{ cm}) = 2.5$ cm^3, which is 0.3 cm^3 smaller than the average value. The uncertainty is ± 0.3 cm^3, and we express the volume as $V = 2.8 \pm 0.3$ cm^3.

b) The ratio of the average diameter to the average thickness is 8.50 cm/0.050 cm = 170. By taking the largest possible value of the diameter and the smallest possible thickness we get the largest possible value for this ratio: 8.52 cm/0.045 cm = 190. The smallest possible value of the ratio is 8.48/0.055 = 150. Thus the uncertainty is ± 20 and we write the ratio as 170 ± 20.

EVALUATE: The thickness is uncertain by 10% and the percentage uncertainty in the diameter is much less, so the percentage uncertainty in the volume and in the ratio should be about 10%.

1.23 IDENTIFY: Estimation Problem.

SET UP: Estimate your scalp area and the number of hairs per mm^2 of scalp.

EXECUTE: I estimate that my scalp's area is about that of a 10 in. diameter circle: $d = 10$ in.(2.54 cm/1 in.)(10 mm/1 cm) = 250 mm. The estimated area is thus $A = \pi r^2$ with $r = d/2 = 125$ mm: $A = \pi(125 \text{ mm})^2 = 5 \times 10^4$ mm^2.

I further **estimate** that on my scalp there are 5 hairs per mm^2. The number of hairs on my head is thus estimated to be about (5 hairs/mm^2)(5×10^4 mm^2) $= 2 \times 10^5$ hairs.

EVALUATE: This is 200,000. Anything much less, like 20,000, would seem to be too small.

1.25 IDENTIFY: Estimation problem

SET UP: Estimate that the pile is 18 in. × 18 in. × 5 ft 8 in.. Use the density of gold to calculate the mass of gold in the pile and from this calculate the dollar value.

EXECUTE:
The volume of gold in the pile is $V =$18 in. × 18 in. × 68 in. = 22,000 in.^{3}. Convert to cm^3: $V = 22,000$ in.^{3}(1000 cm^3/61.02 in.^{3}) $= 3.6 \times 10^5$ cm^3.

From Example 1.4 the density of gold is 19.3 g/cm^3, so the mass of this volume of gold is $m = (19.3 \text{ g/cm}^3)(3.6 \times 10^5 \text{ cm}^3) = 7 \times 10^6$ g. Convert to ounces: $m = 7 \times 10^6$ g (1 ounce/30 g) $= 2 \times 10^5$ ounces.

Also from Example 1.4, the monetary value of one ounce is \$400, so the gold has a value of (\$400/ounce)(2×10^5 ounces) $= \$8 \times 10^7$, or about \$100×10^6 (one hundred million dollars).

EVALUATE: This is quite a large pile of gold, so such a large monetary value is reasonable.

1.35 **IDENTIFY:** The target variables are the x- and y-components of each vector.

SET UP: Use Eq.(1.7). Be sure to use the angles given in the sketch to find the angles each vector makes with the $+x$-axis and use these angles in the equations for the components.

EXECUTE:

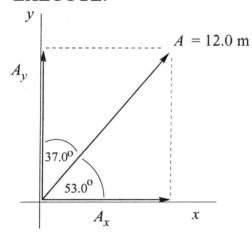

$A_x = A \cos 53.0°$
$A_x = (12.0 \text{ m}) \cos 53.0° = +7.22 \text{ m}$

$A_y = A \sin 53.0°$
$A_y = (12.0 \text{ m}) \sin 53.0° = +9.58 \text{ m}$

EVALUATE:
The sketch shows that A_x and A_y should both be positive, in agreement with our calculation.

EXECUTE:

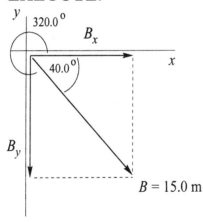

$B_x = B \cos 320.0°$
$B_x = (15.0 \text{ m}) \cos 320.0° = +11.5 \text{ m}$

$B_y = B \sin 320.0°$
$B_y = (15.0 \text{ m}) \sin 320.0° = -9.64 \text{ m}$

EVALUATE:
The sketch shows that B_x is positive and B_y is negative, in agreement with our calculation.

EXECUTE:

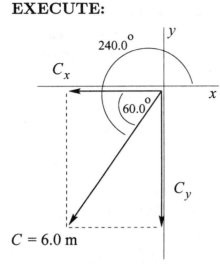

$C_x = C \cos 240.0°$
$C_x = (6.0 \text{ m}) \cos 240.0° = -3.0 \text{ m}$

$C_y = C \sin 240.0°$
$C_y = (6.0 \text{ m}) \sin 240.0° = -5.2 \text{ m}$

EVALUATE:
The sketch shows that both C_x and C_y are negative, in agreement with our calculation.

1.37 IDENTIFY: Find the vector sum of the two forces.

SET UP: Use components to add the two forces. Take the $+x$-direction to be forward and the $+y$-direction to be upward.

EXECUTE: The second force has components $F_{2x} = F_2 \cos 32.4° = 433$ N and $F_{2y} = F_2 \sin 32.4° = 275$ N. The first force has components $F_{1x} = 725$ N and $F_{1y} = 0$.

$F_x = F_{1x} + F_{2x} = 1158$ N and $F_y = F_{1y} + F_{2y} = 275$ N

The resultant force is 1190 N in the direction 13.4° above the forward direction.

EVALUATE: Since the two forces are not in the same direction the magnitude of their vector sum is less than the sum of their magnitudes.

1.39 IDENTIFY: The target variables are the magnitude and direction (angle θ) for each of the vector sums or differences.

SET UP: Use Eq.(1.7) to find the components of $\vec{A}$ and $\vec{B}$. Eq.(1.10) then gives the components of the sum and using Eqs.(1.8) and (1.9) we can find the magnitude and direction.

EXECUTE:

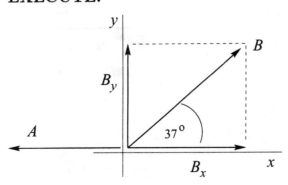

$A_x = -12.0 \text{ m}, \quad A_y = 0$

$B_x = B \cos 37°$
$B_x = (18.0 \text{ m}) \cos 37° = 14.38 \text{ m}$
$B_y = B \sin 37°$
$B_y = (18.0 \text{ m}) \sin 37° = 10.83 \text{ m}$

Note that both B_x and B_y are positive. A_x is negative because because $\vec{A}$ is in the negative x-direction.

a) Let $\vec{R} = \vec{A} + \vec{B}$.

$R_x = A_x + B_x = -12.0 \text{ m} + 14.38 \text{ m} = +2.38 \text{ m}.$

$R_y = A_y + B_y = 0 + 10.83 \text{ m} = +10.83 \text{ m}.$

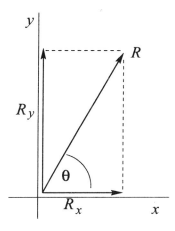

$$R = \sqrt{R_x^2 + R_y^2}$$
$$= \sqrt{(2.38 \text{ m})^2 + (10.83 \text{ m})^2} = 11.1 \text{ m}$$

$$\tan\theta = \frac{R_y}{R_x} = \frac{10.83 \text{ m}}{2.38 \text{ m}} = 4.550$$

$\theta = 77.6°$, measured counterclockwise from the $+x$-axis.

b) $\vec{B} + \vec{A} = \vec{A} + \vec{B}$, so the vector sum is the same as in part (a).

c) Now let $\vec{R} = \vec{A} - \vec{B}$.

$R_x = A_x - B_x = -12.0 \text{ m} - 14.38 \text{ m} = -26.38 \text{ m}.$

$R_y = A_y - B_y = 0 - 10.83 \text{ m} = -10.83 \text{ m}.$

Now both R_x and R_y are negative.

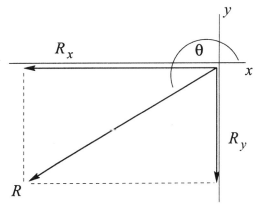

$$R = \sqrt{R_x^2 + R_y^2}$$
$$= \sqrt{(-26.38 \text{ m})^2 + (-10.83 \text{ m})^2} = 28.5 \text{ m}$$

$$\tan\theta = \frac{R_y}{R_x} = \frac{-10.83 \text{ m}}{-26.38 \text{ m}} = 0.4105$$

$\theta = 202°$, measured counterclockwise from the $+x$-axis.

Note: My calculator gives $\arctan(+0.4105) = 22.3°$. But the sketch shows that $\vec{R}$ is in the third quadrant. The angle $22.3° + 180° = 202°$ also has a tangent of $+0.4105$ and the sketch shows this is the correct answer.

d) $\vec{R} = \vec{B} - \vec{A}$. Then $R_x = B_x - A_x = -(A_x - B_x) = +26.38 \text{ m}$ and $R_y = B_y - A_y = -(A_y - B_y) = +10.83 \text{ m}$, using the results of part (b). R_x and R_y now have the same magnitudes but opposite signs from part (c), so now $\vec{R}$ has the same magnitude and opposite direction as in part (c).

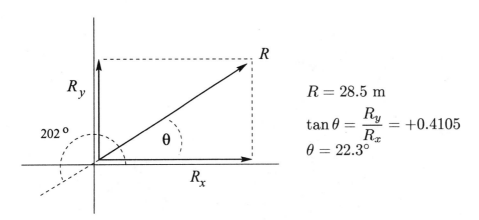

$R = 28.5$ m

$\tan \theta = \dfrac{R_y}{R_x} = +0.4105$

$\theta = 22.3°$

EVALUATE: $\vec{A} + \vec{B} = \vec{B} + \vec{A}$ and $\vec{A} - \vec{B} = -(\vec{B} - \vec{A})$. For each calculation the vector $\vec{R}$ we calculated with components agrees with the vector addition diagram.

1.41 IDENTIFY: Vector addition problem. We are given the magnitude and direction of three vectors and are asked to find their sum.

SET UP:

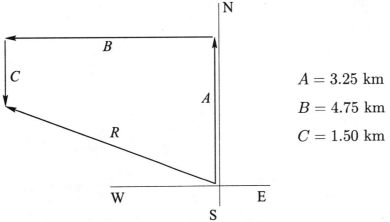

$A = 3.25$ km

$B = 4.75$ km

$C = 1.50$ km

Select a coordinate system where $+x$ is east and $+y$ is north. Let $\vec{A}$, $\vec{B}$ and $\vec{C}$ be the three displacements of the professor. Then the resultant displacement $\vec{R}$ is given by $\vec{R} = \vec{A} + \vec{B} + \vec{C}$. By the method of components, $R_x = A_x + B_x + C_x$ and $R_y = A_y + B_y + C_y$. Find the x and y components of each vector; add them to find the components of the resultant. Then the magnitude and direction of the resultant can be found from its x and y components that we have calculated. As always, it is essential to draw a sketch.

EXECUTE:

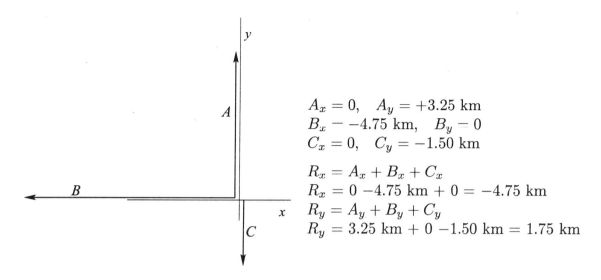

$A_x = 0, \quad A_y = +3.25$ km
$B_x - -4.75$ km, $\quad B_y - 0$
$C_x = 0, \quad C_y = -1.50$ km

$R_x = A_x + B_x + C_x$
$R_x = 0 - 4.75$ km $+ 0 = -4.75$ km
$R_y = A_y + B_y + C_y$
$R_y = 3.25$ km $+ 0 - 1.50$ km $= 1.75$ km

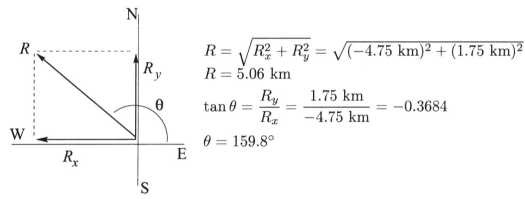

$R = \sqrt{R_x^2 + R_y^2} = \sqrt{(-4.75 \text{ km})^2 + (1.75 \text{ km})^2}$
$R = 5.06$ km

$\tan\theta = \dfrac{R_y}{R_x} = \dfrac{1.75 \text{ km}}{-4.75 \text{ km}} = -0.3684$

$\theta = 159.8°$

The angle θ measured counterclockwise from the $+x$-axis. In terms of compass directions, the resultant displacement is 20.2° N of W.

EVALUATE: $R_x < 0$ and $R_y > 0$, so $\vec{R}$ is in 2nd quadrant. This agrees with the vector addition diagram.

1.43 IDENTIFY: Vector addition problem. $\vec{A} - \vec{B} = \vec{A} + (-\vec{B})$.

SET UP: Find the x- and y-components of $\vec{A}$ and $\vec{B}$. Then the x- and y-components of the vector sum are calculated from the x- and y-components of $\vec{A}$ and $\vec{B}$.

EXECUTE:

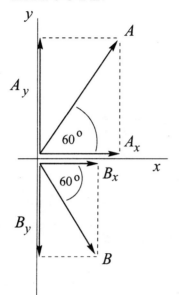

$A_x = A \cos(60.0°)$
$A_x = (2.80 \text{ cm}) \cos(60.0°) = +1.40 \text{ cm}$
$A_y = A \sin(60.0°)$
$A_y = (2.80 \text{ cm}) \sin(60.0°) = +2.425 \text{ cm}$

$B_x = B \cos(-60.0°)$
$B_x = (1.90 \text{ cm}) \cos(-60.0°) = +0.95 \text{ cm}$
$B_y = B \sin(-60.0°)$
$B_y = (1.90 \text{ cm}) \sin(-60.0°) = -1.645 \text{ cm}$

Note that the signs of the components correspond to the directions of the component vectors.

a) Now let $\vec{R} = \vec{A} + \vec{B}$.

$R_x = A_x + B_x = +1.40 \text{ cm} + 0.95 \text{ cm} = +2.35 \text{ m}.$

$R_y = A_y + B_y = +2.425 \text{ cm} - 1.645 \text{ cm} = +0.78 \text{ cm}.$

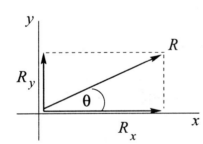

$R = \sqrt{R_x^2 + R_y^2} = \sqrt{(2.35 \text{ cm})^2 + (0.78 \text{ cm})^2}$
$R = 2.48 \text{ km}$

$\tan \theta = \dfrac{R_y}{R_x} = \dfrac{+0.78 \text{ cm}}{+2.35 \text{ cm}} = +0.3319$

$\theta = 18.4°$

EVALUATE: The vector addition diagram for $\vec{R} = \vec{A} + \vec{B}$ is

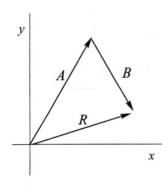

$\vec{R}$ is in the 1st quadrant, with $|R_y| < |R_x|$, in agreement with our calculation.

b) EXECUTE: Now let $\vec{R} = \vec{A} - \vec{B}$.

$R_x = A_x - B_x = +1.40 \text{ cm} - 0.95 \text{ cm} = +0.45 \text{ cm}.$

$R_y = A_y - B_y = +2.425$ cm $+ 1.645$ cm $= +4.070$ cm.

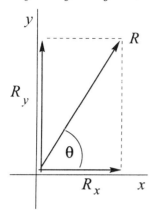

$R = \sqrt{R_x^2 + R_y^2} = \sqrt{(0.45 \text{ cm})^2 + (4.070 \text{ cm})^2}$

$R = 4.09$ cm

$\tan\theta = \dfrac{R_y}{R_x} = \dfrac{4.070 \text{ cm}}{0.45 \text{ cm}} = +9.044$

$\theta = 83.7°$

EVALUATE: The vector addition diagram for $\vec{R} = \vec{A} + (-\vec{B})$ is

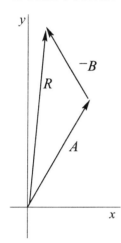

$\vec{R}$ is in the 1st quadrant, with $|R_x| < |R_y|$, in agreement with our calculation.

c) EXECUTE:

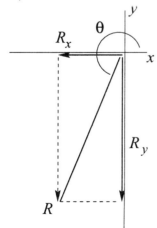

$\vec{B} - \vec{A} = -(\vec{A} - \vec{B})$
$\vec{B} - \vec{A}$ and $\vec{A} - \vec{B}$ are equal in magnitude and opposite in direction.

$R = 4.09$ cm and $\theta = 83.7° + 180° = 264°$

EVALUATE: The vector addition diagram for $\vec{R} = \vec{B} + (-\vec{A})$ is

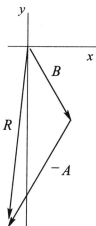

$\vec{R}$ is in the 3rd quadrant, with $|R_x| < |R_y|$, in agreement with our calculation.

1.45 IDENTIFY: Write the vectors in the form of Eq.(1.14).

SET UP: We can use the x- and y-components of each vector as calculated in Exercise 1.35.

EXECUTE:

$\vec{A} = (+7.22 \text{ m})\hat{i} + (+9.58 \text{ m})\hat{j}$

$\vec{B} = (+11.5 \text{ m})\hat{i} + (-9.64 \text{ m})\hat{j}$

$\vec{C} = (-3.0 \text{ m})\hat{i} + (-5.2 \text{ m})\hat{j}$

EVALUATE: Sketching each of these vectors agrees with Fig.(1.28).

1.47 IDENTIFY: Find A and B.

SET UP: Deduce the x- and y-components and use Eq.(1.8).

EXECUTE:

a) $\vec{A} = 4.00\hat{i} + 3.00\hat{j}$; $A_x = +4.00$; $A_y = +3.00$

$A = \sqrt{A_x^2 + A_y^2} = \sqrt{(4.00)^2 + (3.00)^2} = 5.00$

$\vec{B} = 5.00\hat{i} - 2.00\hat{j}$; $B_x = +5.00$; $B_y = -2.00$

$B = \sqrt{B_x^2 + B_y^2} = \sqrt{(5.00)^2 + (-2.00)^2} = 5.39$

EVALUATE: Note that the magnitudes of $\vec{A}$ and $\vec{B}$ are each larger than either of their components.

EXECUTE:

b) $\vec{A} - \vec{B} = 4.00\hat{i} + 3.00\hat{j} - (5.00\hat{i} - 2.00\hat{j}) = (4.00 - 5.00)\hat{i} + (3.00 + 2.00)\hat{j}$

$\vec{A} - \vec{B} = -1.00\hat{i} + 5.00\hat{j}$

c) Let $\vec{R} = \vec{A} - \vec{B} = -1.00\hat{i} + 5.00\hat{j}$. Then $R_x = -1.00$, $R_y = 5.00$.

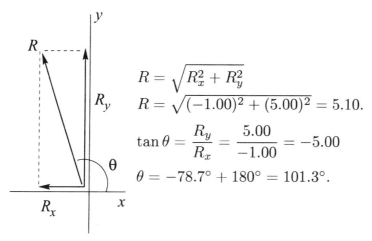

$$R = \sqrt{R_x^2 + R_y^2}$$
$$R = \sqrt{(-1.00)^2 + (5.00)^2} = 5.10.$$
$$\tan\theta = \frac{R_y}{R_x} = \frac{5.00}{-1.00} = -5.00$$
$$\theta = -78.7° + 180° = 101.3°.$$

EVALUATE: $R_x < 0$ and $R_y > 0$, so $\vec{R}$ is in the 2nd quadrant.

1.51 IDENTIFY: Target variables are $\vec{A} \cdot \vec{B}$ and the angle ϕ between the two vectors.
SET UP: We are given $\vec{A}$ and $\vec{B}$ in unit vector form and can take the scalar product using Eq.(1.19). The angle ϕ can then be found from Eq.(1.18).
EXECUTE:
a) $\vec{A} = 4.00\hat{i} + 3.00\hat{j}$, $\vec{B} = 5.00\hat{i} - 2.00\hat{j}$; $A = 5.00$, $B = 5.39$
$\vec{A} \cdot \vec{B} = (4.00\hat{i} + 3.00\hat{j}) \cdot (5.00\hat{i} - 2.00\hat{j}) = (4.00)(5.00) + (3.00)(-2.00) = 20.0 - 6.0 = +14.0$.

b) $\cos\phi = \dfrac{\vec{A} \cdot \vec{B}}{AB} = \dfrac{14.0}{(5.00)(5.39)} = 0.519$; $\phi = 58.7°$.

EVALUATE: The component of $\vec{B}$ along $\vec{A}$ is in the same direction as $\vec{A}$, so the scalar product is positive and the angle ϕ is less than 90°.

1.55 IDENTIFY: Target variable is the vector $\vec{A}\mathbf{X}\vec{B}$, expressed in terms of unit vectors.
SET UP: We are given $\vec{A}$ and $\vec{B}$ in unit vector form and can take the scalar product using Eq.(1.24).
EXECUTE:
$\vec{A} = 4.00\hat{i} + 3.00\hat{j}$, $\vec{B} = 5.00\hat{i} - 2.00\hat{j}$

$\vec{A}\mathbf{X}\vec{B} = (4.00\hat{i} + 3.00\hat{j})\mathbf{X}(5.00\hat{i} - 2.00\hat{j}) =$
$20.0\ \hat{i}\mathbf{X}\hat{i} - 8.00\ \hat{i}\mathbf{X}\hat{j} + 15.0\ \hat{j}\mathbf{X}\hat{i} - 6.00\ \hat{j}\mathbf{X}\hat{j}$

But $\hat{i}\mathbf{X}\hat{i} = \hat{j}\mathbf{X}\hat{j} = \mathbf{0}$ and $\hat{i}\mathbf{X}\hat{j} = \hat{k}$, $\hat{j}\mathbf{X}\hat{i} = -\hat{k}$, so
$\vec{A}\mathbf{X}\vec{B} = -8.00\ \hat{k} + 15.0\ (-\hat{k}) = -23.0\ \hat{k}$.
The magnitude of $\vec{A}\mathbf{X}\vec{B}$ is 23.0.

EVALUATE: Sketch the vectors $\vec{A}$ and $\vec{B}$ in a coordinate system where the xy-plane is in the plane of the paper and the z-axis is directed out toward you.

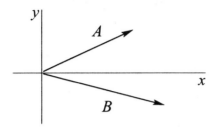

By the right-hand rule $\vec{A} \times \vec{B}$ is directed into the plane of the paper, in the $-z$-direction. This agrees with the above calculation that used unit vectors.

Problems

1.59 IDENTIFY and **SET UP:** Unit conversion.

EXECUTE:

a) $f = 1.420 \times 10^9$ cycles/s, so $\dfrac{1}{1.420 \times 10^9}$ s $= 7.04 \times 10^{-10}$ s for one cycle.

b) $\dfrac{3600 \text{ s/h}}{7.04 \times 10^{-10} \text{ s/cycle}} = 5.11 \times 10^{12}$ cycles/h

c) Calculate the number of seconds in 4600 million years $= 4.6 \times 10^9$ y and divide by the time for 1 cycle:

$$\frac{(4.6 \times 10^9 \text{ y})(3.156 \times 10^7 \text{ s/y})}{7.04 \times 10^{-10} \text{ s/cycle}} = 2.1 \times 10^{26} \text{ cycles}$$

d) The clock is off by 1 s in 100,000 y $= 1 \times 10^5$ y, so in 4.60×10^9 y it is off by

$$(1 \text{ s}) \left(\frac{4.60 \times 10^9}{1 \times 10^5} \right) = 4.6 \times 10^4 \text{ s (about 13 h)}.$$

EVALUATE: In each case the units in the calculation combine algebraically to give the correct units for the answer.

1.67 IDENTIFY: $\vec{A} + \vec{B} = \vec{C}$ (or $\vec{B} + \vec{A} = \vec{C}$). The target variable is vector $\vec{A}$.

SET UP: Use components and Eq.(1.10) to solve for the components of $\vec{A}$. Find the magnitude and direction of $\vec{A}$ from its components.

EXECUTE:

a)

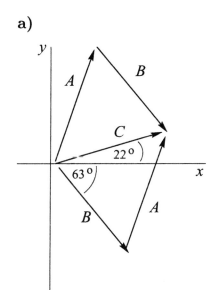

$C_x = A_x + B_x,$ so $A_x = C_x - B_x$
$C_y = A_y + B_y,$ so $A_y = C_y - B_y$

$C_x = C \cos 22.0° = (6.40 \text{ cm}) \cos 22.0°$
$C_x = +5.934 \text{ cm}$
$C_y = C \sin 22.0° = (6.40 \text{ cm}) \sin 22.0°$
$C_y = +2.397 \text{ cm}$

$B_x = B \cos(360° - 63.0°) = (6.40 \text{ cm}) \cos 297.0°$
$B_x = +2.906 \text{ cm}$
$B_y = B \sin 297.0° = (6.40 \text{ cm}) \sin 297.0°$
$B_y = -5.702 \text{ cm}$

b) $A_x = C_x - B_x = +5.934 \text{ cm} - 2.906 \text{ cm} = +3.03 \text{ cm}$
$A_y = C_y - B_y = +2.397 \text{ cm} - (-5.702) \text{ cm} = +8.10 \text{ cm}$

c)

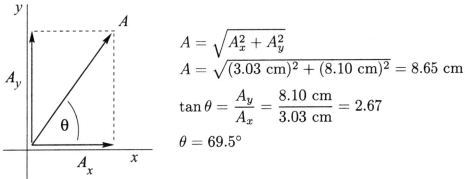

$A = \sqrt{A_x^2 + A_y^2}$
$A = \sqrt{(3.03 \text{ cm})^2 + (8.10 \text{ cm})^2} = 8.65 \text{ cm}$

$\tan \theta = \dfrac{A_y}{A_x} = \dfrac{8.10 \text{ cm}}{3.03 \text{ cm}} = 2.67$

$\theta = 69.5°$

EVALUATE: The $\vec{A}$ we calculated agrees qualitatively with vector $\vec{A}$ in the vector addition diagram in part (a).

1.69 IDENTIFY: Vector addition. Target variable is the 4th displacement.

SET UP: Use a coordinate system where east is in the $+x$-direction and north is in the $+y$-direction.

Let $\vec{A}$, $\vec{B}$, and $\vec{C}$ be the three displacements that are given and let $\vec{D}$ be the fourth unmeasured displacement. Then the resultant displacement is $\vec{R} = \vec{A} + \vec{B} + \vec{C} + \vec{D}$. And since she ends up back where she started, $\vec{R} = \mathbf{0}$.

$\mathbf{0} = \vec{A} + \vec{B} + \vec{C} + \vec{D}$, so $\vec{D} = -(\vec{A} + \vec{B} + \vec{C})$
$D_x = -(A_x + B_x + C_x)$ and $D_y = -(A_y + B_y + C_y)$

EXECUTE:

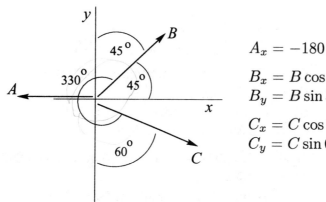

$A_x = -180$ m, $A_y = 0$

$B_x = B\cos 315° = (210 \text{ m})\cos 315° = +148.5$ m
$B_y = B\sin 315° = (210 \text{ m})\sin 315° = -148.5$ m

$C_x = C\cos 60° = (280 \text{ m})\cos 60° = +140$ m
$C_y = C\sin 60° = (280 \text{ m})\sin 60° = +242.5$ m

$D_x = -(A_x + B_x + C_x) = -(-180 \text{ m} + 148.5 \text{ m} + 140 \text{ m}) = -108.5$ m
$D_y = -(A_y + B_y + C_y) = -(0 - 148.5 \text{ m} + 242.5 \text{ m}) = -94.0$ m

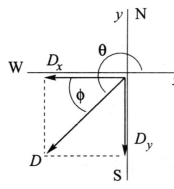

$D = \sqrt{D_x^2 + D_y^2}$
$D = \sqrt{(-108.5 \text{ m})^2 + (-94.0 \text{ m})^2} = 144$ m

$\tan\theta = \dfrac{D_y}{D_x} = \dfrac{-94.0 \text{ m}}{-108.5 \text{ m}} = 0.8664$

$\theta = 180° + 40.9° = 220.9°$
($\vec{D}$ is in the third quadrant since both D_x and D_y are negative.)

The direction of $\vec{D}$ can also be specified in terms of $\phi = \theta - 180° = 40.9°$; $\vec{D}$ is 41° south of west.

EVALUATE: The vector addition diagram, approximately to scale, is

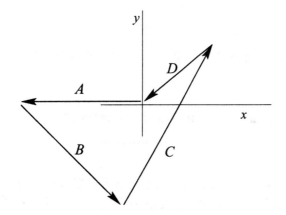

Vector $\vec{D}$ in this diagram agrees qualitatively with our calculation using components.

1.71 IDENTIFY: Vector addition. Target variable is the magnitude of the vector sum of the three displacements.

SET UP:

a) Let the three given displacements be called $\vec{A}$, $\vec{B}$, and $\vec{C}$.

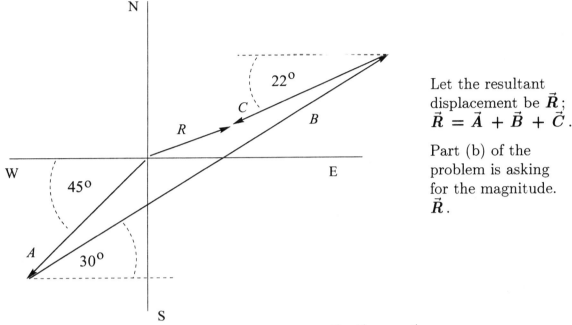

Let the resultant displacement be $\vec{R}$; $\vec{R} = \vec{A} + \vec{B} + \vec{C}$.

Part (b) of the problem is asking for the magnitude. $\vec{R}$.

Find R_x and R_y by adding the components of $\vec{A}$, $\vec{B}$, and $\vec{C}$.

b) EXECUTE: Use a coordinate system where the x-direction is east and the y-direction is north.

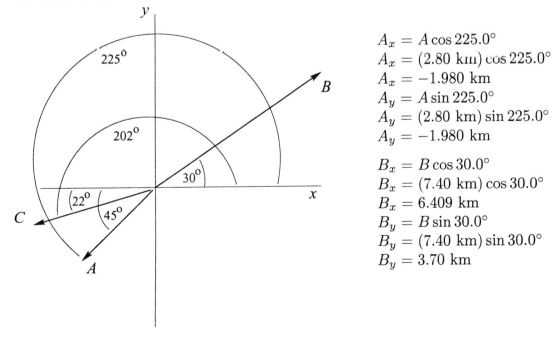

$$A_x = A\cos 225.0°$$
$$A_x = (2.80\text{ km})\cos 225.0°$$
$$A_x = -1.980\text{ km}$$
$$A_y = A\sin 225.0°$$
$$A_y = (2.80\text{ km})\sin 225.0°$$
$$A_y = -1.980\text{ km}$$

$$B_x = B\cos 30.0°$$
$$B_x = (7.40\text{ km})\cos 30.0°$$
$$B_x = 6.409\text{ km}$$
$$B_y = B\sin 30.0°$$
$$B_y = (7.40\text{ km})\sin 30.0°$$
$$B_y = 3.70\text{ km}$$

$C_x = C \cos 202.0° = (3.30 \text{ km}) \cos 202.0° = -3.060 \text{ km}$

$C_y = C \sin 202.0° = (3.30 \text{ km}) \sin 202.0° = -1.236 \text{ km}$

Note that in each case the signs of the components correspond to the directions of the component vectors.

$R_x = A_x + B_x + C_x = -1.980 \text{ km} + 6.409 \text{ km} - 3.060 \text{ km} = +1.369 \text{ km}$

$R_y = A_y + B_y + C_y = -1.980 \text{ km} + 3.70 \text{ km} - 1.236 \text{ km} = +0.484 \text{ km}$

$R = \sqrt{R_x^2 + R_y^2} = \sqrt{(+1.369 \text{ km})^2 + (0.484 \text{ km})^2} = 1.45 \text{ km}.$

EVALUATE: The calculated R_x and R_y agree qualitatively with the vector addition diagram in part (a).

1.75   IDENTIFY: Vector addition. One vector and the sum are given; find the second vector (magnitude and direction).

SET UP: Let $+x$ be east and $+y$ be north. Let $\vec{A}$ be the displacement 285 km at 40.0° north of west and let $\vec{B}$ be the unknown displacement.

$\vec{A} + \vec{B} = \vec{R}$, where $\vec{R} = 115$ km, east

$\vec{B} = \vec{R} - \vec{A}$

$B_x = R_x - A_x,\ B_y = R_y - A_y$

EXECUTE:

$A_x = -A \cos 40.0° = -218.3 \text{ km},\ A_y = +A \sin 40.0° = +183.2 \text{ kg}$

$R_x = 115 \text{ km},\ R_y = 0$

Then $B_x = 333.3 \text{ km},\ B_y = -183.2 \text{ km}.$

$B = \sqrt{B_x^2 + B_y^2} = 380 \text{ km};$

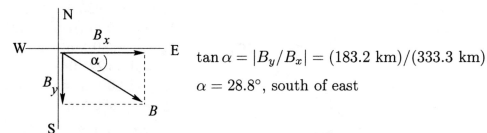

$\tan \alpha = |B_y / B_x| = (183.2 \text{ km})/(333.3 \text{ km})$

$\alpha = 28.8°$, south of east

EVALUATE: The southward component of $\vec{B}$ cancels the northward component of $\vec{A}$. The eastward component of $\vec{B}$ must be 115 km larger than the magnitude of the westward component of $\vec{A}$.

1.77   IDENTIFY: Vector addition. One force and the vector sum are given; find the second force.

SET UP: Use components. Let $+y$ be upward.

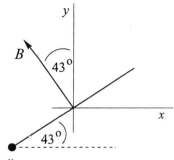

$\vec{B}$ is the force the biceps exerts.

elbow

$\vec{E}$ is the force the elbow exerts.

$\vec{E} + \vec{B} = \vec{R}$, where $R = 132.5$ N and is upward.

$E_x = R_x - B_x, \qquad E_y = R_y - B_y$

EXECUTE:

$B_x = -B \sin 43° = -158.2$ N, $B_y = +B \cos 43° = +169.7$ N

$R_x = 0, R_y = +132.5$ N

Then $E_x = +158.2$ N, $E_y = -37.2$ N

$E = \sqrt{E_x^2 + E_y^2} = 160$ N;

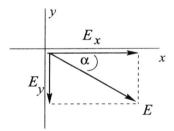

$\tan \alpha = |E_y / E_x| = 37.2/158.2$

$\alpha = 13°$, below horizontal

EVALUATE: The x-component of $\vec{E}$ cancels the x-component of $\vec{B}$. The resultant upward force is less than the upward component of $\vec{B}$, so E_y must be downward.

1.79 IDENTIFY: Vector addition. Three vectors add to zero. The target variables are the magnitudes of two of the vectors.

SET UP: Use components.

Let the displacement from your camp to the store be $\vec{A}$; $A = 240$ m, $32°$ south of east

$\vec{B}$ is $32°$ south of west and $\vec{C}$ is $62°$ south of west

Let $+x$ be east and $+y$ be north.

$\vec{A} + \vec{B} + \vec{C} = 0$

EXECUTE:

$A_x + B_x + C_x = 0$, so $A \cos 32° - B \cos 48° - C \cos 62° = 0$

$A_y + B_y + C_y = 0$, so $-A \sin 32° + B \sin 48° - C \sin 62° = 0$

A is known so we have two equations in the two unknowns B and C.

$0.6691B + 0.4695C = 203.5$ and $0.7431B - 0.8829C = 127.2$

$B = 1.188C + 171.2$; substituting this into the first equation and solving gives

$B = 255$ m and $C = 70$ m.

EVALUATE: The store is 240 m from your camp. On the return trip you traveled more than 240 m since you did not travel in a straight line back to camp.

1.85 IDENTIFY: $\vec{A}$ and $\vec{B}$ are given in unit vector form.
Find A, B and the vector difference $\vec{A} - \vec{B}$.

SET UP:

$\vec{A} = -2.00\hat{i} + 3.00\hat{j} + 4.00\hat{k}$, $\vec{B} = 3.00\hat{i} + 1.00\hat{j} - 3.00\hat{k}$

Use Eq.(1.8) to find the magnitudes of the vectors.

EXECUTE:

a) $A = \sqrt{A_x^2 + A_y^2 + A_z^2} = \sqrt{(-2.00)^2 + (3.00)^2 + (4.00)^2} = 5.38$

$B = \sqrt{B_x^2 + B_y^2 + B_z^2} = \sqrt{(3.00)^2 + (1.00)^2 + (-3.00)^2} = 4.36$

b) $\vec{A} - \vec{B} = (-2.00\hat{i} + 3.00\hat{j} + 4.00\hat{k}) - (3.00\hat{i} + 1.00\hat{j} - 3.00\hat{k})$

$\vec{A} - \vec{B} = (-2.00 - 3.00)\hat{i} + (3.00 - 1.00)\hat{j} + (4.00 - (-3.00))\hat{k} = -5.00\hat{i} + 2.00\hat{j} + 7.00\hat{k}$.

c) Let $\vec{C} = \vec{A} - \vec{B}$, so $C_x = -5.00$, $C_y = +2.00$, $C_z = +7.00$

$C = \sqrt{C_x^2 + C_y^2 + C_z^2} = \sqrt{(-5.00)^2 + (2.00)^2 + (7.00)^2} = 8.83$

$\vec{B} - \vec{A} = -(\vec{A} - \vec{B})$, so $\vec{A} - \vec{B}$ and $\vec{B} - \vec{A}$ have the same magnitude but opposite directions.

EVALUATE: A, B and C are each larger than any of their components.

1.89 IDENTIFY: Find the angle between specified pairs of vectors.

SET UP: Use $\cos\phi = \dfrac{\vec{A} \cdot \vec{B}}{AB}$

EXECUTE:

a) $\vec{A} = \hat{k}$ (along line ab)

$\vec{B} = \hat{i} + \hat{j} + \hat{k}$ (along line ad)

$A = 1$, $B = \sqrt{1^2 + 1^2 + 1^2} = \sqrt{3}$

$\vec{A} \cdot \vec{B} = \hat{k} \cdot (\hat{i} + \hat{j} + \hat{k}) = 1$

So $\cos\phi = \dfrac{\vec{A} \cdot \vec{B}}{AB} = 1/\sqrt{3}$; $\phi = 54.7°$

b) $\vec{A} = \hat{i} + \hat{j} + \hat{k}$ (along line ad)

$\vec{B} = \hat{j} + \hat{k}$ (along line ac)

$A = \sqrt{1^2 + 1^2 + 1^2} = \sqrt{3}; \quad B = \sqrt{1^2 + 1^2} = \sqrt{2}$

$\vec{A} \cdot \vec{B} = (\hat{i} + \hat{j} + \hat{k}) \cdot (\hat{i} + \hat{j}) = 1 + 1 = 2$

So $\cos\phi = \dfrac{\vec{A} \cdot \vec{B}}{AB} = \dfrac{2}{\sqrt{3}\sqrt{2}} = \dfrac{2}{\sqrt{6}}; \quad \phi = 35.3°$

EVALUATE: Each angle is computed to be les than 90°, in agreement with what is deduced from Fig.(1.35).

1.91 **IDENTIFY** and **SET UP:** The target variables are the components of $\vec{C}$. We are given $\vec{A}$ and $\vec{B}$. We also know $\vec{A} \cdot \vec{C}$ and $\vec{B} \cdot \vec{C}$, and this gives us two equations in the two unknowns C_x and C_y.

EXECUTE:

$\vec{A}$ and $\vec{C}$ are perpendicular, so $\vec{A} \cdot \vec{C} = 0$. $A_x C_x + A_y C_y = 0$, which gives $5.0C_x - 6.5C_y = 0$.

$\vec{B} \cdot \vec{C} = 15.0$, so $-3.5C_x + 7.0C_y = 15.0$

We have two equations in two unknowns C_x and C_y. Solving gives

$C_x = 8.0$ and $C_y = 6.1$.

EVALUATE: We can check that our result does give us a vector $\vec{C}$ that satisfies the two equations $\vec{A} \cdot \vec{C} = 0$ and $\vec{B} \cdot \vec{C} = 15.0$.

1.93 **a) IDENTIFY:** Prove that $\vec{A} \cdot (\vec{B} \times \vec{C}) = (\vec{A} \times \vec{B}) \cdot \vec{C}$.

SET UP: Express the scalar and vector products in terms of components.

EXECUTE:

$\vec{A} \cdot (\vec{B} \times \vec{C}) = A_x(\vec{B} \times \vec{C})_x + A_y(\vec{B} \times \vec{C})_y + A_z(\vec{B} \times \vec{C})_z$

$\vec{A} \cdot (\vec{B} \times \vec{C}) = A_x(B_y C_z - B_z C_y) + A_y(B_z C_x - B_x C_z) + A_z(B_x C_y - B_y C_x)$

$(\vec{A} \times \vec{B}) \cdot \vec{C} = (\vec{A} \times \vec{B})_x C_x + (\vec{A} \times \vec{B})_y C_y + (\vec{A} \times \vec{B})_z C_z$

$(\vec{A} \times \vec{B}) \cdot \vec{C} = (A_y B_z - A_z B_y)C_x + (A_z B_x - A_x B_z)C_y + (A_x B_y - A_y B_x)C_z$

Comparison of the expressions for $\vec{A} \cdot (\vec{B} \times \vec{C})$ and $(\vec{A} \times \vec{B}) \cdot \vec{C}$ shows they contain the same terms, so $\vec{A} \cdot (\vec{B} \times \vec{C}) = (\vec{A} \times \vec{B}) \cdot \vec{C}$.

b) IDENTIFY: Calculate $(\vec{A} \times \vec{B}) \cdot \vec{C}$, given the magnitude and direction of $\vec{A}$, $\vec{B}$, and $\vec{C}$.

SET UP: Use Eq.(1.22) to find the magnitude and direction of $\vec{A} \times \vec{B}$. Then we know the components of $\vec{A} \times \vec{B}$ and of $\vec{C}$ and can use an expression like Eq.(1.21) to find the scalar product in terms of components.

EXECUTE:

$A = 5.00$, $\theta_A = 26.0°$; $B = 4.00$, $\theta_B = 63.0°$

$|\vec{A} \times \vec{B}| = AB \sin \phi$.

The angle ϕ between $\vec{A}$ and $\vec{B}$ is equal to $\phi = \theta_B - \theta_A = 63.0° - 26.0° = 37.0°$.

So $|\vec{A} \times \vec{B}| = (5.00)(4.00) \sin 37.0° = 12.04$, and by the right hand-rule $\vec{A} \times \vec{B}$ is in the $+z$-direction.

Thus $(\vec{A} \times \vec{B}) \cdot \vec{C} = (12.04)(6.00) = 72.2$

EVALUATE: $\vec{A} \times \vec{B}$ is a vector, so taking its scalar product with $\vec{C}$ is a legitimate vector operation. $(\vec{A} \times \vec{B}) \cdot \vec{C}$ is a scalar product between two vectors so the result is a scalar.

CHAPTER 2
MOTION ALONG A STRAIGHT LINE

Exercises 3, 7, 9, 13, 15, 19, 21, 31, 33, 37, 39, 43, 45, 47, 51, 53
Problems 57, 59, 65, 67, 69, 73, 75, 79, 81, 83, 85, 89, 93

Exercises

2.3 **IDENTIFY:** Target variable is the time Δt it takes to make the trip in heavy traffic. Use Eq.(2.2) that relates the average velocity to the displacement and average time.

SET UP: $v_{av-x} = \dfrac{\Delta x}{\Delta t}$ so $\Delta x = v_{av-x} \Delta t$ and $\Delta t = \dfrac{\Delta x}{v_{av-x}}$.

EXECUTE:

Use the information given for normal driving conditions to calculate the distance between the two cities:

$\Delta x = v_{av-x} \Delta t = (105 \text{ km/h})(1 \text{ h/60 min})(140 \text{ min}) = 245 \text{ km}$.

Now use v_{av-x} for heavy traffic to calculate Δt; Δx is the same as before:

$\Delta t = \dfrac{\Delta x}{v_{av-x}} = \dfrac{245 \text{ km}}{70 \text{ km/h}} = 3.50 \text{ h} = 3 \text{ h and } 30 \text{ min}$.

The trip takes an additional 1 hour and 10 minutes.

EVALUATE: The time is inversely proportional to the average speed, so the time in traffic is $(105/70)(140 \text{ min}) - 210 \text{ min}$.

2.7 **IDENTIFY:** Use Eq.(2.2) to calculate the magnitude of the average velocity.

a) SET UP: $v_{av} = \dfrac{\Delta x}{\Delta t}$. We can use this equation to calculate Δx for any time interval for which we know v_{av-x}.

EXECUTE:

$\Delta x = (8.0 \text{ m/s})(60 \text{ s}) + (20.0 \text{ m/s})(60 \text{ s}) = 1680 \text{ m}$

$v_{av-x} = \dfrac{1680 \text{ m}}{120 \text{ s}} = 14 \text{ m/s}$

b) SET UP: $v_{av-x} = \dfrac{\Delta x}{\Delta t}$

We know that the total displacement is 240 m + 240 m = 480 m, but we must calculate the total elapsed time by calculating the elapsed time for each segment.

EXECUTE:

$\Delta t = \dfrac{\Delta x}{v_{av-x}}$, so $\Delta t = \dfrac{240 \text{ m}}{8.0 \text{ m/s}} + \dfrac{240 \text{ m}}{20.0 \text{ m/s}} = 30 \text{ s} + 12 \text{ s} = 42 \text{ s}$.

Then $v_{av-x} = \dfrac{480 \text{ m}}{42 \text{ m}} = 11.4 \text{ m/s}.$

c) In part (a) the numerical average of the two speeds is $\dfrac{(8.0 \text{ m/s} + 20.0 \text{ m/s})}{2} =$
14 m/s, which does equal v_{av-x}.

In part (b) the numerical average of the two speeds again is 14 m/s, but this does not equal v_{av-x}.

EVALUATE: The numerical average of the two speeds equals v_{av-x} in part (a) since the two speeds are maintained for equal amounts of time; this is not the case in part (b).

2.9 a) IDENTIFY: Calculate the average velocity using Eq.(2.2).

SET UP: $v_{av-x} = \dfrac{\Delta x}{\Delta t}$ so use $x(t)$ to find the displacement Δx for this time interval.

EXECUTE:

$t = 0$: $x = 0$

$t = 10.0$ s: $x = (2.40 \text{ m/s}^2)(10.0 \text{ s})^2 - (0.120 \text{ m/s}^3)(10.0 \text{ s})^3 = 240 \text{ m} - 120 \text{ m} = 120 \text{ m}.$

Then $v_{av-x} = \dfrac{\Delta x}{\Delta t} = \dfrac{120 \text{ m}}{10.0 \text{ s}} = 12.0 \text{ m/s}.$

b) IDENTIFY: Use Eq.(2.3) to calculate $v_x(t)$ and evaluate this expression at each specified t.

SET UP: $v_x = \dfrac{dx}{dt} = 2bt - 3ct^2.$

EXECUTE:

(i) $t = 0$: $v_x = 0$

(ii) $t = 5.0$ s: $v_x = 2(2.40 \text{ m/s}^2)(5.0 \text{ s}) - 3(0.120 \text{ m/s}^3)(5.0 \text{ s})^2 =$
24.0 m/s $-$ 9.0 m/s $=$ 15.0 m/s.

(iii) $t = 10.0$ s: $v_x = 2(2.40 \text{ m/s}^2)(10.0 \text{ s}) - 3(0.120 \text{ m/s}^3)(10.0 \text{ s})^2 =$
48.0 m/s $-$ 36.0 m/s $=$ 12.0 m/s.

c) IDENTIFY: Find the value of t when $v_x(t)$ from part (b) is zero.

SET UP: $v_x = 2bt - 3ct^2$

$v_x = 0$ at $t = 0$.

$v_x = 0$ next when $2bt - 3ct^2 = 0$

EXECUTE: $2b = 3ct$ so $t = \dfrac{2b}{3c} = \dfrac{2(2.40 \text{ m/s}^2)}{3(0.120 \text{ m/s}^3)} = 13.3 \text{ s}$

EVALUATE: $v_x(t)$ for this motion says the car starts from rest, speeds up, and then slows down again.

2.13   IDENTIFY: Calculate the instantaneous acceleration from a graph of v_x versus t.

SET UP: The acceleration a_x is the slope of the v_x versus t curve.

EXECUTE:

a) a_x has its most positive value when the curve has its largest positive slope; this occurs between approximately 4 s and 7 s.

b) a_x has its most negative value when the curve has its most negative slope; this ocurs between approximately 30 s to 40 s.

c) At $t = 20$ s, the v_x versus t curve is a horizontal straight line with zero slope, so $a_x = 0$.

d) Between 30 s and 40 s the v_x versus t curve is a horizontal straight line with slope

$$\frac{0 - 60 \text{ km/h}}{40 \text{ s} - 30 \text{ s}} = \frac{(-60 \text{ km/h})(10^3 \text{ m/1 km})(1 \text{ h/3600 s})}{10 \text{ s}} = -1.7 \text{ m/s}^2.$$

The acceleration is constant with this value in this time interval, so at $t = 35$ s it is $a_x = -1.7 \text{ m/s}^2$.

e) $\underline{t = 5 \text{ s}}$

The average velocity for $t = 0$ to $t = 5$ s is approximately 15 km/h or 4.1 m/s, so $\Delta x = v_{\text{av}-x}\Delta t = 20$ m; if the car is at $x_0 = 0$ at $t = 0$ then it is at $x = 20$ m at $t = 5$ s.

The velocity at $t = 5$ s is 30 km/h = 8 m/s.

The acceleration at $t = 5$ s is approximately

$$\left(\frac{60 \text{ km/h} - 30 \text{ km/h}}{10 \text{ s} - 5 \text{ s}}\right)\left(\frac{10^3 \text{ m}}{1 \text{ km}}\right)\left(\frac{1 \text{ h}}{3600 \text{ s}}\right) = +1.7 \text{ m/s}^2.$$

$\underline{t = 15 \text{ s}}$

In the first 10 s the car has displacement $\Delta x \approx (8 \text{ m/s})(10 \text{ s}) = 80$ m. From 10 s to 15 s the car has constant speed 60 km/h = 17 m/s, so $\Delta x = (17 \text{ m/s})(5 \text{ s}) = 85$ m. At $t = 15$ s the car is at $x = 80$ m + 85 m = 165 m. The velocity at $t = 15$ s is 17 m/s. The acceleration is zero.

$\underline{t = 25 \text{ s}}$

For the time interval 10 s to 25 s the car has constant speed 17 m/s, so $\Delta x = (17 \text{ m/s})(15 \text{ s}) = 255$ m. At $t = 25$ s the car is at $x = 80$ m + 255 m = 335 m. The velocity is 17 m/s. The acceleration is zero.

$\underline{t = 35 \text{ s}}$

For the time interval 10 s to 30 s, $\Delta x = (17 \text{ m/s})(20 \text{ s}) = 340$ m. From $t = 30$ s to 35 s the average velocity is 45 km/h = 12 m/s, so $\Delta x = (12 \text{ m/s})(5 \text{ s}) = 60$ m. At $t = 35$ s the car is at $x = 80$ m + 340 m + 60 m = 480 m. The velocity at $t = 35$ s is $v_x = 30$ km/h = 8 m/s. From part (d), $a_x = -1.7 \text{ m/s}^2$.

These results allow construction of the motion diagrams.

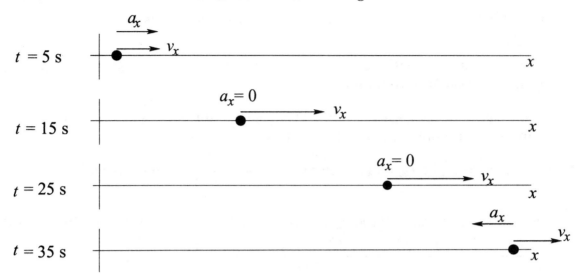

EVALUATE: The motion diagram shows that v_x is increasing when v_x and a_x are in the same direction, v_x is constant when $a_x = 0$, and v_x is decreasing when v_x and a_x are in opposite directions

2.15   IDENTIFY and **SET UP:** Use $v_x = \dfrac{dx}{dt}$ and $a_x = \dfrac{dv_x}{dt}$ to calculate $v_x(t)$ and $a_x(t)$.

EXECUTE:

$v_x = \dfrac{dx}{dt} = 2.00 \text{ cm/s} - (0.125 \text{ cm/s}^2)t$

$a_x = \dfrac{dv_x}{dt} = -0.125 \text{ cm/s}^2$

a) At $t = 0$, $x = 50.0$ cm, $v_x = 2.00$ cm/s, $a_x = -0.125$ cm/s^2.

b) Set $v_x = 0$ and solve for t: $t = 16.0$ s.

c) Set $x = 50.0$ cm and solve for t. This gives $t = 0$ and $t = 32.0$ s. The turtle returns to the starting point after 32.0 s.

d) Turtle is 10.0 cm from starting point when $x = 60.0$ cm or $x = 40.0$ cm.

Set $x = 60.0$ cm and solve for t: $t = 6.20$ s and $t = 25.8$ s.

At $t = 6.20$ s, $v_x = +1.23$ cm/s.

At $t = 25.8$ s, $v_x = -1.23$ cm/s.

Set $x = 40.0$ cm and solve for t: $t = 36.4$ s (other root to the quadratic equation is negative and hence nonphysical).

At $t = 36.4$ s, $v_x = -2.55$ cm/s.

e)

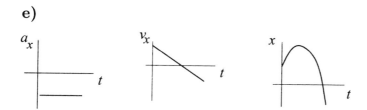

EVALUATE: The acceleration is constant and negative. v_x is linear in time. It is initially positive, decreases to zero, and then becomes negative with increasing magnitude. The turtle initially moves farther away from the origin but then stops and moves in the $-x$-direction.

2.19 a) IDENTIFY and **SET UP:** v_x is the slope of the x versus t curve and a_x is the slope of the v_x versus t curve.

EXECUTE:

$t = 0$ to $t = 5$ s: x versus t is a parabola so a_x is a constant. The curvature is positive so a_x is positive. v_x versus t is a straight line with positive slope. $v_{0x} = 0$.

$t = 5$ s to $t = 15$ s: x versus t is a straight line so v_x is constant and $a_x = 0$. The slope of x versus t is positive so v_x is positive.

$t = 15$ s to $t = 25$ s: x versus t is a parabola with negative curvature, so a_x is constant and negative. v_x versus t is a straight line with negative slope. The velocity is zero at 20 s, positive for 15 s to 20 s, and negative for 20 s to 25 s.

$t = 25$ s to $t = 35$ s: x versus t is a parabola with positive curvature, so a_x is constant and positive. v_x versus t is a straight line with positive slope. The velocity reaches zero at $t = 40$ s.

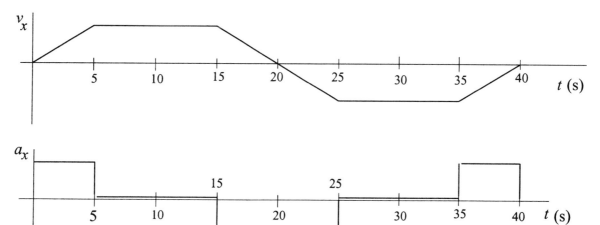

b) The motions diagrams are

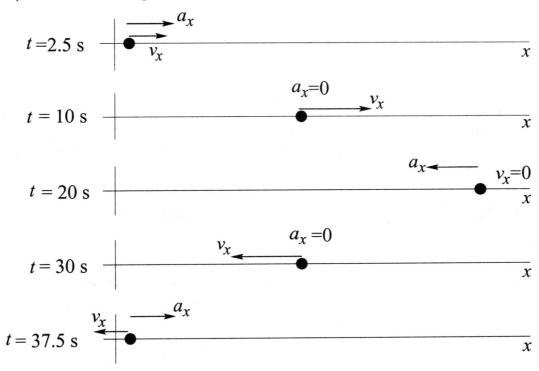

EVALUATE: The spider speeds up for the first 5 s, since v_x and a_x are both positive. Starting at $t = 15$ s the spider starts to slow down, stops momentarily at $t = 20$ s, and then moves in the opposite direction. At $t = 35$ s the spider starts to slow down again and stops at $t = 40$ s.

2.21 IDENTIFY: Use the constant acceleration equations to find v_{0x} and a_x.

a) SET UP:

$x - x_0 = 70.0$ m
$t = 7.00$ s
$v_x = 15.0$ m/s
$v_{0x} = ?$

EXECUTE: Use $x - x_0 = \left(\dfrac{v_{0x} + v_x}{2}\right) t$, so $v_{0x} = \dfrac{2(x - x_0)}{t} - v_x = \dfrac{2(70.0 \text{ m})}{7.00 \text{ s}} - 15.0$ m/s $= 5.0$ m/s.

b) Use $v_x = v_{0x} + a_x t$, so $a_x = \dfrac{v_x - v_{0x}}{t} = \dfrac{15.0 \text{ m/s} - 5.0 \text{ m/s}}{7.00 \text{ s}} = 1.43$ m/s^2.

EVALUATE: The average velocity is $(70.0 \text{ m})/(7.00 \text{ s}) = 10.0$ m/s. The final velocity is larger than this, so the antelope must be speeding up during the time interval; $v_{0x} < v_x$ and $a_x > 0$.

2.31 a) IDENTIFY and **SET UP:** The acceleration a_x at time t is the slope of the tangent to the v_x versus t curve at time t.

EXECUTE:

At $t = 3$ s, the v_x versus t curve is a horizontal straight line, with zero slope. Thus $a_x = 0$.

At $t = 7$ s, the v_x versus t curve is a straight-line segment with slope $\dfrac{45 \text{ m/s} - 20 \text{ m/s}}{9 \text{ s} - 5 \text{ s}} = 6.3 \text{ m/s}^2$. Thus $a_x = 6.3 \text{ m/s}^2$.

At $t = 11$ s the curve is again a straight-line segment, now with slope $\dfrac{-0 - 45 \text{ m/s}}{13 \text{ s} - 9 \text{ s}} = -11.2 \text{ m/s}^2$. Thus $a_x = -11.2 \text{ m/s}^2$.

EVALUATE: $a_x = 0$ when v_x is constant, $a_x > 0$ when v_x is positive and the speed is increasing, and $a_x < 0$ when v_x is positive and the speed is decreasing.

b) IDENTIFY: Calculate the displacement during the specified time interval.

SET UP: We can use the constant acceleration equations only for time intervals during which the acceleration is constant. If necessary, break the motion up into constant acceleration segments and apply the constant acceleration equations for each segment. For the time interval $t = 0$ to $t = 5$ s the acceleration is constant and equal to zero. For the time interval $t = 5$ s to $t = 9$ s the acceleration is constant and equal to 6.25 m/s^2. For the interval $t = 0$ s to $t = 13$ s the acceleration is constant and equal to -11.2 m/s^2.

EXECUTE:

During the first 5 seconds the acceleration is constant, so the constant acceleration kinematic formuals can be used.

$v_{0x} = 20$ m/s $a_x = 0$ $t = 5$ s $x - x_0 = ?$

$x - x_0 = v_{0x}t$ ($a_x = 0$ so no $\frac{1}{2}a_x t^2$ term)

$x - x_0 = (20 \text{ m/s})(5 \text{ s}) = 100$ m; this is the distance the officer travels in the first 5 seconds.

During the interval $t = 5$ s to 9 s the acceleration is again constant. The constant acceleration formulas can be applied to this 4 second interval. It is convenient to restart our clock so the interval starts at time $t = 0$ and ends at time $t = 5$ s. (Note that the acceleration is __not__ constant over the entire $t = 0$ to $t = 9$ s interval.)

$v_{0x} = 20$ m/s $a_x = 6.25$ m/s^2 $t = 4$ s $x_0 = 100$ m $x - x_0 = ?$

$x - x_0 = v_{0x}t + \frac{1}{2}a_x t^2$

$x - x_0 = (20 \text{ m/s})(4 \text{ s}) + \frac{1}{2}(6.25 \text{ m/s}^2)(4 \text{ s})^2 = 80 \text{ m} + 50 \text{ m} = 130$ m.

Thus $x = x_0 + 130$ m $= 100$ m $+ 130$ m $= 230$ m.

At $t = 9$ s the officer is at $x = 230$ m, so she has traveled 230 m in the first 9 seconds.

During the interval $t = 9$ s to $t = 13$ s the acceleration is again constant. The constant acceleration formulas can be applied for this 4 second interval but $\underline{not}$ for the whole $t = 0$ to $t = 13$ s interval. To use the equations restart our clock so this interval begins at time $t = 0$ and ends at time $t = 4$ s.

$v_{0x} = 45$ m/s (at the start of this time interval)

$a_x = -11.2$ m/s^2 $t = 4$ s $x_0 = 230$ m $x - x_0 = ?$

$x - x_0 = v_{0x}t + \frac{1}{2}a_x t^2$

$x - x_0 = (45 \text{ m/s})(4 \text{ s}) + \frac{1}{2}(-11.2 \text{ m/s}^2)(4 \text{ s})^2 = 180 \text{ m} - 89.6 \text{ m} = 90.4 \text{ m}.$

Thus $x = x_0 + 90.4$ m $= 230$ m $+ 90.4$ m $= 320$ m.

At $t = 13$ s the officer is at $x = 320$ m, so she has traveled 320 m in the first 13 seconds.

EVALUATE: The velocity v_x is always positive so the displacement is always positive and displacement and distance traveled are the same. The average velocity for time interval Δt is $v_{av-x} = \Delta x/\Delta t$. For $t = 0$ to 5 s, $v_{av-x} = 20$ m/s. For $t = 0$ to 9 s, $v_{av-x} = 26$ m/s. For $t = 0$ to 13 s, $v_{av-x} = 25$ m/s. These results are consistent with Fig.2.33.

2.33 a) IDENTIFY: The maximum speed occurs at the end of the initial acceleration period.

SET UP: $a_x = 20.0$ m/s^2 $t = 15.0$ min $= 900$ s $v_{0x} = 0$ $v_x = ?$

$v_x = v_{0x} + a_x t$

EXECUTE: $v_x = 0 + (20.0 \text{ m/s}^2)(900 \text{ s}) = 1.80 \times 10^4$ m/s

b) IDENTIFY: Use constant acceleration formulas to find the displacement Δx. The motion consists of three constant acceleration intervals. In the middle segment of the trip $a_x = 0$ and $v_x = 1.80 \times 10^4$ m/s, but we can't directly find the distance traveled during this part of the trip because we don't know the time. Instead, find the distance traveled in the first part of the trip (where $a_x = +20.0$ m/s^2) and in the last part of the trip (where $a_x = -20.0$ m/s^2). Subtract these two distances from the total distance of 3.84×10^8 m to find the distance traveled in the middle part of the trip (where $a_x = 0$).

first segment

SET UP: $x - x_0 = ?$ $t = 15.0$ min $= 900$ s $a_x = +20.0$ m/s^2 $v_{0x} = 0$

$x - x_0 = v_{0x}t + \frac{1}{2}a_x t^2$

EXECUTE: $x - x_0 = 0 + \frac{1}{2}(20.0 \text{ m/s}^2)(900 \text{ s})^2 = 8.10 \times 10^6$ m $= 8.10 \times 10^3$ km

second segment

SET UP: $x - x_0 = ?$ $t = 15.0$ min $= 900$ s $a_x = -20.0$ m/s^2

$v_{0x} = 1.80 \times 10^4$ s

$x - x_0 = v_{0x}t + \frac{1}{2}a_x t^2$

EXECUTE: $x - x_0 = (1.80 \times 10^4 \text{ s})(900 \text{ s}) + \frac{1}{2}(-20.0 \text{ m/s}^2)(900 \text{ s})^2 = 8.10 \times 10^6 \text{ m} = 8.10 \times 10^3 \text{ km}$ (The same distance as traveled as in the first segment.)

Therefore, the distance traveled at constant speed is $3.84 \times 10^8 \text{ m} - 8.10 \times 10^6 \text{ m} - 8.10 \times 10^6 \text{ m} = 3.678 \times 10^8 \text{ m} = 3.678 \times 10^5 \text{ km}$.

The fraction this is of the total distance is $\dfrac{3.678 \times 10^8 \text{ m}}{3.84 \times 10^8 \text{ m}} = 0.958$.

c) IDENTIFY: We know the time for each acceleration period, so find the time for the constant speed segment.

SET UP: $x - x_0 = 3.678 \times 10^8 \text{ m} \qquad v_x = 1.80 \times 10^4 \text{ m/s} \qquad a_x = 0 \qquad t = ?$

$x - x_0 = v_{0x}t + \frac{1}{2}a_x t^2$

EXECUTE: $t = \dfrac{x - x_0}{v_{0x}} = \dfrac{3.678 \times 10^8 \text{ m}}{1.80 \times 10^4 \text{ m/s}} = 2.043 \times 10^4 \text{ s} = 340.5 \text{ min}$.

The total time for the whole trip is thus $15.0 \text{ min} + 340.5 \text{ min} + 15.0 \text{ min} = 370$ min.

EVALUATE: If the speed was a constant $1.80 \times 10^4 \text{ m/s}$ for the entire trip, the trip would take $(3.84 \times 10^8 \text{ m})/(1.80 \times 10^4 \text{ m/s}) = 356 \text{ min}$. The trip actually takes a bit longer than this since the average is less than $1.80 \times 10^8 \text{ m/s}$ during the relatively brief acceleration phases.

2.37 IDENTIFY: Apply the constant acceleration equations to the motion of the car and of the motorcycle.

SET UP:

a) For the car v_x is constant and $a_x = 0$, so x versus t is a straight line with positive slope. For the motorcycle $v_{0x} = 0$ and a_x is constant and positive, so x versus t is a parabola with positive curvature and zero slope at $t = 0$. Let both vehicles be at $x_0 = 0$ at $t = 0$.

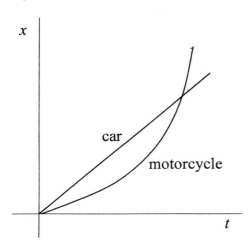

Let T be the time t when the motorcycle overtakes the car and let d be the displacement of the car (and motorcycle) when this occurs

<u>car</u>

$v_{0x} = v_C$ $a_x = 0$ $t = T$ $x - x_0 = d$

Putting these values into $x - x_0 = v_{0x}t + \frac{1}{2}a_x t^2$ gives $\underline{d = v_C T}$.

<u>motorcycle</u>

$v_{0x} = 0$ $a_x = a_M$ $t = T$ $x - x_0 = d$

Putting these values into $x - x_0 = v_{0x}t + \frac{1}{2}a_x t^2$ gives $\underline{d = a_M T^2/2}$.

EXECUTE: Combine these two equations to eliminate d:

$v_C T = \frac{1}{2}a_M T^2$

$T = \dfrac{2v_C}{a_M}$

Then $v_x = v_{0x} + a_x t$ for the motorcycle gives $v_M = 0 + a_M \left(\dfrac{2v_C}{a_M} \right) = 2v_C$, when the motorcycle has overtaken the car.

b) From part (a), $d = v_C T = v_C \left(\dfrac{2v_C}{a_M} \right) = 2v_C^2/a_M$ and $v_C^2 = \frac{1}{2}a_M d$.

Apply $v_x^2 = v_{0x}^2 + 2a_x(x - x_0)$ to the motorcycle. This gives $v_M^2 = 2a_M(x - x_0)$. Set $v_M = v_C$; then $x - x_0$ is the distance the motorcycle has traveled when its velocity equals the velocity of the car. $x - x_0 = \dfrac{v_M^2}{2a_M} = \dfrac{v_C^2}{2a_M} = \dfrac{1}{2a_M}\left(\dfrac{a_M d}{2} \right) = \dfrac{d}{4}$.

The motorcycle has traveled a distance $d/4$ when its speed equals v_C.

EVALUATE: The x-t graph in part (a) shows that the two curves intersect, the motorcycle overtakes the car, no matter what the slope (velocity) for the car is. When the velocities are equal the slopes of the two curves are equal, and the graph shows this occurs well before the intersection of the two curves.

2.39 IDENTIFY: Apply the constant acceleration equations to the motion of the flea. After the flea leaves the ground, $a_y = g$, downward. Take the origin at the ground and the positive direction to be upward.

a) SET UP: At the maximum height v_y.

$v_y = 0$ $y - y_0 = 0.440$ m $a_y = -9.80$ m/s^2 $v_{0y} = ?$

$v_y^2 = v_{0y}^2 + 2a_y(y - y_0)$

EXECUTE: $v_{0y} = \sqrt{-2a_y(y - y_0)} = \sqrt{-2(-9.80 \text{ m/s}^2)(0.440 \text{ m})} = 2.94$ m/s

b) SET UP: When the flea has returned to the ground $y - y_0 = 0$.

$y - y_0 = 0$ $v_{0y} = +2.94$ m/s $a_y = -9.80$ m/s^2 $t = ?$

$y - y_0 = v_{0y}t + \frac{1}{2}a_y t^2$

EXECUTE: With $y - y_0 = 0$ this gives $t = -\dfrac{2v_{0y}}{a_y} = -\dfrac{2(2.94 \text{ m/s})}{-9.80 \text{ m/s}^2} = 0.600$ s.

EVALUATE: We can use $v_y = v_{0y} + a_y t$ to show that with $v_{0y} = 2.94$ m/s, $v_y = 0$ after 0.600 s.

2.43 IDENTIFY: The ring has constant acceleration $a_y = g$, downward.

a) SET UP: $v_{\text{av}-y} = \dfrac{v_{0y} + v_y}{2}$

Need to use the constant acceleration formulas to solve for the velocity v_y of the ring just before it strikes the ground. Take the origin of coordinates at the roof and take the positive y-direction to be upward.

$v_y = ?$ $v_{0y} = +5.00$ m/s $a_y = -9.80$ m/s^2 $y - y_0 = -12.0$ m (When the ring is at the ground its displacement is 12.0 m downward.)

$v_y^2 = v_{0y}^2 + 2a_y(y - y_0)$

EXECUTE: $v_y = -\sqrt{(5.00 \text{ m/s})^2 + 2(-9.80 \text{ m/s}^2)(-12.0 \text{ m})} = -16.13$ m/s

Then $v_{\text{av}-y} = \dfrac{v_{0y} + v_y}{2} = \dfrac{+5.00 \text{ m/s} - 16.13 \text{ m/s}}{2} = -5.56$ m/s. The minus sign indicates that the average velocity is downward.

b) SET UP and **EXECUTE:** The acceleration is constant and equal to -9.80 m/s^2, so the average value must equal this value: $a_{\text{av}-y} = -9.80$ m/s^2.

c) SET UP: $t = ?$ $v_y = -16.13$ m/s $a_y = -9.80$ m/s^2, $v_{0y} = +5.00$ m/s

$v_y = v_{0y} + a_y t$

EXECUTE: $t = \dfrac{v_y - v_{0y}}{a_y} = \dfrac{-16.13 \text{ m/s} - 5.00 \text{ m/s}}{-9.80 \text{ m/s}^2} = 2.16$ s

d) We found in part (a) that $v_y = -16.1$ m/s just before the ring strikes the ground.

e)

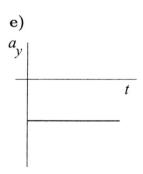
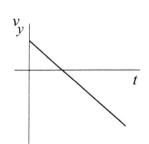
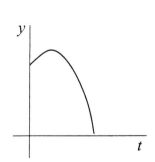

EVALUATE: We can check our results by using $y - y_0 = (v_{\mathrm{av}-y})t$ to calculate $y - y_0 = (-5.56 \text{ m/s})(2.16 \text{ s}) = -12.0 \text{ m}$, which checks.

2.45 IDENTIFY: The balloon has constant acceleration $a_y = g$, downward.

a) SET UP: Take the $+y$ direction to be upward.

$t = 2.00 \text{ s}, \quad v_{0y} = -6.00 \text{ m/s}, \quad a_y = -9.80 \text{ m/s}^2, \quad v_y = ?$

EXECUTE: $v_y = v_{0y} + a_y t = -6.00 \text{ m/s} + (-9.80 \text{ m/s}^2)(2.00 \text{ s}) = -25.5 \text{ m/s}$

b) SET UP: $y - y_0 = ?$

EXECUTE: $y - y_0 = v_{0y}t + \frac{1}{2}a_y t^2 = (-6.00 \text{ m/s})(2.00 \text{ s}) + \frac{1}{2}(-9.80 \text{ m/s}^2)(2.00 \text{ s})^2 = -31.6 \text{ m}$

c) SET UP: $y - y_0 = -10.0 \text{ m}, \quad v_{0y} = -6.00 \text{ m/s}, \quad a_y = -9.80 \text{ m/s}^2, \quad v_y = ?$

$v_y^2 = v_{0y}^2 + 2a_y(y - y_0)$

EXECUTE: $v_y = -\sqrt{2a_y(y - y_0)} = -\sqrt{(-6.00 \text{ m/s})^2 + 2(-9.80 \text{ m/s}^2)(-10.0 \text{ m})} = -15.2 \text{ m/s}$

d)

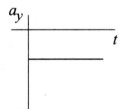 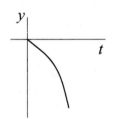

EVALUATE: The speed of the balloon increases steadily since the acceleration and velocity are in the same direction. $|v_y| = 25.5 \text{ m/s}$ when $|y - y_0| = 31.6 \text{ m}$, so $|v_y|$ is less than this (15.2 m/s) when $|y - y_0|$ is less (10.0 m).

2.47 IDENTIFY: Use the constant acceleration equations to calculate a_x and $x - x_0$.

a) SET UP: $v_x = 224 \text{ m/s}, \quad v_{0x} = 0, \quad t = 0.900 \text{ s}, \quad a_x = ?$

$v_x = v_{0x} + a_x t$

EXECUTE: $a_x = \dfrac{v_x - v_{0x}}{t} = \dfrac{224 \text{ m/s} - 0}{0.900 \text{ s}} = 249 \text{ m/s}^2$

b) $a_x/g = (249 \text{ m/s}^2)/(9.80 \text{ m/s}^2) = 25.4$

c) $x - x_0 = v_{0x}t + \frac{1}{2}a_x t^2 = 0 + \frac{1}{2}(249 \text{ m/s}^2)(0.900 \text{ s})^2 = 101 \text{ m}$

d) SET UP: Calculate the acceleration, assuming it is constant:

$t = 1.40 \text{ s}, \quad v_{0x} = 283 \text{ m/s}, \quad v_x = 0 \text{ (stops)}, \quad a_x = ?$

$$v_x = v_{0x} + a_x t$$

EXECUTE: $a_x = \dfrac{v_x - v_{0x}}{t} = \dfrac{0 - 283 \text{ m/s}}{1.40 \text{ s}} = -202 \text{ m/s}^2$

$a_x/g = (-202 \text{ m/s}^2)/(9.80 \text{ m/s}^2) = -20.6; \quad a_x = -20.6g$

If the acceleration while the sled is stopping is constant then the magnitude of the acceleration is only 20.6g. But if the acceleration is not constant it is certainly possible that at some point the instantaneous acceleration could be as large as 40g.

EVALUATE: It is reasonable that for this motion the acceleration is much larger than g.

2.51 $a_x = At - Bt^2$ with $A = 1.50 \text{ m/s}^3$ and $B = 0.120 \text{ m/s}^4$

a) IDENTIFY: Integrate $a_x(t)$ to find $v_x(t)$ and then integrate $v_x(t)$ to find $x(t)$.

SET UP: $v_x = v_{0x} + \int_0^t a_x \, dt$

EXECUTE: $v_x = v_{0x} + \int_0^t (At - Bt^2) \, dt = v_{0x} + \frac{1}{2}At^2 - \frac{1}{3}Bt^3$

At rest at $t = 0$ says that $v_{0x} = 0$, so

$v_x = \frac{1}{2}At^2 - \frac{1}{3}Bt^3 = \frac{1}{2}(1.50 \text{ m/s}^3)t^2 - \frac{1}{3}(0.120 \text{ m/s}^4)t^3$

$v_x = (0.75 \text{ m/s}^3)t^2 - (0.040 \text{ m/s}^4)t^3$

SET UP: $x = x_0 + \int_0^t v_x \, dt$

EXECUTE: $x = x_0 + \int_0^t (\frac{1}{2}At^2 - \frac{1}{3}Bt^3) \, dt = x_0 + \frac{1}{6}At^3 - \frac{1}{12}Bt^4$

At the origin at $t = 0$ says that $x_0 = 0$, so

$x = \frac{1}{6}At^3 - \frac{1}{12}Bt^4 = \frac{1}{6}(1.50 \text{ m/s}^3)t^3 - \frac{1}{12}(0.120 \text{ m/s}^4)t^4$

$x = (0.25 \text{ m/s}^3)t^3 - (0.010 \text{ m/s}^4)t^4$

EVALUATE: We can check our results by using them to verify that $v_x(t) = \dfrac{da_x}{dt}$ and $a_x(t) = \dfrac{dv_x}{dt}$.

b) IDENTIFY and **SET UP:** At time t, when v_x is a maximum, $\dfrac{dv_x}{dt} = 0$. (Since $a_x = \dfrac{dv_x}{dt}$, the maximum velocity is when $a_x = 0$. For earlier times a_x is positive so v_x is still increasing. For later times a_x is negative and v_x is decreasing.)

EXECUTE:

$a_x = \dfrac{dv_x}{dt} = 0$ so $At - Bt^2 = 0$

One root is $t = 0$, but at this time $v_x = 0$ and not a maximum.

The other root is $t = \dfrac{A}{B} = \dfrac{1.50 \text{ m/s}^3}{0.120 \text{ m/s}^4} = 12.5$ s

At this time $v_x = (0.75 \text{ m/s}^3)t^2 - (0.040 \text{ m/s}^4)t^3$ gives

$v_x = (0.75 \text{ m/s}^3)(12.5 \text{ s})^2 - (0.040 \text{ m/s}^4)(12.5 \text{ s})^3 = 117.2 \text{ m/s} - 78.1 \text{ m/s} = 39.1 \text{ m/s}.$

EVALUATE: For $t < 12.5$ s, $a_x > 0$ and v_x is increasing. For $t > 12.5$ s, $a_x < 0$ and v_x is decreasing.

2.53 a) IDENTIFY and **SET UP:** The change in speed is the area under the a_x versus t curve between vertical lines at $t = 2.5$ s and $t = 7.5$ s.

EXECUTE: This area is

$\frac{1}{2}(4.00 \text{ cm/s}^2 + 8.00 \text{ cm/s}^2)(7.5 \text{ s} - 2.5 \text{ s}) = 30.0 \text{ cm/s}$

This acceleration is positive so the change in velocity is positive.

b) Slope of v_x versus t is positive and increasing with t.

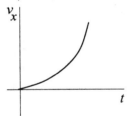

EVALUATE: The calculation in part (a) is equivalent to $\Delta v_x = (a_{\text{av}-x})\,\Delta t$. Since a_x is linear in t, $a_{\text{av}-x} = (a_{0x} + a_x)/2$. Thus $a_{\text{av}-x} = \frac{1}{2}(4.00 \text{ cm/s}^2 + 8.00 \text{ cm/s}^2)$ for the time interval $t = 2.5$ s to $t = 7.5$ s.

Problems

2.57 IDENTIFY: Use information about displacement and time to calculate average speed and average velocity. Take the origin to be at Seward and the positive direction to be west.

a) SET UP: average speed $= \dfrac{\text{distance traveled}}{\text{time}}$

EXECUTE: The distance traveled (different from the net displacement $(x - x_0)$) is 76 km + 34 km = 110 km.

Find the total elapsed time by using $v_{\text{av}-x} = \dfrac{\Delta x}{\Delta t} = \dfrac{x - x_0}{t}$ to find t for each leg of the journey.

Seward to Auora: $t = \dfrac{x - x_0}{v_{\text{av}-x}} = \dfrac{76 \text{ km}}{88 \text{ km/h}} = 0.8636 \text{ h}$

Auora to York: $t = \dfrac{x - x_0}{v_{\text{av}-x}} = \dfrac{-34 \text{ km}}{-72 \text{ km/h}} = 0.4722 \text{ h}$

Total $t = 0.8636 \text{ h} + 0.4722 \text{ h} = 1.336 \text{ h}$.

Then average speed $= \dfrac{110 \text{ km}}{1.336 \text{ h}} = 82 \text{ km/h}$.

b) SET UP: $v_{\text{av}-x} = \dfrac{\Delta x}{\Delta t}$, where Δx is the displacement, not the total distance traveled.

For the whole trip he ends up $76 \text{ km} - 34 \text{ km} = 42 \text{ km}$ west of his starting point.

$v_{\text{av}-x} = \dfrac{42 \text{ km}}{1.336 \text{ h}} = 31 \text{ km/h}$.

EVALUATE: The motion is not uniformly in the same direction so the displacement is less than the distance traveled and the magnitude of the average velocity is less than the average speed.

2.59 a) IDENTIFY: Calculate the average acceleration using $a_{\text{av}-x} = \dfrac{\Delta v_x}{\Delta t} = \dfrac{v_x - v_{0x}}{t}$
Use the information about the time and total distance to find his maximum speed.

SET UP: $v_{0x} = 0$ since the runner starts from rest.

$t = 4.0$ s, but we need to calculate v_x, the speed of the runner at the end of the acceleration period.

EXECUTE:

For the last $9.1 \text{ s} - 4.0 \text{ s} = 5.1 \text{ s}$ the acceleration is zero and the runner travels a distance of $d_1 = (5.1 \text{ s})v_x$ (obtained using $x - x_0 = v_{0x}t + \frac{1}{2}a_x t^2$)

During the acceleration phase of 4.0 s, where the velocity goes from 0 to v_x, the runner travels a distance

$d_2 = \left(\dfrac{v_{0x} + v_x}{2} \right) t = \dfrac{v_x}{2}(4.0 \text{ s}) = (2.0 \text{ s})v_x$

The total distance traveled is 100 m, so $d_1 + d_2 = 100$ m. This gives $(5.1 \text{ s})v_x + (2.0 \text{ s})v_x = 100$ m.

$v_x = \dfrac{100 \text{ m}}{7.1 \text{ s}} = 14.08 \text{ m/s}$.

Now we can calculate $a_{\text{av}-x}$: $a_{\text{av}-x} = \dfrac{v_x - v_{0x}}{t} = \dfrac{14.08 \text{ s} - 0}{4.0 \text{ s}} = 3.5 \text{ m/s}^2$.

b) For this time interval the velocity is constant, so $a_{\text{av}-x} = 0$.

EVALUATE: Now that we have v_x we can calculate $d_1 = (5.1 \text{ s})(14.08 \text{ m/s}) =$

71.9 m and $d_2 = (4.0 \text{ s})(14.08 \text{ m/s}) = 28.2$ m. So, $d_1 + d_2 = 100$ m, which checks.

c) IDENTIFY and **SET UP:** $a_{\text{av}-x} = \dfrac{v_x - v_{0x}}{t}$, where now the time interval is the full 9.1 s of the race.

We have calculated the final speed to be 14.08 m/s, so

$$a_{\text{av}-x} = \frac{14.08 \text{ m/s}}{9.1 \text{ s}} = 1.5 \text{ m/s}^2.$$

EVALUATE: The acceleration is zero for the last 5.1 s, so it makes sense for the answer in part (c) to be less than half the answer in part (a).

d) The runner spends different times moving with the average accelerations of parts (a) and (b).

2.65 IDENTIFY and SET UP: Apply constant acceleration equations.
Find the velocity at the start of the second 5.0 s; this is the velocity at the end of the first 5.0 s. Then find $x - x_0$ for the first 5.0 s.

EXECUTE:

For the first 5.0 s of the motion, $v_{0x} = 0$, $t = 5.0$ s.

$v_x = v_{0x} + a_x t$ gives $v_x = a_x(5.0 \text{ s})$.

This is the initial speed for the second 5.0 s of the motion. For the second 5.0 s:

$v_{0x} = a_x(5.0 \text{ s})$, $t = 5.0$ s, $x - x_0 = 150$ m.

$x - x_0 = v_{0x} t + \frac{1}{2} a_x t^2$ gives $150 \text{ m} = (25 \text{ s}^2) a_x + (12.5 \text{ s}^2) a_x$ and $a_x = 4.0 \text{ m/s}^2$

Use this a_x and consider the first 5.0 s of the motion:

$x - x_0 = v_{0x} t + \frac{1}{2} a_x t^2 = 0 + \frac{1}{2}(4.0 \text{ m/s}^2)(5.0 \text{ s})^2 = 50.0$ m.

EVALUATE: The ball is speeding up so it travels farther in the second 5.0 s interval than in the first. In fact, $x - x_0$ is proportional to t^2 since it starts from rest. If it goes 50.0 m in 5.0 s, in twice the time (10.0 s) it should go four times as far. In 10.0 s we calculated it went 50 m + 150 m = 200 m, which is four times 50 m.

2.67 IDENTIFY: Apply constant acceleration equations to the motion of the two objects, you and the cockroach. You catch up with the roach when both objects are at the same place at the same time. Let T be the time when you catch up with the cockroach.

SET UP:

Take $x = 0$ to be at the $t = 0$ location of the roach and positive x to be in the direction of motion of the two objects.

roach:

$v_{0x} = 1.50 \text{ m/s}, \quad a_x = 0, \quad x_0 = 0, \quad x = 1.20 \text{ m}, \quad t = T$

you:

$$v_{0x} = 0.80 \text{ m/s}, \quad x_0 = -0.90 \text{ m}, \quad x = 1.20 \text{ m}, \quad t = T, \quad a_x = ?$$

Apply $x - x_0 = v_{0x}t + \frac{1}{2}a_x t^2$ to both objects:

EXECUTE:

roach: $1.20 \text{ m} = (1.50 \text{ m/s})T$, so $T = 0.800$ s.

you: $1.20 \text{ m} - (-0.90 \text{ m}) = (0.80 \text{ m/s})T + \frac{1}{2}a_x T^2$

$2.10 \text{ m} = (0.80 \text{ m/s})(0.800 \text{ s}) + \frac{1}{2}a_x(0.800 \text{ s})^2$

$2.10 \text{ m} = 0.64 \text{ m} + (0.320 \text{ s}^2)a_x$

$a_x = 4.6 \text{ m/s}^2$.

EVALUATE: Your final velocity is $v_x = v_{0x} + a_x t = 4.48$ m/s. Then $x - x_0 = \left(\dfrac{v_{0x} + v_x}{2}\right)t = 2.10$ m, which checks. You have to accelerate to a speed greater than that of the roach so you will travel the extra 0.90 m you are initially behind.

2.69 IDENTIFY: Apply constant acceleration equations to each object.
Take the origin of coordinates to be at the initial position of the truck.

Let d be the distance that the auto initially is behind the truck, so $x_0(\text{auto}) = -d$ and $x_0(\text{truck}) = 0$. Let T be the time it takes the auto to catch the truck. Thus at time T the truck has undergone a displacement $x - x_0 = 40.0$ m, so is at $x = x_0 + 40.0$ m $= 40.0$ m. The auto has caught the truck so at time T it is also at $x = 40.0$ m.

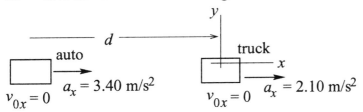

a) SET UP: Use the motion of the truck to calculate T:

$x - x_0 = 40.0 \text{ m}, \quad v_{0x} = 0 \text{ (starts from rest)}, \quad a_x = 2.10 \text{ m/s}^2, \quad t = T$

$x - x_0 = v_{0x}t + \frac{1}{2}a_x t^2$

Since $v_{0x} = 0$, this gives $t = \sqrt{\dfrac{2(x - x_0)}{a_x}}$

EXECUTE: $T = \sqrt{\dfrac{2(40.0 \text{ m})}{2.10 \text{ m/s}^2}} = 6.17$ s

b) SET UP: Use the motion of the auto to calculate d:

$x - x_0 = 40.0 \text{ m} + d, \quad v_{0x} = 0, \quad a_x = 3.40 \text{ m/s}^2, \quad t = 6.17 \text{ s}$

$x - x_0 = v_{0x}t + \frac{1}{2}a_x t^2$

EXECUTE:

$d + 40.0 \text{ m} = \frac{1}{2}(3.40 \text{ m/s}^2)(6.17 \text{ s})^2$

$d = 64.8 \text{ m} - 40.0 \text{ m} = 24.8 \text{ m}$

c) auto: $v_x = v_{0x} + a_x t = 0 + (3.40 \text{ m/s}^2)(6.17 \text{ s}) = 21.0 \text{ m/s}$

truck: $v_x = v_{0x} + a_x t = 0 + (2.10 \text{ m/s}^2)(6.17 \text{ s}) = 13.0 \text{ m/s}$

d)

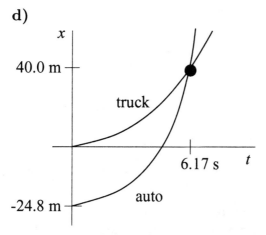

EVALUATE: In part (c) we found that the auto was traveling faster than the truck when they come abreast. The graph in part (a) agrees with this: at the intersection of the two curves the slope of the x-t curve for the auto is greater than that of the truck. The auto must have an average velocity greater than that of the truck since it must travel farther in the same time interval.

2.73 IDENTIFY: Apply constant acceleration equations to each vehicle.

SET UP:

a) It is very convenient to work in coordinates attached to the truck.

Note that these coordinates move at constant velocity relative to the earth. In these coordinates the truck is at rest, and the initial velocity of the car is $v_{0x} = 0$. Also, the car's acceleration in these coordinates is the same as in coordinates fixed to the earth.

EXECUTE:

First, let's calculate how far the car must travel relative to the truck:

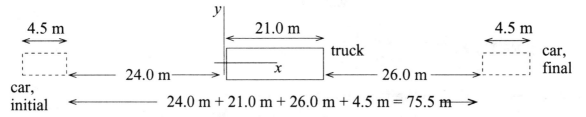

The car goes from $x_0 = -24.0$ m to $x = 51.5$ m. So $x - x_0 = 75.5$ m for the car. Calculate the time it takes the car to travel this distance:

$a_x = 0.600 \text{ m/s}^2, \quad v_{0x} = 0, \quad x - x_0 = 75.5 \text{ m}, \quad t = ?$

$x - x_0 = v_{0x}t + \frac{1}{2}a_x t^2$

$t = \sqrt{\dfrac{2(x - x_0)}{a_x}} = \sqrt{\dfrac{2(75.5 \text{ m})}{0.600 \text{ m/s}^2}} = 15.86 \text{ s}$

It takes the car 15.9 s to pass the truck.

b) Need how far the car travels relative to the earth, so go now to coordinates fixed to the earth. In these coordinates $v_{0x} = 20.0$ m/s for the car. Take the origin to be at the initial position of the car.

$v_{0x} = 20.0 \text{ m/s}, \quad a_x = 0.600 \text{ m/s}^2, \quad t = 15.86 \text{ s}, \quad x - x_0 = ?$

$x - x_0 = v_{0x}t + \frac{1}{2}a_x t^2 = (20.0 \text{ m/s})(15.86 \text{ s}) + \frac{1}{2}(0.60 \text{ m/s}^2)(15.86 \text{ s})^2$

$x - x_0 = 317.2 \text{ m} + 75.5 \text{ m} = 393 \text{ m}.$

c) In coordinates fixed to the earth:

$v_x = v_{0x} + a_x t = 20.0 \text{ m/s} + (0.600 \text{ m/s}^2)(15.86 \text{ s}) = 29.5 \text{ m/s}$

EVALUATE: In 15.9 s the truck travels $x - x_0 = (20.0 \text{ m/s})(15.86 \text{ s}) = 317.2$ m. The car travels 392.7 m $-$ 317.2 m $= 75$ m farther than the truck,which checks with part (a). In coordinates attached to the truck, for the car $v_{0x} = 0$, $v_x = 9.5$ m/s and in 15.86 s the car travels $x - x_0 = \left(\dfrac{v_{0x} + v_x}{2} \right) t = 75$ m, whihc checks with part (a).

2.75 $a(t) = \alpha + \beta t$, with $\alpha = -2.00 \text{ m/s}^2$ and $\beta = 3.00 \text{ m/s}^3$

a) IDENTIFY and **SET UP:** Integrate $a_x(t)$ to find $v_x(t)$ and then integrate $v_x(t)$ to find $x(t)$.

EXECUTE: $v_x = v_{0x} + \int_0^t a_x\,dt = v_{0x} + \int_0^t (\alpha + \beta t)\,dt = v_{0x} + \alpha t + \frac{1}{2}\beta t^2$

$x = x_0 + \int_0^t v_x\,dt = x_0 + \int_0^t (v_{0x} + \alpha t + \frac{1}{2}\beta t^2)\,dt = x_0 + v_{0x}t + \frac{1}{2}\alpha t^2 + \frac{1}{6}\beta t^3$

At $t = 0$, $x = x_0$.

To have $x = x_0$ at $t_1 = 4.00$ s requires that $v_{0x}t_1 + \frac{1}{2}\alpha t_1^2 + \frac{1}{6}\beta t_1^3 = 0$.

Thus $v_{0x} = -\frac{1}{6}\beta t_1^2 - \frac{1}{2}\alpha t_1 = -\frac{1}{6}(3.00 \text{ m/s}^3)(4.00 \text{ s})^2 - \frac{1}{2}(-2.00 \text{ m/s}^2)(4.00 \text{ s}) = -4.00$ m/s.

b) With v_{0x} as calculated in part (a) and $t = 4.00$ s,

$v_x = v_{0x} + \alpha t + \frac{1}{2}\beta t^2 = -4.00 \text{ s} + (-2.00 \text{ m/s}^2)(4.00 \text{ s}) + \frac{1}{2}(3.00 \text{ m/s}^3)(4.00 \text{ s})^2 = +12.0$ m/s.

EVALUATE: $a_x = 0$ at $t = 0.67$ s. For $t > 0.67$ s, $a_x > 0$. At $t = 0$, the particle is moving in the $-x$-direction and is speeding up. After $t = 0.67$ s, when the acceleration is positive, the object slows down and then starts to move in the $+x$-direction with increasing speed.

2.79 **IDENTIFY:** Use constant acceleration equations, with $a_y = g$, downward, to calculate the speed of the diver when she reaches the water.

SET UP:

Take the origin of coordinates to be at the platform, and take the $+y$-direction to be downward.

$y - y_0 = +21.3$ m, $a_y = +9.80$ m/s^2, $v_{0y} = 0$ (since diver just steps off), $v_y = ?$

$v_y^2 = v_{0y}^2 + 2a_y(y - y_0)$

EXECUTE: $v_y = +\sqrt{2a_y(y - y_0)} = +\sqrt{2(9.80 \text{ m/s}^2)(21.3 \text{ m})} = +20.4$ m/s.

We know that v_y is positive because the diver is traveling downward when she reaches the water.

The announcer has exaggerated the speed of the diver.

EVALUATE: We could also use $y - y_0 = v_{0y}t + \frac{1}{2}a_y t^2$ to find $t = 2.085$ s. The diver gains 9.80 m/s of speed each second, so has $v_y = (9.80 \text{ m/s}^2)(2.085 \text{ s}) = 20.4$ m/s when she reaches the water, which checks

b) IDENTIFY: Calculate the initial upward velocity needed to give the diver a speed of 25.0 m/s when she reaches the water. Use the same coordinates as in part (a).

SET UP: $v_{0y} = ?$, $v_y = +25.0$ m/s, $a_y = +9.80$ m/s^2, $y - y_0 = +21.3$ m

$v_y^2 = v_{0y}^2 + 2a_y(y - y_0)$

EXECUTE: $v_{0y} = -\sqrt{v_y^2 - 2a_y(y - y_0)} =$

$-\sqrt{(25.0 \text{ m/s})^2 - 2(9.80 \text{ m/s}^2)(21.3 \text{ m})} = -14.4$ m/s.

(v_{0y} is negative since the direction of the initial velocity is upward.)

EVALUATE: One way to decide if this speed is reasonable is to calculate the maximum height above the platform it would produce:

$v_{0y} = -14.4$ m/s, $v_y = 0$ (at maximum height), $a_y = +9.80$ m/s^2, $y - y_0 = ?$

$v_y^2 = v_{0y}^2 + 2a_y(y - y_0)$

$y - y_0 = \dfrac{v_y^2 - v_{0y}^2}{2a_y} = \dfrac{0 - (-14.4 \text{ s})^2}{2(+9.80 \text{ m/s})} = -10.6$ m

This is not physically attainable; a vertical leap of 10.6 m upward is not possible.

2.81 **IDENTIFY:** Let $y = 0$ at the ground and let positive y be upward.

Apply constant acceleration equations to the motion of the football, with $a_y = -g$.

a) SET UP: Let $t = 0$ when the football is at the window. Consider the motion from the window to the maximum height.

$v_{0y} = 5.00$ m/s, $v_y = 0$ (at maximum height), $a_y = -9.80$ m/s^2, $y_0 = +12.0$ m, $y = ?$

$v_y^2 = v_{0y}^2 + 2a_y(y - y_0)$

EXECUTE: $y - y_0 = \dfrac{v_y^2 - v_{0y}^2}{2a_y} = \dfrac{0 - (5.00 \text{ s})^2}{2(-9.80 \text{ m/s}^2)} = 1.28$ m.

$y = y_0 + 1.28$ m $= 12.0$ m $+ 1.28$ m $= 13.3$ m (maximum height above ground).

b) SET UP: Now let $t = 0$ be when the football is on the ground. Use the motion from the ground to the window to find v_{0y}, the velocity of the football as it leaves the ground.

$v_{0y} = ?$, $a_y = -9.80$ m/s^2, $v_y = +5.00$ s, $y - y_0 = +12.0$ m

$v_y^2 = v_{0y}^2 + 2a_y(y - y_0)$

EXECUTE: $v_{0y} = +\sqrt{v_y^2 - 2a_y(y - y_0)} =$

$\sqrt{(5.00 \text{ m/s})^2 - 2(-9.80 \text{ m/s}^2)(+12.0 \text{ m})} = +16.13$ m/s

SET UP: Now consider the motion from the ground to the maximum height.

$v_{0y} = +16.13$ m/s, $t = ?$, $v_y = 0$ (at maximum height), $a_y = -9.80$ m/s^2

$v_y = v_{0y} + a_y t$

EXECUTE: $t = \dfrac{v_y - v_{0y}}{a_y} = \dfrac{0 - 16.13 \text{ m/s}}{-9.80 \text{ m/s}^2} = +1.65$ s

EVALUATE: $y - y_0 = v_{0y}t + \frac{1}{2}a_y t^2 = (16.13 \text{ m/s})(1.65 \text{ s}) + \frac{1}{2}(-9.80 \text{ m/s}^2)(1.65 \text{ s})^2$ $= 13.3$ m. This agrees with the maximum height calculated in part (a), so checks.

2.83 Take positive y to be upward.

a) IDENTIFY: Consider the motion from when he applies the acceleration to when the shot leaves his hand.

SET UP: $v_{0y} = 0$, $v_y = ?$, $a_y = 45.0$ m/s^2, $y - y_0 = 0.640$ m

$v_y^2 = v_{0y}^2 + 2a_y(y - y_0)$

EXECUTE: $v_y = \sqrt{2a_y(y - y_0)} = \sqrt{2(45.0 \text{ m/s}^2)(0.640 \text{ m})} = 7.59$ m/s

b) IDENTIFY: Consider the motion of the shot from the point where he releases it to its maximum height, where $v = 0$. Take $y = 0$ at the ground.

SET UP: $y_0 = 2.20$ m, $y = ?$, $a_y = -9.80$ m/s^2 (free fall), $v_{0y} = 7.59$ m/s

(from part (a), $v_y = 0$ (at maximum height)
$$v_y^2 = v_{0y}^2 + 2a_y(y - y_0)$$

EXECUTE: $y - y_0 = \dfrac{v_y^2 - v_{0y}^2}{2a_y} = \dfrac{0 - (7.59 \text{ m/s})^2}{2(-9.80 \text{ m/s}^2)} = 2.94 \text{ m}$

$y = 2.20 \text{ m} + 2.94 \text{ m} = 5.14 \text{ m}.$

c) IDENTIFY: Consider the motion of the shot from the point where he releases it to when it returns to the height of his head. Take $y = 0$ at the ground.

SET UP: $y_0 = 2.20 \text{ m}$, $y = 1.83 \text{ m}$, $a_y = -9.80 \text{ m/s}^2$, $v_{0y} = +7.59 \text{ m/s}$,

$t = ?$ $y - y_0 = v_{0y}t + \frac{1}{2}a_y t^2$

EXECUTE:

$1.83 \text{ m} - 2.20 \text{ m} = (7.59 \text{ m/s})t + \frac{1}{2}(-9.80 \text{ m/s}^2)t^2 - 0.37 \text{ m}$
$= (7.59 \text{ m/s})t - (4.90 \text{ m/s}^2)t^2$
$4.90t^2 - 7.59t - 0.37 = 0$, with t in seconds.

Use the quadratic formula to solve for t:

$t = \frac{1}{9.80}(7.59 \pm \sqrt{(7.59)^2 - 4(4.90)(-0.37)}) = 0.774 \pm 0.822$

t must be positive, so $t = 0.774 \text{ s} + 0.822 \text{ s} = 1.60 \text{ s}$

EVALUATE: Calculate the time to the maximum height: $v_y = v_{0y} + a_y t$, so $t = (v_y - v_{0y})/a_y = -(7.59 \text{ m/s})/(-9.80 \text{ m/s}^2) = 0.77 \text{ s}$. It also takes 0.77 s to return to 2.2 m above the ground, for a total time of 1.54 s. His head is a little lower than 2.20 m, so it is reasonable for the shot to reach the level of his head a little later than 1.54 s after being thrown; the answer of 1.60 s in part (c) makes sense.

2.85 **IDENTIFY** and **SET UP:** Let $+y$ be upward. Each ball moves with constant acceleration $a_y = -9.80 \text{ m/s}^2$. In parts (c) and (d) require that the two balls be at the same height at the same time.

EXECUTE:

a) At ceiling, $v_y = 0$, $y - y_0 = 3.0 \text{ m}$, $a_y = -9.80 \text{ m/s}^2$. Solve for v_{0y}.
$v_y^2 = v_{0y}^2 + 2a_y(y - y_0)$ gives $v_{0y} = 7.7 \text{ m/s}$.

b) $v_y = v_{0y} + a_y t$ with the information from part (a) gives $t = 0.78 \text{ s}$.

c) Let the first ball travel downward a distance d in time t. It starts from its maximum height, so $v_{0y} = 0$.

$y - y_0 = v_{0y}t + \frac{1}{2}a_y t^2$ gives $d = (4.9 \text{ m/s}^2)t^2$

The second ball has $v_{0y} = \frac{2}{3}(7.7 \text{ m/s}) = 5.1 \text{ m/s}$. In time t it must travel upward $3.0 \text{ m} - d$ to be at the same place as the first ball.

$y - y_0 = v_{0y}t + \frac{1}{2}a_y t^2$ gives $3.0 \text{ m} - d = (5.1 \text{ m/s})t - (4.9 \text{ m/s}^2)t^2.$

We have two equations in two unknowns, d and t. Solving gives $t = 0.59$ s and $d = 1.7$ m.

d) 3.0 m $-d = 1.3$ m

EVALUATE: In 0.59 s the first ball falls $d = (4.9 \text{ m/s}^2)(0.59 \text{ s})^2 = 1.7$ m, so is at the same height as the second ball.

2.89 a) IDENTIFY: Let $+y$ be upward. The can has constant acceleration $a_y = -g$. The initial upward velocity of the can equals the upward velocity of the scaffolding; first find this speed.

SET UP: $y - y_0 = -15.0$ m, $t = 3.25$ s, $a_y = -9.80$ m/s^2, $v_{0y} = ?$

EXECUTE: $y - y_0 = v_{0y}t + \frac{1}{2}a_yt^2$ gives $v_{0y} = 11.31$ m/s

Use this v_{0y} in $v_y = v_{0y} + a_yt$ to solve for v_y: $v_y = -20.5$ m/s

b) IDENTIFY: Find the maximum height of the can, above the point where it falls from the scaffolding:

SET UP: $v_y = 0$, $v_{0y} = +11.31$ m/s, $a_y = -9.80$ m/s^2, $y - y_0 = ?$

EXECUTE: $v_y^2 = v_{0y}^2 + 2a_y(y - y_0)$ gives $y - y_0 = 6.53$ m

The can will pass the location of the other painter. Yes, he gets a chance.

EVALUATE: Relative to the ground the can is initially traveling upward, so it moves upward before stopping momentarily and starting to fall back down.

2.93 IDENTIFY and **SET UP:** Use $v_x = dx/dt$ and $a_x = dv_x/dt$ to calculate $v_x(t)$ and $a_x(t)$ for each car. Use these equations to answer the questions about the motion.

EXECUTE:

$$x_A = \alpha t + \beta t^2, \quad v_{Ax} = \frac{dx_A}{dt} = \alpha + 2\beta t, \quad a_{Ax} = \frac{dv_{Ax}}{dt} = 2\beta$$

$$x_B = \gamma t^2 - \delta t^3, \quad v_{Bx} = \frac{dx_B}{dt} = 2\gamma t - 3\delta t^2, \quad a_{Bx} = \frac{dv_{Bx}}{dt} = 2\gamma - 6\delta t$$

a) IDENTIFY and **SET UP:** The car that initially moves ahead is the one that has the larger v_{0x}.

EXECUTE: At $t = 0$, $v_{Ax} = \alpha$ and $v_{Bx} = 0$. So initially car A moves ahead.

b) IDENTIFY and **SET UP:** Cars at the same point implies $x_A = x_B$.

$\alpha t + \beta t^2 = \gamma t^2 - \delta t^3$

EXECUTE: One solution is $t = 0$, which says that they start from the same point. To find the other solutions, divide by t: $\alpha + \beta t = \gamma t - \delta t^2$

$\delta t^2 + (\beta - \gamma)t + \alpha = 0$

$t = \frac{1}{2\delta}(-(\beta - \gamma) \pm \sqrt{(\beta - \gamma)^2 - 4\delta\alpha}) = \frac{1}{0.40}(+1.60 \pm \sqrt{(1.60)^2 - 4(0.20)(2.60)}) = 4.00 \text{ s} \pm 1.73 \text{ s}$

So $x_A = x_B$ for $t = 0$, $t = 2.27$ s and $t = 5.73$ s.

EVALUATE: Car A has constant, positive a_x. Its v_x is positive and increasing. Car B has $v_{0x} = 0$ and a_x that is initially positive but then becomes negative. Car B initially moves in the $+x$-direction but then slows down and finally reverses direction. At $t = 2.27$ s car B has overtaken car A and then passes it. At $t = 5.73$ s, car B is moving in the $-x$-direction as it passes car A again.

c) IDENTIFY: The distance from A to B is $x_B - x_A$. The rate of change of this distance is $\dfrac{d(x_B - x_A)}{dt}$. If this distance is not changing, $\dfrac{d(x_B - x_A)}{dt} = 0$. But this says $v_{Bx} - v_{Ax} = 0$. (The distance between A and B is neither decreasing nor increasing at the instant when they have the same velocity.)

SET UP: $v_{Ax} = v_{Bx}$ requires $\alpha + 2\beta t = 2\gamma t - 3\delta t^2$

EXECUTE: $3\delta t^2 + 2(\beta - \gamma)t + \alpha = 0$

$t = \frac{1}{6\delta}(-2(\beta - \gamma) \pm \sqrt{4(\beta - \gamma)^2 - 12\delta\alpha}) = \frac{1}{1.20}(-3.20 \pm \sqrt{4(-1.60)^2 - 12(0.20)(2.60)})$

$t = 2.667$ s ± 1.667 s, so $v_{Ax} = v_{Bx}$ for $t = 1.00$ s and $t = 4.33$ s.

EVALUATE: At $t = 1.00 s$, $v_{Ax} = v_{Bx} = 5.00$ m/s. At $t = 4.33$ s, $v_{Ax} = v_{Bx} = 13.0$ m/s. Now car B is slowing down while A continues to speed up, so their velocities aren't ever equal again.

d) IDENTIFY and **SET UP** $a_{Ax} = a_{Bx}$ requires $2\beta = 2\gamma - 6\delta t$

EXECUTE: $t = \dfrac{\gamma - \beta}{3\delta} = \dfrac{2.80 \text{ m/s}^2 - 1.20 \text{ m/s}^2}{3(0.20 \text{ m/s}^3)} = 2.67$ s.

EVALUATE: At $t = 0$, $a_{Bx} > a_{Ax}$, but a_{Bx} is decreasing while a_{Ax} is constant. They are equal at $t = 2.67$ s but for all times after that $a_{Bx} < a_{Ax}$.

CHAPTER 3
MOTION IN TWO OR THREE DIMENSIONS

Exercises 1, 3, 5, 7, 9, 11, 19, 21, 23, 25, 33, 37, 41, 43
Problems 45, 51, 53, 55, 57, 61, 63, 65, 69, 71, 73, 75, 79, 81, 85

Exercises

3.1 **IDENTIFY** and **SET UP:** Use Eq.(3.2), in component form.

EXECUTE:

$$(v_{av})_x = \frac{\Delta x}{\Delta t} = \frac{x_2 - x_1}{t_2 - t_1} = \frac{5.3 \text{ m} - 1.1 \text{ m}}{3.0 \text{ s} - 0} = 1.4 \text{ m/s}$$

$$(v_{av})_y = \frac{\Delta y}{\Delta t} = \frac{y_2 - y_1}{t_2 - t_1} = \frac{-0.5 \text{ m} - 3.4 \text{ m}}{3.0 \text{ s} - 0} = -1.3 \text{ m/s}$$

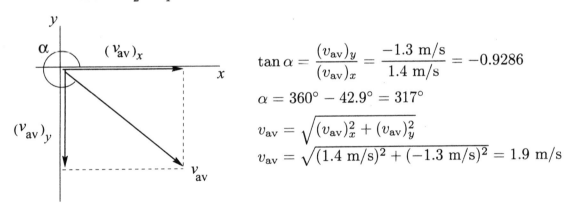

$$\tan \alpha = \frac{(v_{av})_y}{(v_{av})_x} = \frac{-1.3 \text{ m/s}}{1.4 \text{ m/s}} = -0.9286$$

$$\alpha = 360° - 42.9° = 317°$$

$$v_{av} = \sqrt{(v_{av})_x^2 + (v_{av})_y^2}$$

$$v_{av} = \sqrt{(1.4 \text{ m/s})^2 + (-1.3 \text{ m/s})^2} = 1.9 \text{ m/s}$$

EVALUATE: Our calculation gives that $\vec{v}_{av}$ is in the 4th quadrant. This corresponds to increasing x and decreasing y.

3.3 **a) IDENTIFY** and **SET UP:** From $\vec{r}$ we can calculate x and y for any t. Then use Eq.(3.2), in component form.

EXECUTE:

$\vec{r} = [4.0 \text{ cm} + (2.5 \text{ cm/s}^2)t^2]\hat{i} + (5.0 \text{ cm/s})t\hat{j}$

At $t = 0$, $\vec{r} = (4.0 \text{ cm})\hat{i}$.

At $t = 2.0$ s, $\vec{r} = (14.0 \text{ cm})\hat{i} + (10.0 \text{ cm})\hat{j}$.

$$(v_{av})_x = \frac{\Delta x}{\Delta t} = \frac{10.0 \text{ cm}}{2.0 \text{ s}} = 5.0 \text{ cm/s}.$$

$$(v_{av})_y = \frac{\Delta y}{\Delta t} = \frac{10.0 \text{ cm}}{2.0 \text{ s}} = 5.0 \text{ cm/s}.$$

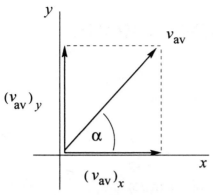

$$v_{av} = \sqrt{(v_{av})_x^2 + (v_{av})_y^2} = 7.1 \text{ cm/s}$$

$$\tan \alpha = \frac{(v_{av})_y}{(v_{av})_x} = 1.00$$

$$\theta = 45°.$$

EVALUATE: Both x and y increase, so $\vec{v}_{av}$ is in the 1st quadrant.

b) IDENTIFY and **SET UP:** Calculate $\vec{r}$ by taking the time derivative of $\vec{r}(t)$.

EXECUTE: $\vec{v} = \dfrac{d\vec{r}}{dt} = ([5.0 \text{ cm/s}^2]t)\hat{i} + (5.0 \text{ cm/s})\hat{j}$

$\underline{t = 0}$: $v_x = 0$, $v_y = 5.0 \text{ cm/s}$; $v = 5.0 \text{ cm/s}$ and $\theta = 90°$

$\underline{t = 1.0 \text{ s}}$: $v_x = 5.0 \text{ cm/s}$, $v_y = 5.0 \text{ cm/s}$; $v = 7.1 \text{ cm/s}$ and $\theta = 45°$

$\underline{t = 2.0 \text{ s}}$: $v_x = 10.0 \text{ cm/s}$, $v_y = 5.0 \text{ cm/s}$; $v = 11 \text{ cm/s}$ and $\theta = 27°$

c) The trajectory is a graph of y versus x.

$x = 4.0 \text{ cm} + (2.5 \text{ cm/s}^2)t^2$, $y = (5.0 \text{ cm/s})t$

For values of t between 0 and 2.0 s, calculate x and y and plot y versus x.

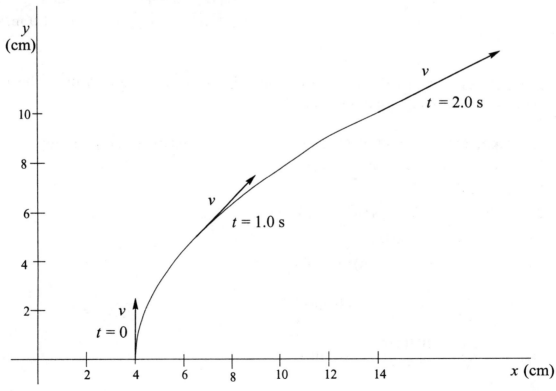

EVALAUTE: The sketch shows that the instantaneous velocity at any t is tangent to the trajectory.

3.5　**IDENTIFY** and **SET UP:** Use Eq.(3.8) in component form to calculate $(a_{\text{av}})_x$ and $(a_{\text{av}})_y$.

EXECUTE:

a) The velocity vectors at $t_1 - 0$ and $t_2 = 30.0$ s are

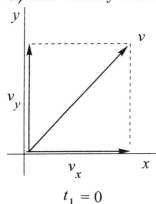 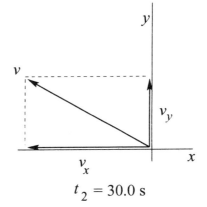

$t_1 = 0$　　　　　　　　　　$t_2 = 30.0$ s

b)　$(a_{\text{av}})_x = \dfrac{\Delta v_x}{\Delta t} = \dfrac{v_{2x} - v_{1x}}{t_2 - t_1} = \dfrac{-170 \text{ m/s} - 90 \text{ m/s}}{30.0 \text{ s}} = -8.67 \text{ m/s}^2$

$(a_{\text{av}})_y = \dfrac{\Delta v_y}{\Delta t} = \dfrac{v_{2y} - v_{1y}}{t_2 - t_1} = \dfrac{40 \text{ m/s} - 110 \text{ m/s}}{30.0 \text{ s}} = -2.33 \text{ m/s}^2$

c)

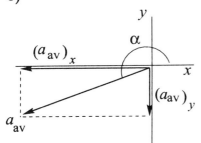

$a = \sqrt{(a_{\text{av}})_x^2 + (a_{\text{av}})_y^2} = 8.98 \text{ m/s}^2$

$\tan \alpha = \dfrac{(a_{\text{av}})_y}{(a_{\text{av}})_x} = \dfrac{-2.33 \text{ m/s}^2}{8.67 \text{ m/s}^2} = 0.269$

$\alpha = 15° + 180° = 195°$

EVALUATE: The changes in v_x and v_y are both in the negative x or y direction, so both components of $\vec{a}_{\text{av}}$ are in the 3rd quadrant.

3.7　**IDENTIFY** and **SET UP:** Use Eqs.(3.4)and (3.12) to find v_x, v_y, a_x, and a_y as functions of time. The magnitude and direction of $\vec{r}$ and $\vec{a}$ can be found once we know their components.

EXECUTE:

a) Calculate x and y for t values in the range 0 to 2.0 s and plot y versus x

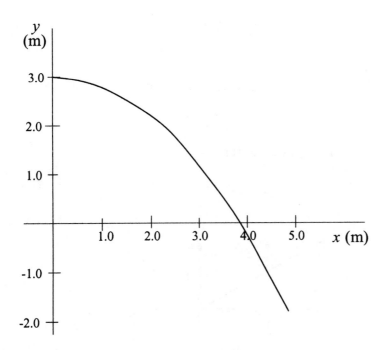

b) $v_x = \dfrac{dx}{dt} = \alpha \qquad v_y = \dfrac{dy}{dt} = -2\beta t$

$a_y = \dfrac{dv_x}{dt} = 0 \qquad a_y = \dfrac{dv_y}{dt} = -2\beta$

Thus $\vec{v} = \alpha\hat{i} - 2\beta t\hat{j} \qquad \vec{a} = -2\beta\hat{j}$

c) <u>velocity:</u> At $t = 2.0$ s, $v_x = 2.4$ m/s, $\quad v_y = -2(1.2 \text{ m/s}^2)(2.0 \text{ s}) = -4.8$ m/s

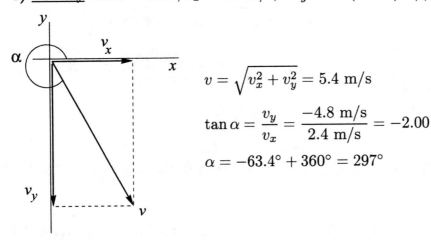

$v = \sqrt{v_x^2 + v_y^2} = 5.4$ m/s

$\tan\alpha = \dfrac{v_y}{v_x} = \dfrac{-4.8 \text{ m/s}}{2.4 \text{ m/s}} = -2.00$

$\alpha = -63.4° + 360° = 297°$

<u>acceleration:</u> At $t = 2.0$ s, $a_x = 0, \quad a_y = -2(1.2 \text{ m/s}^2) = -2.4 \text{ m/s}^2$

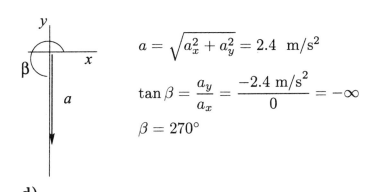

$$a = \sqrt{a_x^2 + a_y^2} = 2.4 \ \text{m/s}^2$$

$$\tan \beta = \frac{a_y}{a_x} = \frac{-2.4 \ \text{m/s}^2}{0} = -\infty$$

$$\beta = 270°$$

d)

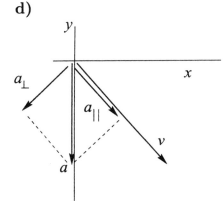

EVALUATE:

$\vec{a}$ has a component $a_{\parallel}$ in the same direction as $\vec{v}$, so we know that v is increasing (the bird is speeding up.)

$\vec{a}$ also has a component $a_{\perp}$ perpendicular to $\vec{v}$, so that the direction of $\vec{v}$ is changing; the bird is turning toward the $-y$-direction (toward the right)

$\vec{v}$ is always tangent to the path; $\vec{v}$ at $t = 2.0$ s shown in part (c) is tangent to the path at this t, conforming to this general rule. $\vec{a}$ is constant and in the $-y$-dirrection; the direction of $\vec{v}$ is turning toward the $-y$-direction.

3.9 **IDENTIFY:** The book moves in projectile motion once it leaves the table top. Its initial velocity is horizontal.

SET UP: Take the positive y-direction to be upward. Take the origin of coordinates at the initial position of the book, at the point where it leaves the table top.

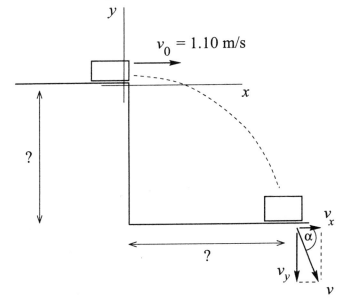

$v_0 = 1.10$ m/s

x-component:
$a_x = 0$, $v_{0x} = 1.10$ m/s,
$t = 0.350$ s

y-component:
$a_y = -9.80$ m/s^2, $v_{0y} = 0$,
$t = 0.350$ s

Use constant acceleration equations for the x and y components of the motion, with $a_x = 0$ and $a_y = -g$.

EXECUTE:

a) $y - y_0 = ?$

$y - y_0 = v_{0y}t + \frac{1}{2}a_y t^2 = 0 + \frac{1}{2}(-9.80 \text{ m/s}^2)(0.350 \text{ s})^2 = -0.600 \text{ m}$. The table top is 0.600 m above the floor.

b) $x - x_0 = ?$

$x - x_0 = v_{0x}t + \frac{1}{2}a_x t^2 = (1.10 \text{ m/s})(0.350 \text{ s}) + 0 = 0.385 \text{ m}$.

c) $v_x = v_{0x} + a_x t = 1.10 \text{ m/s}$ (The x-component of the velocity is constant, since $a_x = 0$.)

$v_y = v_{0y} + a_y t = 0 + (-9.80 \text{ m/s}^2)(0.350 \text{ s}) = -3.43 \text{ m/s}$

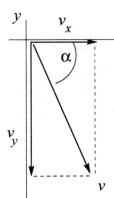

$v = \sqrt{v_x^2 + v_y^2} = 3.60 \text{ m/s}$

$\tan\alpha = \dfrac{v_y}{v_x} = \dfrac{-3.43 \text{ m/s}}{1.10 \text{ m/s}} = -3.118$

$\alpha = -72.2°$

Direction of $\vec{v}$ is 72.2° below the horizontal.

d)

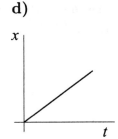 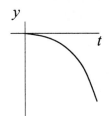 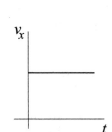 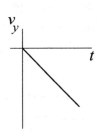

EVALUATE: In the x-direction, $a_x = 0$ and v_x is constant. In the y-direction, $a_y = -9.80 \text{ m/s}^2$ and v_y is downward and increasing in magnitude since a_y and v_y are in the same directions. The x and y motions occur independently, connected only by the time. The time it takes the book to fall 0.600 m is the time it travels horizontally.

3.11   IDENTIFY: Both objects move with projectile motion. Use constant acceleration equations for the x- and y-components.

SET UP: Take $+y$ to be upward.

Use Chirpy's vertical motion to find the height of the cliff.

$v_{0y} = 0$, $a_y = -9.80$ m/s^2, $y - y_0 = -h$, $t = 3.50$ s

EXECUTE: $y - y_0 = v_{0y}t + \frac{1}{2}a_yt^2$ gives $h = 60.0$ m

SET UP: Milada: Use vertical motion to find time in the air.

$y - y_0 = v_{0y}t + \frac{1}{2}a_yt^2$ gives $h = 60.0$ m

EXECUTE: $y - y_0 = v_{0y}t + \frac{1}{2}a_yt^2$ gives -60.0 m $= (0.503$ m/s$)t - (4.90$ m/s$^2)t^2$.

Solving for t and taking the positive root gives $t = 3.55$ s.

Then $v_{0x} = v_0 \cos 32.0°$, $a_x = 0$, $y - y_0 = -h$, $t = 3.55$ s gives $x - x_0 = 2.86$ m.

EVALUATE: We did not need to know Chirpy's initial speed since it is horizontal and doesn't affect his vertical motion. Milada is in the air slightly longer than Chirpy because she has a small upward component of initial velocity.

3.19　IDENTIFY: The baseball moves in projectile motion. In part (c) first calculate the components of the velocity at this point and then get the resultant velocity form its components.

SET UP:

First find the x- and y-components of the initial velocity. Use coordinates where the $+y$-direction is upward, the $+x$-direction is to the right and the origin is at the point where the baseball leaves the bat.

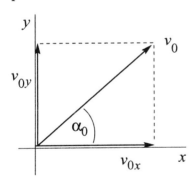

$v_{0x} = v_0 \cos \alpha_0 = (30.0$ m/s$) \cos 36.9° = 24.0$ m/s
$v_{0y} = v_0 \sin \alpha_0 = (30.0$ m/s$) \sin 36.9° = 18.0$ m/s

Use constant acceleration equations for the x and y motions, with $a_x = 0$ and $a_y = -g$.

EXECUTE:

a) *y-component* (vertical motion):

$y - y_0 = +10.0$ m/s,　$v_{0y} = 18.0$ m/s,　$a_y = -9.80$ m/s^2,　$t = ?$

$y - y_0 = v_{0y} + \frac{1}{2}a_yt^2$

10.0 m $= (18.0$ m/s$)t - (4.90$ m/s$^2)t^2$

$(4.90$ m/s$^2)t^2 - (18.0$ m/s$)t + 10.0$ m $= 0$

Apply the quadratic formula: $t = \frac{1}{9.80}[18.0 \pm \sqrt{(-18.0)^2 - 4(4.90)(10.0)}]$ s $= (1.837 \pm 1.154)$ s

The ball is at a height of 10.0 above the point where it left the bat at $t_1 = 0.683$ s and at $t_2 = 2.99$ s. At the earlier time the ball passes through a height of 10.0 m as its way up and at the later time it passes through 10.0 m on its way down.

b) $v_x = v_{0x} = +24.0$ m/s, at all times since $a_x = 0$.

$v_y = v_{0y} + a_y t$

$\underline{t_1 = 0.683\ s}$: $v_y = +18.0$ m/s $+ (-9.80$ m/s$^2)(0.683$ s$) = +11.3$ m/s. (v_y is positive means that the ball is traveling upward at this point.)

$\underline{t_2 = 2.99\ s}$: $v_y = +18.0$ m/s $+ (-9.80$ m/s$^2)(2.99$ s$) = -11.3$ m/s. (v_y is negative means that the ball is traveling downward at this point.)

c) $v_x = v_{0x} = 24.0$ m/s

Solve for v_y:

$v_y = ?$, $y - y_0 = 0$ (when ball returns to height where motion started),

$a_y = -9.80$ m/s^2, $v_{0y} = +18.0$ m/s

$v^2 = v_{0y}^2 + 2a_y(y - y_0)$

$v_y = -v_{0y} = -18.0$ m/s (negative, since the baseball must be traveling downward at this point)

Now that have the components can solve for the magnitude and direction of $\vec{v}$.

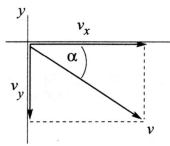

$v = \sqrt{v_x^2 + v_y^2}$

$v = \sqrt{(24.0\ \text{m/s})^2 + (-18.0\ \text{m/s})^2} = 30.0$ m/s

$\tan \alpha = \dfrac{v_y}{v_x} = \dfrac{-18.0\ \text{m/s}}{24.0\ \text{m/s}}$

$\alpha = -36.9°$,   36.9° below the horizontal

The velocity of the ball when it returns to the level where it left the bat has magnitude 30.0 m/s and is directed at an angle of 36.9° below the horizontal.

EVALUATE: The discussion in parts (a)and (b) explains the significance of two values of t for which $y - y_0 = +10.0$ m. When the ball returns to its initial height, our results give that its speed is the same as its initial speed and the angle of its velocity below the horizontal is equal to the angle of its initial velocity above the horizontal; both of these are general results.

3.21 IDENTIFY: Take the origin of coordinates at the point where the quarter leaves your hand and take positive y to be upward. The quarter moves in projectile motion, with $a_x = 0$, and $a_y = -g$. It travels vertically for the time it takes it to travel horizontally 2.1 m.

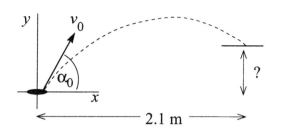

$$v_{0x} = v_0 \cos \alpha_0 = (6.4 \text{ m/s}) \cos 60°$$
$$v_{0x} = 3.20 \text{ m/s}$$

$$v_{0y} = v_0 \sin \alpha_0 = (6.4 \text{ m/s}) \sin 60°$$
$$v_{0y} = 5.54 \text{ m/s}$$

a) SET UP: Use the horizontal (x-component) of motion to solve for t, the time the quarter travels through the air:

$$t = ?, \quad x - x_0 = 2.1 \text{ m}, \quad v_{0x} = 3.2 \text{ m/s}, \quad a_x = 0$$

$$x - x_0 = v_{0x}t + \tfrac{1}{2}a_x t^2 = v_{0x}t, \text{ since } a_x = 0$$

EXECUTE: $t = \dfrac{x - x_0}{v_{0x}} = \dfrac{2.1 \text{ m}}{3.2 \text{ m/s}} = 0.656 \text{ s}$

SET UP: Now find the vertical displacement of the quarter after this time:

$$y - y_0 = ?, \quad a_y = -9.80 \text{ m/s}^2, \quad v_{0y} = +5.54 \text{ m/s}, \quad t = 0.656 \text{ s}$$

$$y - y_0 + v_{0y}t + \tfrac{1}{2}a_y t^2$$

EXECUTE: $y - y_0 = (5.54 \text{ m/s})(0.656 \text{ s}) + \tfrac{1}{2}(-9.80 \text{ m/s}^2)(0.656 \text{ s})^2 =$
3.63 m − 2.11 m =1.5 m.

b) SET UP: $v_y =, \quad t = 0.656 \text{ s}, \quad a_y = -9.80 \text{ m/s}^2, \quad v_{0y} = +5.54 \text{ m/s}$

$$v_y = v_{0y} + a_y t$$

EXECUTE: $v_y = 5.54 \text{ m/s} + (-9.80 \text{ m/s}^2)(0.656 \text{ s}) = -0.89 \text{ m/s}.$

EVALUATE: The minus sign for v_y indicates that the y-component of $\vec{v}$ is downward. At this point the quarter has passed through the highest point in its path and is on its way down. From Eq.(3.20), the horizontal range if it returned to its original height (it doesn't!) would be 3.6 m. It reaches its maximum height after traveling horizontally 1.8 m, so at $x - x_0 = 2.1$ m it is on its way down.

3.23 IDENTIFY: Take the origin of coordinates at the roof and let the $+y$-direction be upward. The rock moves in projectile motion, with $a_x = 0$ and $a_y = -g$. Apply constant acceleration equations for the x and y components of the motion.
SET UP:

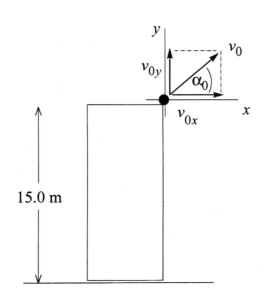

$$v_{0x} = v_0 \cos \alpha_0 = 25.2 \text{ m/s}$$

$$v_{0y} = v_0 \sin \alpha_0 = 16.3 \text{ m/s}$$

a) At the maximum height $v_y = 0$.

$a_y = -9.80 \text{ m/s}^2, \quad v_y = 0, \quad v_{oy} = +16.3 \text{ m/s}, \quad y - y_0 = ?$

$v_y^2 = v_{0y}^2 + 2a_y(y - y_0)$

EXECUTE: $y - y_0 = \dfrac{v_y^2 - v_{0y}^2}{2a_y} = \dfrac{0 - (16.3 \text{ m/s})^2}{2(-9.80 \text{ m/s}^2)} = +13.6 \text{ m}$

b) SET UP: Find the velocity by solving for its x and y components.

$v_x = v_{0x} = 25.2 \text{ m/s}$ (since $a_x = 0$)

$v_y = ?, \quad a_y = -9.80 \text{ m/s}^2, \quad y - y_0 = -15.0 \text{ m}$ (negative because at the ground the rock is below its initial position), $\quad v_{0y} = 16.3 \text{ m/s}$

$v_y^2 = v_{0y}^2 + 2a_y(y - y_0)$

$v_y = -\sqrt{v_{0y}^2 + 2a_y(y - y_0)}$ (v_y is negative because at the ground the rock is traveling downward.)

EXECUTE: $v_y = -\sqrt{(16.3 \text{ m/s})^2 + 2(-9.80 \text{ m/s}^2)(-15.0 \text{ m})} = -23.7 \text{ m/s}$

Then $v = \sqrt{v_x^2 + v_y^2} = \sqrt{(25.2 \text{ m/s})^2 + (-23.7 \text{ m/s})^2} = 34.6 \text{ m/s}$.

c) SET UP: Use the vertical motion (y-component) to find the time the rock is in the air:

$t = ?, \quad v_y = -23.7 \text{ m/s}$ (from part (b)), $\quad a_y = -9.80 \text{ m/s}^2, \quad v_{0y} = +16.3 \text{ m/s}$

EXECUTE: $t = \dfrac{v_y - v_{0y}}{a_y} = \dfrac{-23.7 \text{ m/s} - 16.3 \text{ m/s}}{-9.80 \text{ m/s}^2} = +4.08 \text{ s}$

SET UP: Can use this t to calculate the horizontal range:

$t = 4.08 \text{ s}, \quad v_{0x} = 25.2 \text{ m/s}, \quad a_x = 0, \quad x - x_0 = ?$

EXECUTE: $x - x_0 = v_{0x}t + \frac{1}{2}a_x t^2 = (25.2 \text{ m/s})(4.08 \text{ s}) + 0 = 103 \text{ m}$

d) Graphs of x versus t, y versus t, v_x versus t, and v_y versus t:

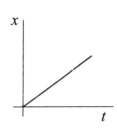 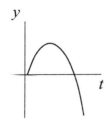 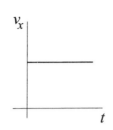 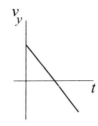

EVALUATE: The time it takes the rock to travel vertically to the ground is the time it has to travel horizontally. With $v_{0y} = +16.3 \text{ m/s}$ the time it takes the rock to return to the level of the roof ($y = 0$) is $t = 2v_{0y}/g = 3.33 \text{ s}$. The time in the air is greater than this because the rock travels an additional 15.0 m to the ground.

3.25 **IDENTIFY** and **SET UP:** The stone moves in projectile motion. Its initial velocity is the same as that of the balloon. Use constant acceleration equations for the x and y components of its motion. Take $+y$ to be upward.

EXECUTE:

a) Use the vertical motion of the rock to find the initial height.

$t = 6.00 \text{ s}$, $v_{0y} = +20.0 \text{ s}$, $a_y = +9.80 \text{ m/s}^2$, $y - y_0 = ?$

$y - y_0 = v_{0y}t + \frac{1}{2}a_y t^2$ gives $y - y_0 = 296 \text{ m}$

b) In 6.00 s the balloon travels downward a distance $y - y_0 = (20.0 \text{ s})(6.00 \text{ s}) = 120$ m. So, its height above ground when the rock hits is 296 m − 120 m = 176 m.

c) The horizontal distance the rock travels in 6.00 s is 90.0 m. The vertical component of the distance between the rock and the basket is 176 m, so the rock is $\sqrt{(176 \text{ m})^2 + (90 \text{ m})^2} = 198 \text{ m}$ from the basket when it hits the ground.

d) (i) The basket has no horizontal velocity, so the rock has horizontal velocity 15.0 m/s relative to the basket.

Just before the rock hits the ground, its vertical component of velocity is $v_y = v_{0y} + a_y t = 20.0 \text{ s} + (9.80 \text{ m/s}^2)(6.00 \text{ s}) = 78.8 \text{ m/s}$, downward, relative to the ground. The basket is moving downward at 20.0 m/s, so relative to the basket the rock has downward component of velocity 58.8 m/s.

e) horizontal: 15.0 m/s; vertical: 78.8 m/s

EVALUATE: The rock has a constant horizontal velocity and accelerates downward

3.33 IDENTIFY: Uniform circular motion.

SET UP: Since the magnitude of $\vec{v}$ is constant, $v_{\text{tan}} = \dfrac{d\,|\vec{v}\,|}{dt} = 0$ and the resultant acceleration is equal to the radial component. At each point in the motion the radial component of the acceleration is directed in toward the center of the circular path and its magnitude is given by v^2/R.

EXECUTE:

a) $a_{\text{rad}} = \dfrac{v^2}{R} = \dfrac{(7.00 \text{ m/s})^2}{14.0 \text{ m}} = 3.50 \text{ m/s}^2$, upward.

b) The radial acceleration has the same magnitude as in part (a), but now the direction toward the center of the circle is downward. The acceleration at this point in the motion is 3.50 m/s^2, downward.

c) **SET UP:** The time to make one rotation is the period T, and the speed v is the distance for one revolution divided by T:

EXECUTE: $v = \dfrac{2\pi R}{T}$ so $T = \dfrac{2\pi R}{v} = \dfrac{2\pi(14.0 \text{ m})}{7.00 \text{ m/s}} = 12.6 \text{ s}$.

EVALUATE: The radial acceleration is constant in magnitude since v is constant and is at every point in the motion directed toward the center of the circular path. The acceleration is perpendicular to $\vec{v}$ and is nonzero because the direction of $\vec{v}$ changes.

3.37 IDENTIFY: Relative velocity problem. The time to walk the length of the moving sidewalk is the length divided by the velocity of the woman relative to the ground.

SET UP: Let W stand for the woman, G for the ground, and S for the sidewalk. Take the positive direction to be the direction in which the sidewalk is moving.

The velocities are $v_{W/G}$ (woman relative to the ground), $v_{W/S}$ (woman relative to the sidewalk), and $v_{S/G}$ (sidewalk relative to the ground).

Eq.(3.33) becomes $v_{W/G} = v_{W/S} + v_{S/G}$.

The time to reach the other end is given by $t = \dfrac{\text{distance traveled relative to ground}}{v_{W/G}}$.

EXECUTE:

a) $v_{S/G} = 1.0 \text{ m/s}$

$v_{W/S} = +1.5 \text{ m/s}$

$v_{W/G} = v_{W/S} + v_{S/G} = 1.5 \text{ m/s} + 1.0 \text{ m/s} = 2.5 \text{ m/s}$.

$t = \dfrac{35.0 \text{ m}}{v_{W/G}} = \dfrac{35.0 \text{ m}}{2.5 \text{ m}} = 14 \text{ s}$.

b) $v_{S/G} = 1.0$ m/s

$v_{W/S} = -1.5$ m/s

$v_{W/G} = v_{W/S} + v_{S/G} = -1.5$ m/s $+ 1.0$ m/s $= -0.5$ m/s. (Since $v_{W/G}$ now is negative, she must get on the moving sidewalk at the opposite end from in part (a).)

$$t = \frac{-35.0 \text{ m}}{v_{W/G}} = \frac{-35.0 \text{ m}}{-0.5 \text{ m}} = 70 \text{ s.}$$

EVALUATE: Her speed relative to the ground is much greater in part (a) when she walks with the motion of the sidewalk.

3.41 IDENTIFY: Relative velocity problem in two dimensions. His motion relative to the earth (time, displacement) depends on his velocity relative to the earth so we must solve for this velocity.

a) SET UP: View the motion from above.

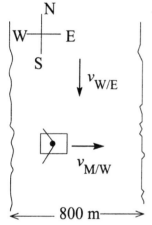

The velocity vectors in the problem are:
$\vec{v}_{M/E}$, the velocity of the man relative to the earth
$\vec{v}_{W/E}$, the velocity of the water relative to the earth
$\vec{v}_{M/W}$, the velocity of the man relative to the water

The rule for adding these velocities is
$\vec{v}_{M/E} = \vec{v}_{M/W} + \vec{v}_{W/E}$

The problem tells us that $\vec{v}_{W/E}$ has magnitude 2.0 m/s and direction due south. It also tells us that $\vec{v}_{M/W}$ has magnitude 4.2 m/s and direction due east.

The vector addition diagram is then

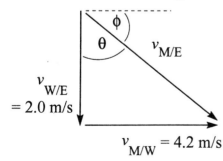

This diagram shows the vector addition
$\vec{v}_{M/E} = \vec{v}_{M/W} + \vec{v}_{W/E}$
and also has $\vec{v}_{M/W}$ and $\vec{v}_{W/E}$ in their specified directions. Note that the vector diagram forms a right triangle.

The Pythagorean theorem applied to the vector addition diagram gives
$v_{M/E}^2 = v_{M/W}^2 + v_{W/E}^2$.

EXECUTE: $v_{M/E} = \sqrt{v_{M/W}^2 + v_{W/E}^2} = \sqrt{(4.2 \text{ m/s})^2 + (2.0 \text{ m/s})^2} = 4.7$ m/s

$\tan\theta = \dfrac{v_{M/W}}{v_{W/E}} = \dfrac{4.2 \text{ m/s}}{2.0 \text{ m/s}} = 2.10; \quad \theta = 25°; \text{ or } \phi = 90° - \theta = 65°.$

The velocity of the man relative to the earth has magnitude 4.7 m/s and direction 25° S of E.

b) This requires careful thought. To cross the river the man must travel 800 m due east relative to the earth. The man's velocity relative to the earth is $\vec{v}_{M/E}$. But, from the vector addition diagram the eastward component of $v_{M/E}$ equals $v_{M/W} = 4.2$ m/s.

Thus $t = \dfrac{x - x_0}{v_x} = \dfrac{800 \text{ m}}{4.2 \text{ m/s}} = 190$ s.

c) The southward component of $\vec{v}_{M/E}$ equals $v_{W/E} = 2.0$ m/s. Therefore, in the 190 s it takes him to cross the river the distance south the man travels relative to the earth is

$y - y_0 = v_y t = (2.0 \text{ m/s})(190 \text{ s}) = 380$ m.

EVALUATE: If there were no currrent he would cross in the same time, (800 m)/(4.2 m/s) = 190 s. The current carries him downstream but doesn't affect his motion in the perpendicular direction, from bank to bank.

3.43 IDENTIFY: Relative velocity problem in two dimensions.

a) SET UP: $\vec{v}_{P/A}$ is the velocity of the plane relative to the air. The problem states that $\vec{v}_{P/A}$ has magnitude 35 m/s and direction south.

$\vec{v}_{A/E}$ is the velocity of the air relative to the earth. The problem states that $\vec{v}_{A/E}$ is to the southwest (45° S of W) and has magnitude 10 m/s.

The relative velocity equation is $\vec{v}_{P/E} = \vec{v}_{P/A} + \vec{v}_{A/E}$.

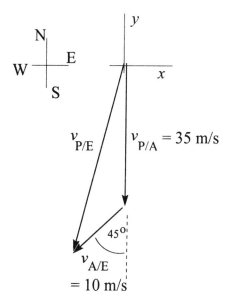

EXECUTE:

b) $(v_{P/A})_x = 0$, $(v_{P/A})_y = -35$ m/s

$(v_{A/E})_x = -(10 \text{ m/s})\cos 45° = -7.07$ m/s,

$(v_{A/E})_y = -(10 \text{ m/s})\sin 45° = -7.07$ m/s

$(v_{P/E})_x = (v_{P/A})_x + (v_{A/E})_x = 0 - 7.07 \text{ m/s} = -7.1$ m/s

$(v_{P/E})_y = (v_{P/A})_y + (v_{A/E})_y = -35 \text{ m/s} - 7.07 \text{ m/s} = -42$ m/s

c)

$$v_{P/E} = \sqrt{(v_{P/E})_x^2 + v_{P/E})_y^2}$$

$$v_{P/E} = \sqrt{(-7.1 \text{ m/s})^2 + (-42 \text{ m/s})^2} = 43 \text{ m/s}$$

$$\tan\phi = \frac{(v_{P/E})_x}{(v_{P/E})_y} = \frac{-7.1}{-42} = 0.169$$

$$\phi = 9.6°; \quad (9.6° \text{ west of south})$$

EVALUATE: The relative velocity addition diagram does not form a right triangle so the vector addition must be done using components. The wind adds both southward and westward components to the velocity of the plane relative to the ground.

Problems

3.45 IDENTIFY: Given $x(t)$, $y(t)$ find the components of the velocity and acceleration by taking derivatives according to Eqs.(3.4) and (3.12).

 SET UP: $x = \alpha t \qquad y = 15.0 \text{ m} - \beta t^2$

a) $v_x = \alpha$ $v_y = -2\beta t$

$a_x = 0$ $a_y = -2\beta$

$\vec{v}$ perpendicular to $\vec{a}$ means that $\vec{v} \cdot \vec{a} = 0$

EXECUTE: $\vec{v} \cdot \vec{a} = v_x a_x + v_y a_y = +4\beta^2 t$

Thus $\vec{v} \cdot \vec{a} = 0$ only for $t = 0$.

EVALUATE: $\vec{a}$ is constant and always in the $-y$-direction. $\vec{v}$ is perpendicular to $\vec{a}$ only when $\vec{v}$ has no y-component, and this is so only at $t = 0$.

b) The speed is instantaneously not changing when there is no component of acceleration in the direction of the velocity, that is, when $\vec{v} \cdot \vec{a} = 0$. From part (a) this happens only at $t = 0$.

c) SET UP: $\vec{v}$ perpendicular to $\vec{r}$ means that $\vec{v} \cdot \vec{r} = 0$

EXECUTE: $\vec{v} \cdot \vec{r} = v_x x + v_y y = \alpha^2 t - 2\beta t(15.0 \text{ m} - \beta t^2) = 0$

$t = 0$ is one root; at $t = 0$, $x = 0$ and $y = 15.0$ m.

The other roots are given by $\alpha^2 - 2\beta(15.0 \text{ m} - \beta t^2) = 0$.

With the numerical values for α and β this equation becomes $0.5t^2 = 13.56 \text{ s}^2$

The positive, physical root is $t = 5.21$ s. At this t, $x = 6.25$ m and $y = 1.44$ m.

EVALUATE: $\vec{r}$ and $\vec{v}$ at this point in the path are shown in the graph of the path in part (e).

d) SET UP: The distance from the origin is given by $r = \sqrt{x^2 + y^2}$. The minumum value of r occurs when $dr/dt = 0$.

EXECUTE:

$$\frac{dr}{dt} = \frac{d\sqrt{x^2 + y^2}}{dt} = (x^2 + y^2)^{-1/2} \left[x\frac{dx}{dt} + y\frac{dy}{dt} \right] = 0.$$

But this says $xv_x + yv_y = 0$, which is the same condition as in part (c). Thus the minimum distance occurs at $t = 5.21$ s and at this t the distance r equals 6.41 m. (Note: For the other solution to this equation, $t = 0$, the distance is 15.0 m, which is larger.)

EVALUATE: At the minimum r there can be no velocity component opposite to the direction of $\vec{r}$; such a component would move the object toward the origin and give a smaller r.

e)

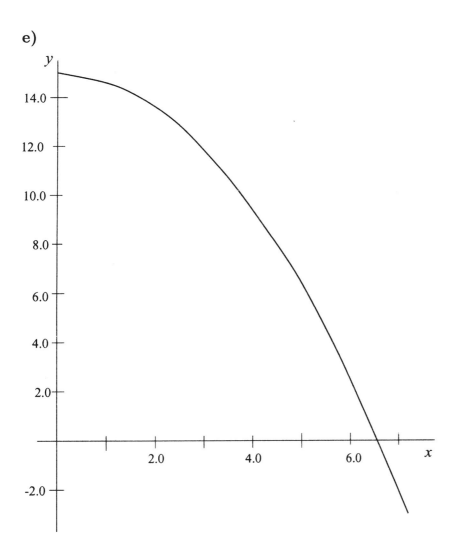

3.51 IDENTIFY: Take $+y$ to be downward. Both objects have the same vertical motion, with v_{0y} and $a_y = +g$. Use constant acceleration equations for the x and y components of the motion.

SET UP: Use the vertical motion to find the time in the air:

$v_{0y} = 0$, $a_y = -9.80$ m/s^2, $y - y_0 = 25$ m, $t = ?$

EXECUTE: $y - y_0 = v_{0y}t + \frac{1}{2}a_y t^2$ gives $t = 2.259$ s

During this time the dart must travel 90 m, so the horizontal component of its velocity must be

$$v_{0x} = \frac{x - x_0}{t} = \frac{90 \text{ m}}{2.25 \text{ s}} = 40 \text{ m/s}$$

EVALUATE: Both objects hit the ground at the same time. The dart hits the monkey for any muzzle velocity greater than 40 m/s.

3.53 IDENTIFY: The cannister moves in projectile motion. Its initial velocity is horizontal. Apply constant acceleration equations for the x and y components of motion.

SET UP:

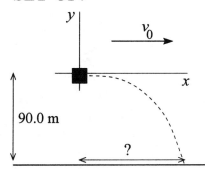

Take the origin of coordinates at the point where the canister is released. Take $+y$ to be upward. The initial velocity of the canister is the velocity of the plane, 64.0 m/s in the $+x$-direction.

Use the vertical motion to find the time of fall:

$t = ?$, $v_{0y} = 0$, $a_y = -9.80 \text{ m/s}^2$, $y - y_0 = -90.0$ m (When the canister reaches the ground it is 90.0 m <u>below</u> the origin.)

$y - y_0 = v_{0y}t + \frac{1}{2}a_y t^2$

EXECUTE: Since $v_{0y} = 0$, $t = \sqrt{\dfrac{2(y - y_0)}{a_y}} = \sqrt{\dfrac{2(-90.0 \text{ m})}{-9.80 \text{ m/s}^2}} = 4.286$ s.

SET UP: Then use the horizontal component of the motion to calculate how far the canister falls in this time:

$x - x_0 = ?$, $a_x = 0$, $v_{0x} = 64.0$ m/s,

EXECUTE: $x - x_0 = v_0 t + \frac{1}{2}at^2 = (64.0 \text{ m/s})(4.286 \text{ s}) + 0 = 274$ m.

EVALUATE: The time it takes the cannister to fall 90.0 m, starting from rest, is the time it travels horizontally at constant speed.

3.55 **IDENTIFY:** Projectile motion problem.

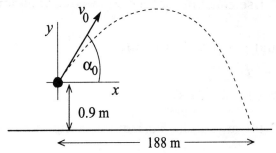

Take the origin of coordinates at the point where the ball leaves the bat, and take $+y$ to be upward.

$v_{0x} = v_0 \cos \alpha_0$
$v_{0y} = v_0 \sin \alpha_0$,
but we don't know v_0.

Write down the equation for the horizontal displacement when the ball hits the ground and the corresponding equation for the vertical displacement. The time t is the same for both components, so this will give us two equations in two unknowns (v_0 and t).

a) SET UP: <u>y-component</u>:

$a_y = -9.80 \text{ m/s}^2$, $y - y_0 = -0.9$ m, $v_{0y} = v_0 \sin 45°$

$y - y_0 = v_{0y}t + \frac{1}{2}a_y t^2$

EXECUTE: $-0.9 \text{ m} = (v_0 \sin 45°)t + \frac{1}{2}(-9.80 \text{ m/s}^2)t^2$

SET UP: x-component:

$a_x = 0, \quad x - x_0 = 188 \text{ m}, \quad v_{0x} = v_0 \cos 45°$

$x - x_0 = v_{0x}t + \frac{1}{2}a_x t^2$

EXECUTE: $t - \dfrac{x - x_0}{v_{0x}} = \dfrac{188 \text{ m}}{v_0 \cos 45°}$

Put the expression for t from the x-component motion into the y-component equation and solve for v_0. (Note that $\sin 45° = \cos 45°$.)

$$-0.9 \text{ m} = (v_0 \sin 45°)\left(\frac{188 \text{ m}}{v_0 \cos 45°}\right) - (4.90 \text{ m/s}^2)\left(\frac{188 \text{ m}}{v_0 \cos 45°}\right)^2$$

$$4.90 \text{ m/s}^2 \left(\frac{188 \text{ m}}{v_0 \cos 45°}\right)^2 = 188 \text{ m} + 0.9 \text{ m} = 188.9 \text{ m}$$

$$\left(\frac{v_0 \cos 45°}{188 \text{ m}}\right)^2 = \frac{4.90 \text{ m/s}^2}{188.9 \text{ m}}, \quad v_0 = \left(\frac{188 \text{ m}}{\cos 45°}\right)\sqrt{\frac{4.90 \text{ m/s}^2}{188.9 \text{ m}}} = 42.8 \text{ m/s}$$

b) Use the horizontal motion to find the time it takes the ball to reach the fence:

SET UP: x-component:

$x - x_0 = 116 \text{ m}, \quad a_x = 0, \quad v_{0x} = v_0 \cos 45° = (42.8 \text{ m/s}) \cos 45° = 30.3 \text{ m/s},$
$t = ?$

$x - x_0 = v_{0x}t + \frac{1}{2}a_x t^2$

EXECUTE: $t = \dfrac{x - x_0}{v_{0x}} = \dfrac{116 \text{ m}}{30.3 \text{ m/s}} = 3.83 \text{ s}$

SET UP: Find the vertical displacement of the ball at this t:

y-component:

$y - y_0 = ?, \quad a_y = -9.80 \text{ m/s}^2, \quad v_{0y} = v_0 \sin 45° = 30.3 \text{ m/s}, \quad t = 3.83 \text{ s}$

$y - y_0 = v_{0y}t + \frac{1}{2}a_y t^2$

EXECUTE: $y - y_0 = (30.3 \text{ s})(3.83 \text{ s}) + \frac{1}{2}(-9.80 \text{ m/s}^2)(3.83 \text{ s})^2$

$y - y_0 = 116.0 \text{ m} - 71.9 \text{ m} = +44.1 \text{ m}$, above the point where the ball was hit. The height of the ball above the ground is 44.1 m + 0.90 m = 45.0 m. It's height then above the top of the fence is 45.0 m − 3.0 m = 42.0 m.

EVALUATE: With $v_0 = 42.8 \text{ m/s}$, $v_{0y} = 30.3 \text{ m/s}$ and it takes the ball 6.18 s to return to the height where it was hit and only slightly longer to reach a point 0.9 m below this height. $t = (188 \text{ m})/(v_0 \cos 45°)$ gives $t = 6.21 \text{ s}$, which agrees with this estimate. The ball reaches its maximum height approximately $(188 \text{ m})/2 = 94$ m from home plate, so at the fence the ball is not far past its maximum height of 47.6 m, so a height of 45.0 m at the fence is reasonable.

3.57 **IDENTIFY:** Projectile motion. Take $+y$ to be upward and the origin of coordinates at the floor. Neither the launch angle α_0 nor the initial speed v_0 are specified. We want the maximum height to be D, so the ball just barely misses hitting the ceiling.

SET UP: First, use the vertical motion to solve for α_0:

$y - y_0 = D,$ $v_{0y} = v_0 \sin \alpha_0 = \sqrt{6gD} \sin \alpha_0,$ $v_y = 0,$ $a_y = -g$

$v_y^2 = v_{0y}^2 + 2a_y(y - y_0)$

EXECUTE: $0 = 6gD \sin^2 \alpha_0 - 2gD$

This gives $\sin \alpha_0 = \sqrt{1/3}$. Note: $\cos^2 \alpha_0 = 1 - \sin^2 \alpha_0 = 2/3$, so $\cos \alpha_0 = \sqrt{2/3}$.

SET UP: Now use the vertical motion to express the time in the air, t in terms of v_0. After time t the ball has returned to the floor and $y - y_0 = 0$.

EXECUTE: $y - y_0 = v_{0y}t + \frac{1}{2}a_yt^2$

$0 = v_0 \sin \alpha_0 - \frac{1}{2}gt,$ so $t = \dfrac{2v_0 \sin \alpha_0}{g} = \dfrac{2v_0}{\sqrt{3}g}.$

SET UP: In the horizontal motion $a_x = 0$ and $v_{0x} = v_0 \cos \alpha_0$.

$x - x_0 = v_{0x}t + \frac{1}{2}a_xt^2 = (v_0 \cos \alpha_0)t = (v_0 \cos \alpha_0)\left(\dfrac{2v_0 \sin \alpha_0}{g}\right)$

EXECUTE: $x - x_0 = \left(\dfrac{2\sin \alpha_0 \cos \alpha_0}{g}\right)(6gD) =$

$(12 \sin \alpha_0 \cos \alpha_0)D = 12(\sqrt{1/3})(\sqrt{2/3})D = 4\sqrt{2}D.$

EVALUATE: $\cos \alpha_0 = \sqrt{2/3}$ so $\sin \alpha_0 = \sqrt{1 - \cos^2 \alpha_0} = 1\sqrt{3}$. Then the horizontal range formula from Example 3.10 gives $x - x_0 = R = (2v_0^2 \sin \alpha_0 \cos \alpha_0)/g = 4\sqrt{2}D$, which checks.

3.61 **a) IDENTIFY** and **SET UP:** Use the equation derived in Example 3.10:

$R = (v_0 \cos \alpha_0)\left(\dfrac{2v_0 \sin \alpha_0}{g}\right)$

Call the range R_1 when the angle is α_0 and R_2 when the angle is $90° - \alpha$.

$R_1 = (v_0 \cos \alpha_0)\left(\dfrac{2v_0 \sin \alpha_0}{g}\right)$

$R_2 = (v_0 \cos(90° - \alpha_0))\left(\dfrac{2v_0 \sin(90° - \alpha_0)}{g}\right)$

The problem asks us to show that $R_1 = R_2$.

EXECUTE: We can use the trig identities in appendix B to show:

$\cos(90° - \alpha_0) = \cos(\alpha_0 - 90°) = \sin \alpha_0$

$\sin(90° - \alpha_0) = -\sin(\alpha_0 - 90°) = -(-\cos \alpha_0) = +\cos \alpha_0$

Thus $R_2 = (v_0 \sin \alpha_0) \left(\dfrac{2v_0 \cos \alpha_0}{g} \right) = (v_0 \cos \alpha_0) \left(\dfrac{2v_0 \sin \alpha_0}{g} \right) = R_1.$

b) $R = \dfrac{v_0^2 \sin 2\alpha_0}{g}$ so $\sin 2\alpha_0 = \dfrac{Rg}{v_0^2} = \dfrac{(0.25 \text{ m})(9.80 \text{ m/s}^2)}{(2.2 \text{ m/s})^2}.$

This gives $\alpha = 15°$ or $75°$.

EVALUATE: $R = (v_0^2 \sin 2\alpha_0)/g$, so the result in part (a) requires that $\sin^2(2\alpha_0) = \sin^2(180° - 2\alpha_0)$, which is true. (Try some values of α_0 and see!)

3.63 a) IDENTIFY: Projectile motion.

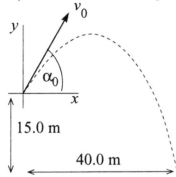

Take the origin of coordinates at the top of the ramp and take $+y$ to be upward. The problem specifies that the object is displaced 40.0 m to the right when it is 15.0 m below the origin.

We don't know t, the time in the air, and we don't know v_0. Write down the equations for the horizontal and vertical displacements. Combine these two equations to eliminate one unknown.

SET UP: y-component:

$y - y_0 = -15.0 \text{ m}, \quad a_y = -9.80 \text{ m/s}^2, \quad v_{0y} = v_0 \sin 53.0°$

$y - y_0 = v_{0y}t + \frac{1}{2}a_y t^2$

EXECUTE: $-15.0 \text{ m} = (v_0 \sin 53.0°)t - (4.90 \text{ m/s}^2)t^2$

SET UP: y-component:

$x - x_0 = 40.0 \text{ m}, \quad a_x = 0, \quad v_{0x} = v_0 \cos 53.0°$

$x - x_0 = v_{0x}t + \frac{1}{2}a_x t^2$

EXECUTE: $40.0 \text{ m} = (v_0 t) \cos 53.0°$

The second equation says $v_0 t = \dfrac{40.0 \text{ m}}{\cos 53.0°} = 66.47 \text{ m}.$

Use this to replace $v_0 t$ in the first equation:

$$-15.0 \text{ m} = (66.47 \text{ m})\sin 53° - (4.90 \text{ m/s}^2)t^2$$

$$t = \sqrt{\frac{(66.46 \text{ m})\sin 53° + 15.0 \text{ m}}{4.90 \text{ m/s}^2}} = \sqrt{\frac{68.08 \text{ m}}{4.90 \text{ m/s}^2}} = 3.727 \text{ s}.$$

Now that we have t we can use the x-component equation to solve for v_0:

$$v_0 = \frac{40.0 \text{ m}}{t\cos 53.0°} = \frac{40.0 \text{ m}}{(3.727 \text{ s})\cos 53.0°} = 17.8 \text{ m/s}.$$

EVALUATE: Using these values of v_0 and t in the $y - y_0 = v_{0y} + \frac{1}{2}a_y t^2$ equation verifies that $y - y_0 = -15.0$ m.

b) IDENTIFY: $v_0 = (17.8 \text{ m/s})/2 = 8.9$ m/s

This is less than the speed required to make it to the other side, so he lands in the river.

Use the vertical motion to find the time it takes him to reach the water:

SET UP: $y - y_0 = -100$ m; $v_{0y} = +v_0 \sin 53.0° = 7.11$ m/s; $a_y = -9.80$ m/s^2

$y - y_0 = v_{0y}t + \frac{1}{2}a_y t^2$ gives $-100 = 7.11t - 4.90t^2$

EXECUTE: $4.90t^2 - 7.11t - 100 = 0$ and $t = \frac{1}{9.80}(7.11 \pm \sqrt{(7.11)^2 - 4(4.90)(-100)})$

$t = 0.726 \text{ s} \pm 4.57$ s so $t = 5.30$ s.

The horizontal distance he travels in this time is

$x - x_0 = v_{0x}t = (v_0 \cos 53.0°)t = (5.36 \text{ m/s})(5.30 \text{ s}) = 28.4$ m.

He lands in the river a horizontal distance of 28.4 m from his launch point.

EVALUATE: He has half the minimum speed and makes it only about halfway across.

3.65 IDENTIFY and **SET UP:** Take $+y$ to be upward. The rocket moves with projectile motion, with $v_{0y} = +40.0$ m/s and $v_{0x} = 30.0$ m/s relative to the ground. The vertical motion of the rocket is unaffected by its horizontal velocity.

EXECUTE:

a) $v_y = 0$ (at maximum height), $v_{0y} = +40.0$ m/s, $a_y = -9.80$ m/s^2, $y - y_0 = ?$

$v_y^2 = v_{0y}^2 + 2a_y(y - y_0)$ gives $y - y_0 = 81.6$ m

b) Both the cart and the rocket have the same constant horizontal velocity, so both travel the same horizontal distance while the rocket is in the air and the rocket lands in the cart.

c) Use the vertical motion of the rocket to find the time it is in the air.

$v_{0y} = 40$ m/s, $a_y = -9.80$ m/s^2, $v_y = -40$ m/s, $t = ?$

$v_y = v_{0y} + a_y t$ gives $t = 8.164$ s

Then $x - x_0 = v_{0x}t = (30.0 \text{ m/s})(8.164 \text{ s}) = 245 \text{ m}$.

d) Relative to the ground the rocket has initial velocity components $v_{0x} = 30.0 \text{ m/s}$ and $v_{0y} = 40.0 \text{ m/s}$, so it is traveling at $53.1°$ above the horizontal.

e) (i)

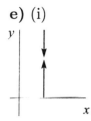

Relative to the cart, the rocket travels straight up and then straight down

(ii)

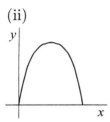

Relative to the ground the rocket travels in a parabola.

EVALUATE: Both the cart and rocket have the same constant horizontal velocity. The rocket lands in the cart.

3.69 IDENTIFY: The shell moves in projectile motion. To find the horizontal distance between the tanks we must find the horizontal velocity of one tank realtive to the other. Take $+y$ to be upward.

a) SET UP: The vertical motion of the shell is unaffected by the horizontal motion of the tank. Use the vertical motion of the shell to find the time the shell is in the air:

$v_{0y} = v_0 \sin \alpha = 43.4 \text{ m/s}$, $a_y = -9.80 \text{ m/s}^2$, $y - y_0 = 0$ (returns to initial height), $t = ?$

EXECUTE: $y - y_0 = v_{0y}t + \frac{1}{2}a_y t^2$ gives $t = 8.86 \text{ s}$

SET UP: Consider the motion of one tank relative to the other.

EXECUTE: Relative to tank #1 the shell has a constant horizontal velocity $v_0 \cos \alpha = 246.2 \text{ m/s}$. Relative to the ground the horizontal velocity component is $246.2 \text{ m/s} + 15.0 \text{ m/s} = 261.2 \text{ m/s}$. Relative to tank #2 the shell has horizontal velocity component $261.2 \text{ m/s} - 35.0 \text{ m/s} = 226.2 \text{ m/s}$. The distance between the tanks when the shell was fired is the $(226.2 \text{ m/s})(8.86 \text{ s}) = 2000 \text{ m}$ that the shell travels relative to tank #2 during the 8.86 s that the shell is in the air.

b) The tanks are initially 2000 m apart. In 8.86 s tank #1 travels 133 m and tank #2 travels 310 m, in the same direction. Therefore, their separation increases by 310 m $-$ 183 m $= 177$ m. So, the separation becomes 2180 m (rounding to 3 significant figures).

EVALUATE: The retreating tank has greater speed than the approaching tank, so they move farther apart while the shell is in the air. We can also calculate the separation in part (b) as the relative speed of the tanks times the time the shell is in the air: $(3.5 \text{ m/s} - 15.0 \text{ m/s})(8.86 \text{ s}) = 177 \text{ m}$.

3.71 IDENTIFY: Projectile Motion.

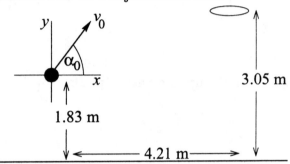

Take the origin of coordinates at the point where the player releases the ball.

$$v_{0x} = v_0 \cos \alpha_0$$
$$v_{0y} = v_0 \sin \alpha_0$$

We are first given v_0 and α_0 and are asked about the subsequent motion.

a) SET UP: y-component:

$v_y = 0$ (at the maximum height), $v_{0y} = v_0 \sin \alpha_0 = (4.88 \text{ m/s}) \sin 35° = 2.80 \text{ m/s}$,
$a_y = -9.80 \text{ m/s}^2$, $y - y_0 = ?$
$v_y^2 = v_{0y}^2 + 2a_y(y - y_0)$

EXECUTE: $y - y_0 = \dfrac{v_y^2 - v_{0y}^2}{2a_y} = \dfrac{0 - (2.80 \text{ m/s})^2}{2(-9.80 \text{ m/s}^2)} = +0.400 \text{ m}$

The height above the floor then is $0.400 \text{ m} + 1.83 \text{ m} = 2.23 \text{ m}$.

b) SET UP: Use the vertical motion to find the time the ball is in the air:

y-component:

$y - y_0 = -1.83 \text{ m}$ (vertical displacement when the ball reaches the floor),
$v_{0y} = 2.80 \text{ m/s}$, $a_y = -9.80 \text{ m/s}^2$, $t = ?$
$y - y_0 = v_{0y}t + \frac{1}{2}a_yt^2$

EXECUTE: $-1.83 \text{ m} = (2.80 \text{ m/s})t - (4.90 \text{ m/s}^2)t^2$
$4.90t^2 - 2.80t - 1.83 = 0$, with t in seconds.

The quadratic formula gives

$t = \dfrac{1}{9.80}\left(2.80 \pm \sqrt{(2.80)^2 + 4(4.90)(1.83)}\right) \text{ s} = 0.286 \text{ s} \pm 0.675 \text{ s}$.

t must be positive, so $t = 0.286 \text{ s} + 0.675 \text{ s} = 0.961 \text{ s}$

SET UP: Find the horizontal displacement for this t:

$x - x_0 = ?$, $v_{0x} = v_0 \cos \alpha_0 = (4.88 \text{ m/s}) \cos 35° = 4.00 \text{ m/s}$, $a_x = 0$,

$t = 0.961$ s

EXECUTE: $x - x_0 = v_{0x}t + \frac{1}{2}a_x t^2 = (4.00 \text{ m/s})(0.961 \text{ s}) = 3.84$ m

EVALUATE: The ball hits the floor before it gets to the basket.

c) IDENTIFY: We don't know either v_0 or the time t for the ball to reach the basket. Write down the equations for the horizontal and vertical displacements to get two equations for these two unknowns. Specify that the ball is 3.05 m above the floor when it has traveled horizontally 4.21 m.

SET UP: y-component:

$a_y = -9.80 \text{ m/s}^2, \quad v_{0y} = v_0 \sin 35°, \quad y - y_0 = 3.05 \text{ m} - 1.83 \text{ m} = 1.22 \text{ m},$

$y - y_0 = v_{0y}t + \frac{1}{2}a_y t^2$

EXECUTE: $1.22 \text{ m} = (v_0 \sin 35°)t - (4.90 \text{ m/s}^2)t^2$

SET UP: x-component:

$a_x = 0, \quad v_{0x} = v_0 \cos 35°, \quad x - x_0 = 4.21 \text{ m}$

$x - x_0 = v_{0x}t + \frac{1}{2}a_x t^2$

EXECUTE: Since $a_x = 0$ this equation gives $v_0 t = \dfrac{x - x_0}{\cos 35°} = \dfrac{4.21 \text{ m}}{\cos 35°} = 5.139$ m

Use this result in the y-component equation:

$1.22 \text{ m} = (5.139 \text{ m}) \sin 35° - (4.90 \text{ m/s}^2)t^2$

$(4.90 \text{ m/s}^2)t^2 = 1.728$ m

$t = \sqrt{\dfrac{1.728 \text{ m}}{4.90 \text{ m/s}^2}} = 0.594$ s

Then $v_0 = \dfrac{5.139 \text{ m}}{t} = \dfrac{5.139 \text{ m}}{0.594 \text{ s}} = 8.65$ m/s.

EVALUATE: This v_0 is considerably larger than v_0 for the failed attempt.

d) IDENTIFY: We know from part (c) that $v_0 = 8.65$ m/s.

Use the y-component motion to find the maximum height above the floor:

SET UP: $v_y = 0$ (at maximum height), $a_y = -9.80 \text{ m/s}^2$,

$v_{0y} = v_0 \sin 35° = (8.65 \text{ m/s}) \sin 35° = 4.96 \text{ m/s}, \quad y - y_0 = ?$

$v_y^2 = v_{0y}^2 + 2a_y(y - y_0)$

EXECUTE: $y - y_0 = \dfrac{v_y^2 - v_{0y}^2}{2a_y} = \dfrac{0 - (4.96 \text{ m/s})^2}{2(-9.80 \text{ m/s}^2)} = +1.26$ m.

The maximum height above the floor is given by 1.83 m + 1.26 m = 3.09 m.

Also use the y-component to find the time t to reach the maximum height. Can then use this t in the x-component equations to find the horizontal displacement at this point.

$$v_y = v_{0y} + a_y t, \text{ so } t = \frac{v_y - v_{0y}}{a_y} = \frac{0 - 4.96 \text{ m/s}}{-9.80 \text{ m/s}^2} = +0.506 \text{ s}.$$

$x - x_0 = ?, \quad a_x = 0, \quad v_{0x} = v_0 \cos 35° = (8.65 \text{ m/s}) \cos 35° = 7.09 \text{ m/s}, \quad t = 0.506 \text{ s}$

$x - x_0 = v_{0x}t + \frac{1}{2}a_x t^2 = (7.09 \text{ m/s})(0.506 \text{ s}) + 0 = 3.59 \text{ m}$

This is the distance of the ball from the point where it was released. Its distance from the basket is $4.21 \text{ m} - 3.59 \text{ m} = 0.62 \text{ m}$.

EVALUATE: The ball passes through the basket just after it passes its maximum height.

3.73 IDENTIFY: There are two stages to the motion: while the engines fire and then the projectile motion after they shut off. The position and velocity at the end of the first stage are the initial values for the second stage.

SET UP: Take $+y$ to be upward. The trajectory of the rocket is:

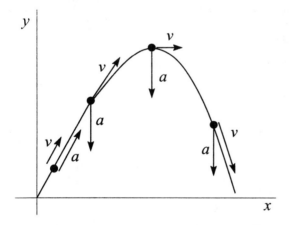

EXECUTE:

a) Find $v_x(t)$ and $v_y(t)$. Note that $\sin \alpha_0 = 4/5$ and $\cos \alpha_0 = 3/5$.

time 0 to time T:

$a_x = g \cos \alpha_0 = (3/5)g \qquad v_{0x} = 0, \quad v_x = v_{0x} = a_x t = (3/5)gt$

$a_y = g \sin \alpha_0 = (4/5)g \qquad v_{0y} = 0, \quad v_y = v_{0y} = a_y t = (4/5)gt$

at time T: $v_x = (3/5)gT \qquad v_y = (4/5)gT$

time T to end:

$a_x = 0 \qquad v_x = (3/5)gT$ (constant during this time interval)

$a_y = -g \qquad v_y = v_{0y} + a_y t = (4/5)gT - g(t - T)$ (Note that t is the time measured from when the rocket is first fired, so $t - T$ is the elapsed time for this part of the motion.)

b) The graphs of $v_x(t)$ and $v_y(t)$ are thus

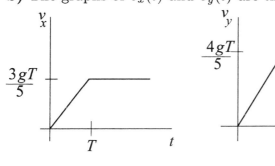

c) **SET UP:** The maximum altitude is reached after the engines shut off, so occurs at a time t that is greater than T.

Set $v_y = 0$ and solve for t: $(4/5)gT - g(t - T) = 0$, $\quad t = (4/5)T + T = (9/5)T$.

The acceleration is constant in the time interval $t = 0$ to $t = T$ and again in the interval $t = T$ to the end. But the acceleration changes at $t = T$ so the constant acceleration equations don't apply for the entire time period of the motion.

EXECUTE: First use $y - y_0 = v_{0y}t + \frac{1}{2}a_y t^2$ in the $t = 0$ to $t = T$ interval to solve for y_T, the altitude at time $t = T$: $y_T = 0 + \frac{1}{2}a_y T^2 = \frac{1}{2}(4/5)gT^2 = (2/5)gT^2$.

For the second part of the motion the initial time is $t = T$ and the constant acceleration equation is $y - y_T = v_{Ty}t + \frac{1}{2}a_y(t - T)^2$. At the maximum height $t - T = (4/5)T$.

$y - (2/5)gT^2 = [(4/5)gT][(4/5)T] - \frac{1}{2}g[(4/5)T]^2$

This gives $y = (18/25)gT^2$.

d) We know that $t = (9/5)T$ when the rocket is at the maximum height and that this maximum height is $y = (18/25)gT^2$. Apply the constant acceleration equations for the y-component of the motion from the maximum height to the ground to find the time t when the rocket reaches the ground.

$y - y_0 = -(18/25)gT^2$, $\quad t = ?$, $\quad a_y = -g$, $\quad v_{0y} = 0$

Since $v_{0y} = 0$ (starts at maximum height), the equation $y - y_0 = v_{0y}t + \frac{1}{2}a_y t^2$ gives

$$t = \sqrt{\frac{2(y - y_0)}{a_y}} = \sqrt{\frac{36T^2}{25}} = (6/5)T.$$

$(6/5)T$ is the time from the maximum height to the ground and $(9/5)T$ is the time from the start of the motion to the maximum height, so the total time in the air is $t = 3T$. The rocket travels for time T in the first constant acceleration segment and for time $2T$ in the second constant acceleration segment.

Find x at $t = T$, at the end of the first segment:

$x - x_0 = v_{0x}t + \frac{1}{2}a_x t^2$ gives $x = 0 + \frac{1}{2}(3/5)gT^2 = (3/10)gT^2$

The horizontal displacement during the second segment, from $t = T$ to $t = 3T$ and where $a_x = 0$, is

$x - x_0 = v_{Tx}t = [(3/5)gT][3T - T] = (6/5)gT^2$.

The total horizontal displacement is the sum, $(3/10)gT^2 + (6/5)gT^2 = (3/2)gT^2$.

EVALUATE: Note that both the maximum height and the horizontal range are proportional to T^2.

3.75 **IDENTIFY** and **SET UP:** Use Eqs. (3.4) and (3.12) to get the velocity and acceleration components from the position components.

EXECUTE: $x = R\cos\omega t, \quad y = R\sin\omega t$

a) $r = \sqrt{x^2 + y^2} = \sqrt{R^2\cos^2\omega t + R^2\sin^2\omega t} = \sqrt{R^2(\sin^2\omega t + \cos^2\omega t)} = \sqrt{R^2} = R$,

since $\sin^2\omega t + \cos^2\omega t = 1$.

b) $v_x = \dfrac{dx}{dt} = -R\omega\sin\omega t, \quad v_y = \dfrac{dy}{dt} = R\omega\cos\omega t$

$\vec{v} \cdot \vec{r} = v_x x + v_y y = (-R\omega\sin\omega t)(R\cos\omega t) + (R\omega\cos\omega t)(R\sin\omega t)$

$\vec{v} \cdot \vec{r} = R^2\omega(-\sin\omega t\cos\omega t + \sin\omega t\cos\omega t) = 0$, so $\vec{v}$ is perpendicular to $\vec{r}$.

c) $a_x = \dfrac{dv_x}{dt} = -R\omega^2\cos\omega t = -\omega^2 x$

$a_y = \dfrac{dv_y}{dt} = -R\omega^2\sin\omega t = -\omega^2 y$

$a = \sqrt{a_x^2 + a_y^2} = \sqrt{\omega^4 x^2 + \omega^4 y^2} = \omega^2\sqrt{x^2 + y^2} = R\omega^2$.

$\vec{a} = a_x\hat{i} + a_y\hat{j} = -\omega^2(x\hat{i} + y\hat{j}) = -\omega^2\vec{r}$.

Since ω^2 is positive this means that the direction of $\vec{a}$ is opposite to the direction of $\vec{r}$.

d) $v = \sqrt{v_x^2 + v_y^2} = \sqrt{R^2\omega^2\sin^2\omega t + R^2\omega^2\cos^2\omega t} = \sqrt{R^2\omega^2(\sin^2\omega t + \cos^2\omega t)}$

$v = \sqrt{R^2\omega^2} = R\omega$.

e) $a = R\omega^2, \quad \omega = v/R, \quad$ so $a = R(v^2/R^2) = v^2/R$.

EVALUATE: The rock moves in uniform circular motion. The position vector is radial, the velocity is tangential, and the acceleration is radially inward.

3.79 **IDENTIFY:** Relative velocity problem. Find the velocity of Harry relative to Larry and use this to find the time it takes for the twins to meet. This is the time the pigeon flies.

SET UP:

The trick is to find the time it takes for the twins to meet. Since the pigeon flies at constant speed, the total distance the pigeon flies is just this speed times the time it flies.

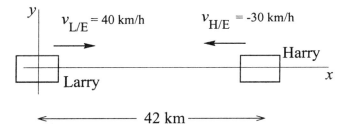

Work in a coordinate system where the velocity of Harry is $v_{H/L}$, the velocity of Harry relative to Larry. Let $v_{L/E} = +40$ km/h be the velocity of Larry relative to the earth and let $v_{H/E} = -30$ km/h be the velocity of Harry relative to the earth (in the opposite direction).

EXECUTE: The relative velocity formula (Eq.(3.36)) gives $v_{H/E} = v_{H/L} + v_{L/E}$, or

$$v_{H/L} = v_{H/E} - v_{L/E} = -30 \text{ km/h} - 40 \text{ km/h} = -70 \text{ km/h}.$$

Then $x - x_0 = vt$ gives $t = \dfrac{x - x_0}{v} = \dfrac{-42 \text{ km}}{-70 \text{ km/h}} = +0.600$ h.

The distance the pigeon flies in this time is

distance = (speed)(time) = (50 km/h)(0.600 h) = 30 km.

EVALUATE: The twins move toward each other, so their speed of approach is the sum of their speeds. Our solution does not tell us where the twins meet and it doesn't tell us how many round trips between the twins the pigeon makes.

3.81 IDENTIFY: Relative velocity problem. The plane's motion relative to the earth is determined by its velocity relative to the earth.

SET UP:

Select a coordinate system where $+y$ is north and $+x$ is east.

The velocity vectors in the problem are:

$\vec{v}_{P/E}$, the velocity of the plane relative to the earth.

$\vec{v}_{P/A}$, the velocity of the plane relative to the air (the magnitude $v_{P/A}$ is the air speed of the plane and the direction of $\vec{v}_{P/A}$ is the compass course set by the pilot).

$\vec{v}_{A/E}$, the velocity of the air relative to the earth (the wind velocity).

The rule for combining relative velocities gives $\vec{v}_{P/E} = \vec{v}_{P/A} + \vec{v}_{A/E}$.

a) We are given the following information about the relative velocities:

$\vec{v}_{P/A}$ has magnitude 220 km/h and its direction is west. In our coordinates it has components $(v_{P/A})_x = -220$ km/h and $(v_{P/A})_y = 0$.

From the displacement of the plane relative to the earth after 0.500 h, we find that $\vec{v}_{P/E}$ has components in our coordinate system of

$$(v_{P/E})_x = -\frac{120 \text{ km}}{0.500 \text{ h}} = -240 \text{ km/h (west)}$$

$$(v_{P/E})_y = -\frac{20 \text{ km}}{0.500 \text{ h}} = -40 \text{ km/h (south)}$$

With this information the diagram corresponding to the velocity addition equation is

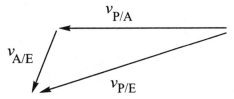

We are asked to find $\vec{v}_{A/E}$, so solve for this vector:

$\vec{v}_{P/E} = \vec{v}_{P/A} + \vec{v}_{A/E}$ gives $\vec{v}_{A/E} = \vec{v}_{P/E} - \vec{v}_{P/A}$.

EXECUTE: The x-component of this equation gives

$(v_{A/E})_x = (v_{P/E})_x - (v_{P/A})_x = -240 \text{ km/h} - (-220 \text{ km/h}) = -20 \text{ km/h}$.

The y-component of this equation gives

$(v_{A/E})_y = (v_{P/E})_y - (v_{P/A})_y = -40 \text{ km/h}$.

Now that we have the components of $\vec{v}_{A/E}$ we can find its magnitude and direction.

$$v_{A/E} = \sqrt{(v_{A/E})_x^2 + (v_{A/E})_y^2}$$

$$v_{A/E} = \sqrt{(-20 \text{ km/h})^2 + (-40 \text{ km/h})^2} = 44.7 \text{ km/h}$$

$$\tan\phi = \frac{40 \text{ km/h}}{20 \text{ km/h}} = 2.00; \quad \phi = 63.4°$$

The direction of the wind velocity is 63.4° S of W, or 26.6° W of S.

EVALUATE: The plane heads west. It goes farther west than it would without wind and also travels south, so the wind velocity has components west and south.

b) SET UP: The rule for combining the relative velocities is still $\vec{v}_{P/E} = \vec{v}_{P/A} + \vec{v}_{A/E}$, but some of these velocities have different values than in part (a).

$\vec{v}_{P/A}$ has magnitude 220 km/h but its direction is to be found.

$\vec{v}_{A/E}$ has magnitude 40 km/h and its direction is due south.

The direction of $\vec{v}_{P/E}$ is west; its magnitude is not given.

The vector diagram for $\vec{v}_{P/E} = \vec{v}_{P/A} + \vec{v}_{A/E}$ and the specified directions for the vectors is

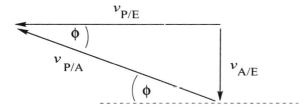

The vector addition diagram forms a right triangle.

EXECUTE: $\sin\phi = \dfrac{v_{A/E}}{v_{P/A}} = \dfrac{40\ \text{km/h}}{220\ \text{km/h}} = 0.1818; \quad \phi = 10.5°.$

The pilot should set her course 10.5° north of west.

EVALUATE: The velocity of the plane relative to the air must have a northward component to counteract the wind and a westward component in order to travel west.

3.85 IDENTIFY: Relative velocity problem.

SET UP: The three relative velocities are:

$\vec{v}_{J/G}$, Juan relative to the ground. This velocity is due north and has magnitude $v_{J/G} = 8.00\ \text{m/s}$.

$\vec{v}_{B/G}$, the ball relative to the ground. This vector is 37.0° east of north and has magnitude $v_{B/G} = 12.0\ \text{m/s}$.

$\vec{v}_{B/J}$, the ball relative to Juan. We are asked to find the magnitude and direction of this vector.

The relative velocity addition equation is $\vec{v}_{B/G} = \vec{v}_{B/J} + \vec{v}_{J/G}$, so $\vec{v}_{B/J} = \vec{v}_{B/G} - \vec{v}_{J/G}$.

The relative velocity addition diagram does not form a right triangle so we must do the vector addition using components.

Take $+y$ to be north and $+x$ to be east.

EXECUTE:

$v_{B/Jx} = +v_{B/G} \sin 37.0° = 7.222\ \text{m/s}$

$v_{B/Jy} = +v_{B/G} \cos 37.0° - v_{J/G} = 1.584\ \text{m/s}$

These two components give $v_{B/J} = 7.39\ \text{m/s}$ at 12.4° north of east.

EVALUATE: Since Juan is running due north, the ball's eastward component of velocity relative to him is the same as its eastward component relative to the earth. The northward component of velocity for Juan and the ball are in the same direction, so the component for the ball relative to Juan is the difference in their components of velocity relative to the ground.

CHAPTER 4
NEWTON'S LAWS OF MOTION

Exercises 5, 11, 13, 15, 17, 21, 23
Problems 31, 33, 35, 37, 39, 41, 43, 45, 49

Exercises

4.5 **IDENTIFY:** Vector addition.

SET UP: Use a coordinate system where the $+x$-axis is in the direction of $\vec{F}_A$, the force applied by dog A.

EXECUTE:

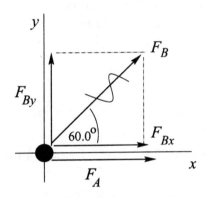

$F_{Ax} = +270 \text{ N}, \quad F_{Ay} = 0$

$F_{Bx} = F_B \cos 60.0° = (300 \text{ N}) \cos 60.0° = +150 \text{ N}$
$F_{By} = F_B \sin 60.0° = (300 \text{ N}) \sin 60.0° = +260 \text{ N}$

$\vec{R} = \vec{F}_A + \vec{F}_B$
$R_x = F_{Ax} + F_{Bx} = +270 \text{ N} + 150 \text{ N} = +420 \text{ N}$
$R_y = F_{Ay} + F_{By} = 0 + 260 \text{ N} = +260 \text{ N}$

$R = \sqrt{R_x^2 + R_y^2}$
$R = \sqrt{(420 \text{ N})^2 + (260 \text{ N})^2} = 494 \text{ N}$

$\tan \theta = \dfrac{R_y}{R_x} = 0.619$

$\theta = 31.8°$

EVALUATE: The forces must be added as vectors. The magnitude of the resultant force is less than the sum of the magnitudes of the two forces and depends on the angle between the two forces.

4.11 **IDENTIFY** and **SET UP:** Use Newton's second law in component form (Eq.4.8) to calculate the acceleration produced by the force. Use constant acceleration equations to calculate the effect of the acceleration on the motion.

EXECUTE:

a) During this time interval the acceleration is constant and equal to

$$a_x = \frac{F_x}{m} = \frac{0.250 \text{ N}}{0.160 \text{ kg}} = 1.562 \text{ m/s}^2$$

We can use the constant acceleration kinematic equations from Chapter 2.

$$x - x_0 = v_{0x}t + \tfrac{1}{2}a_x t^2 = 0 + \tfrac{1}{2}(1.562 \text{ m/s}^2)(2.00 \text{ s})^2,$$

so the puck is at $x = 3.12$ m.

$$v_x = v_{0x} + a_x t = 0 + (1.562 \text{ m/s}^2)(2.00 \text{ s}) = 3.12 \text{ m/s}.$$

b) In the time interval from $t = 2.00$ s + 5.00 s the force has been removed so the acceleration is zero. The speed stays constant at $v_x = 3.12$ m/s. The distance the puck travels is $x - x_0 = v_{0x}t = (3.12 \text{ s})(5.00 \text{ s} - 2.00 \text{ s}) = 9.36$ m. At the end of the interval it is at $x = x_0 + 9.36$ m $= 12.5$ m.

In the time interval from $t = 5.00$ s to 7.00 s the acceleration is again $a_x = 1.562$ m/s^2. At the start of this interval $v_{0x} = 3.12$ m/s and $x_0 = 12.5$ m.

$$x - x_0 = v_{0x}t + \tfrac{1}{2}a_x t^2 = (3.12 \text{ m/s})(2.00 \text{ s}) + \tfrac{1}{2}(1.562 \text{ m/s}^2)(2.00 \text{ s})^2.$$

$$x - x_0 = 6.24 \text{ m} + 3.12 \text{ m} = 9.36 \text{ m}.$$

Therefore, at $t = 7.00$ s the puck is at $x = x_0 + 9.36$ m $= 12.5$ m $+ 9.36$ m $=$ 21.9 m.

$$v_x = v_{0x} + a_x t = 3.12 \text{ m/s} + (1.562 \text{ m/s}^2)(2.00 \text{ s}) = 6.24 \text{ m/s}$$

EVALUATE: The acceleration says the puck gains 1.56 m/s of velocity for every second the force acts. The force acts a total of 4.00 s so the final velocity is (1.56 m/s)(4.00 s) = 6.24 m/s.

4.13 **IDENTIFY** and **SET UP:** A net force produces an acceleration in the direction of the force. An acceleration produces a change of velocity in the direction of the acceleration.

EXECUTE and **EVALAUTE:**

a) Constant velocity implies $\vec{a} = 0$, so $\sum \vec{F} = 0$; the resultant force is zero; the vector sum of all forces on the puck is zero.

b)

Constant $\vec{v}$ says the puck travels in a straight line.

c) The new force produces a constant acceleration that is in the direction perpendicular to the line connecting points A and B. The path is a parabola, just as in projectile motion.

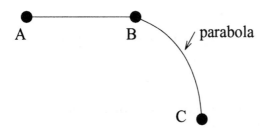

d) Now the acceleration is always perpendicular to the instantaneous velocity $\vec{v}$, so the puck moves in a circular path with constant speed.

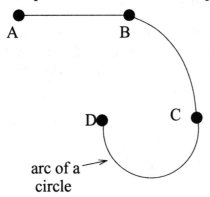

4.15 IDENTIFY and **SET UP:** $F = ma$. We must use $w = mg$ to find the mass of the boulder.

EXECUTE: $m = \dfrac{w}{g} = \dfrac{2400 \text{ N}}{9.80 \text{ m/s}^2} = 244.9 \text{ kg}$

Then $F = ma = (244.9 \text{ kg})(12.0 \text{ m/s}^2) = 2940 \text{ N}$.

EVALUATE: We must use mass in Newton's second law. Mass and weight are proportional.

4.17 IDENTIFY AND SET UP: $w = mg$. The mass of the watermelon is constant, independent of its location. Its weight differs on earth and Jupiter's moon. Use the information about the watermelon's weight on earth to calculate its mass:

EXECUTE: $w = mg$ gives that $m = \dfrac{w}{g} = \dfrac{44.0 \text{ N}}{9.80 \text{ m/s}^2} = 4.49 \text{ kg}$.

On Jupiter's moon, $m = 4.49$ kg, the same as on earth. Thus the weight on Jupiter's moon is $w = mg = (4.49 \text{ kg})(1.81 \text{ m/s}^2) = 8.13 \text{ N}$.

EVALUATE: The weight of the watermelon is less on Io, since g is smaller there.

4.21 IDENTIFY: Identify the forces on the bottle.

SET UP: Classify forces as contact or noncontact forces. The noncontact force is gravity and the contact forces come from things that touch the object. Gravity is always directed downward toward the center of the earth. Air resistance is always directed opposite to the velocity of the object relative to the air.

EXECUTE: a) The free-body diagram for the bottle is

The only forces on the bottle are
gravity (downward) and air resistance (upward).

b)

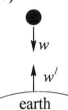

w is the force of gravity that the earth exerts
on the bottle. The reaction to this force is w',
force that the bottle exerts on the earth

Note that these two equal and opposite forces produce very different accelerations because the bottle and the earth have very different masses.

F_{air} is the force that the air exerts on the bottle and is upward. The reaction to this force is a downward force F'_{air} that the bottle exerts on the air. These two forces have equal magnitudes and opposite directions.

EVALUATE: The only thing in contact with the bottle while it is falling is the air. Newton's third law always deals with forces on two different objects.

4.23 IDENTIFY: Apply Newton's second law to the earth.

SET UP: The force of gravity that the earth exerts on her is her weight, $w = mg = (45 \text{ kg})(9.8 \text{ m/s}^2) = 441$ N. By Newton's 3rd law, she exerts an equal and opposite force on the earth.

Apply $\sum \vec{F} = m\vec{a}$ to the earth, with $|\sum \vec{F}| = w = 441$ N, but must use the mass of the earth for m.

EXECUTE: $a = \dfrac{w}{m} = \dfrac{441 \text{ N}}{6.0 \times 10^{24} \text{ kg}} = 7.4 \times 10^{-23} \text{ m/s}^2.$

EVALUATE: This is <u>much</u> smaller than her acceleration of 9.8 m/s^2. The force she exerts on the earth equals in magnitude the force the earth exerts on her, but the acceleration the force produces depends on the mass of the object and her mass is <u>much</u> less than the mass of the earth.

Problems

4.31 IDENTIFY: Vector addition problem. Write the vector addition equation in component form. We know one vector and its resultant and are asked to solve for the other vector.

SET UP: Use coordinates with the $+x$-axis along $\vec{F}_1$ and the $+y$-axis along $\vec{R}$.

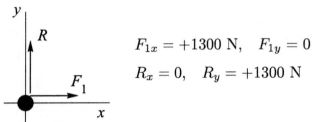

$$F_{1x} = +1300 \text{ N}, \quad F_{1y} = 0$$

$$R_x = 0, \quad R_y = +1300 \text{ N}$$

$$\vec{F}_1 + \vec{F}_2 = \vec{R}, \quad \text{so } \vec{F}_2 = \vec{R} - \vec{F}_1$$

EXECUTE:

$$F_{2x} = R_x = F_{1x} = 0 - 1300 \text{ N} = -1300 \text{ N}$$
$$F_{2y} = R_y - F_{1y} = +1300 \text{ N} - 0 = +1300 \text{ N}$$

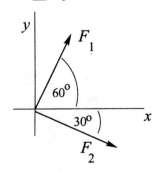

$$F_2 = \sqrt{F_{2x}^2 + F_{2y}^2} = \sqrt{(-1300 \text{ N})^2 + (1300 \text{ N})^2}$$
$$F = 1840 \text{ N}$$

$$\tan \theta = \frac{F_{2y}}{F_{2x}} = \frac{+1300 \text{ N}}{-1300 \text{ N}} = -1.00$$

$$\theta = 135°$$

The magnitude of $\vec{F}_2$ is 1840 N and its direction is 135° counterclockwise from the direction of $\vec{F}_1$.

EVALUATE: $\vec{F}_2$ has a negative x-component to cancel $\vec{F}_1$ and a y-component to equal $\vec{R}$.

4.33 a) IDENTIFY: If the box moves in the $+x$-direction it must have $a_y = 0$, so $\sum F_y = 0$.

The smallest force the child can exert and still produce such motion is a force that makes the y-components of all three forces sum to zero, but that doesn't have any x-component.

SET UP: Let $\vec{F}_3$ be the force exerted by the child.

$\sum F_y = ma_y$ implies $F_{1y} + F_{2y} + F_{3y} = 0$, so $F_{3y} = -(F_{1y} + F_{2y})$.

EXECUTE:

$F_{1y} = +F_1 \sin 60° = (100 \text{ N}) \sin 60° = 86.6 \text{ N}$

$F_{2y} = +F_2 \sin(-30°) = -F_2 \sin 30° = -(140 \text{ N}) \sin 30° = -70.0 \text{ N}$

Then $F_{3y} = -(F_{1y} + F_{2y}) = -(86.6 \text{ N} - 70.0 \text{ N}) = -16.6 \text{ N}$; $F_{3x} = 0$

The smallest force the child can exert has magnitude 17 N and is directed at 90° clockwise from the $+x$-axis shown in the figure.

b) IDENTIFY and SET UP: Apply $\sum F_x = ma_x$. We know the forces and a_x so can solve for m. The force exerted by the child is in the $-y$-direction and has no x-component.

EXECUTE:

$F_{1x} = F_1 \cos 60° = 50 \text{ N}$

$F_{2x} = F_2 \cos 30° = 121.2 \text{ N}$

$\sum F_x = F_{1x} + F_{2x} = 50 \text{ N} + 121.2 \text{ N} = 171.2 \text{ N}$

$m = \dfrac{\sum F_x}{a_x} = \dfrac{171.2 \text{ N}}{2.00 \text{ m/s}^2} = 85.6 \text{ kg}$

Then $w = mg = 840 \text{ N}$.

EVALUATE: In part (b) we don't need to consider the y-component of Newton's second law. $a_y = 0$ so the mass doesn't appear in the $\sum F_y = ma_y$ equation.

4.35 IDENTIFY: We can apply constant acceleration equations to relate the kinematic variables and we can use Newton's second law to relate the forces and acceleration.

a) SET UP: First use the information given about the height of the jump to calculate the speed he has at the instant his feet leave the ground. Use a coordinate system with the $+y$-axis upward and the origin at the position when his feet leave the ground.

$v_y = 0$ (at the maximum height), $v_{0y} = ?$, $a_y = -9.80 \text{ m/s}^2$, $y - y_0 = +1.2 \text{ m}$

$v_y^2 = v_{0y}^2 + 2a_y(y - y_0)$

EXECUTE: $v_{0y} = \sqrt{-2a_y(y - y_0)} = \sqrt{-2(-9.80 \text{ m/s}^2)(1.2 \text{ m})} = 4.85 \text{ m/s}$

b) SET UP: Now consider the acceleration phase, from when he starts to jump until when his feet leave the ground. Use a coordinate system where the $+y$-axis is upward and the origin is at his position when he starts his jump.

EXECUTE: Calculate the average acceleration:

$$(a_{av})_y = \frac{v_y - v_{0y}}{t} = \frac{4.85 \text{ m/s} - 0}{0.300 \text{ s}} = 16.2 \text{ m/s}^2$$

c) SET UP: Finally, find the average upward force that the ground must exert on him to produce this average upward acceleration. (Don't forget about the downward force of gravity.)

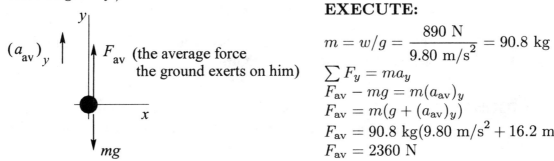

EXECUTE:

$$m = w/g = \frac{890 \text{ N}}{9.80 \text{ m/s}^2} = 90.8 \text{ kg}$$

$$\sum F_y = ma_y$$
$$F_{av} - mg = m(a_{av})_y$$
$$F_{av} = m(g + (a_{av})_y)$$
$$F_{av} = 90.8 \text{ kg}(9.80 \text{ m/s}^2 + 16.2 \text{ m/s}^2)$$
$$F_{av} = 2360 \text{ N}$$

This is the average force exerted on him by the ground. But by Newton's 3rd law, the average force he exerts on the ground is equal and opposite, so is 2360 N, downward.

EVALUATE: In order for him to accelerate upward, the ground must exert an upward force greater than his weight.

4.37 IDENTIFY: Apply Newton's second law to calculate a.

a) SET UP: The free-body diagram for the bucket is

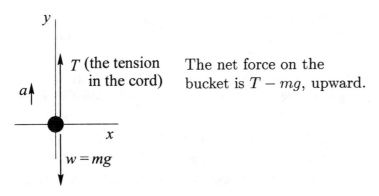

T (the tension in the cord)   The net force on the bucket is $T - mg$, upward.

b) EXECUTE: $\sum F_y = ma_y$ gives $T - mg = ma$

$$a = \frac{T - mg}{m} = \frac{75.0 \text{ N} - (4.80 \text{ kg})(9.80 \text{ m/s}^2)}{4.80 \text{ kg}} = \frac{75.0 \text{ N} - 47.04 \text{ N}}{4.80 \text{ kg}} = 5.82 \text{ m/s}^2.$$

EVALUATE: The weight of the bucket is 47.0 N. The upward force exerted by the cord is larger than this, so the bucket accelerates upward.

4.39 IDENTIFY: Use Newton's 2nd law to relate the acceleration and forces for each crate.

a) SET UP: Since the crates are connected by a rope, they both have the same acceleration, 2.50 m/s^2.

b) Forces on the 4.00 kg crate:

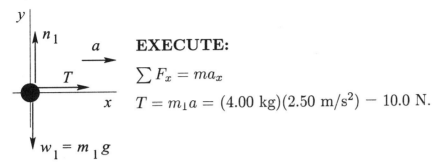

EXECUTE:

$\sum F_x = ma_x$

$T = m_1 a = (4.00 \text{ kg})(2.50 \text{ m/s}^2) - 10.0 \text{ N}.$

c) SET UP: Forces on the 6.00 kg crate:

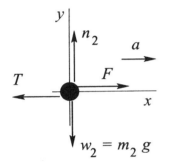

The crate accelerates to the right, so the net force is to the right. F must be larger than T.

d) EXECUTE: $\sum F_x = ma_x$ gives $F - T = m_2 a$

$F = T + m_2 a = 10.0 \text{ N} + (6.00 \text{ kg})(2.50 \text{ m/s}^2) = 10.0 \text{ N} + 15.0 \text{ N} = 25.0 \text{ N}$

EVALUATE: We can also consider the two crates and the rope connecting them as a single object of mass $m = m_1 + m_2 = 10.0$ kg. The free-body diagram is

$\sum F_x = ma_x$

$F = ma = (10.0 \text{ kg})(2.50 \text{ m/s}^2) = 25.0 \text{ N}$

This agrees with our answer in part (d).

4.41 IDENTIFY: and **SET UP:** Take derivatives of $x(t)$ to find v_x and a_x.
Use Newton's second law to relate the acceleration to the net force on the object.

EXECUTE:

a) $x = (9.0 \times 10^3 \text{ m/s}^2)t^2 - (8.0 \times 10^4 \text{ m/s}^3)t^3$

$x = 0$ at $t = 0$

When $t = 0.025$ s, $x = (9.0 \times 10^3 \text{ m/s}^2)(0.025 \text{ s})^2 - (8.0 \times 10^4 \text{ m/s}^3)(0.025\text{s})^3 = 4.4$ m.

The length of the barrel must be 4.4 m.

b) $v_x = \dfrac{dx}{dt} = (18.0 \times 10^3 \text{ m/s}^2)t - (24.0 \times 10^4 \text{ m/s}^3)t^2$

At $t = 0$, $v_x = 0$ (object starts from rest).

At $t = 0.025$ s, when the object reaches the end of the barrel,

$v_x = (18.0 \times 10^3 \text{ m/s}^2)(0.025 \text{ s}) - (24.0 \times 10^4 \text{ m/s}^3)(0.025 \text{ s})^2 = 300$ m/s

c) $\sum F_x = ma_x$, so must find a_x.

$a_x = \dfrac{dv_x}{dt} = 18.0 \times 10^3 \text{ m/s}^2 - (48.0 \times 10^4 \text{ m/s}^3)t$

(i) At $t = 0$, $a_x = 18.0 \times 10^3 \text{ m/s}^2$ and $\sum F_x = (1.50 \text{ kg})(18.0 \times 10^3 \text{ m/s}^2) = 2.7 \times 10^4$ N.

(ii) At $t = 0.025$ s, $a_x = 18.0 \times 10^3 \text{ m/s}^2 - (48.0 \times 10^4 \text{ m/s}^3)(0.025 \text{ s}) = 6.0 \times 10^3 \text{ m/s}^2$ and $\sum F_x = (1.50 \text{ kg})(6.0 \times 10^3 \text{ m/s}^2) = 9.0 \times 10^3$ N.

EVALUATE: The acceleration and net force decrease as the object moves along the barrel.

4.43   **IDENTIFY:** Use Newton's second law, in each case applied to one specific car or group of cars.

a) SET UP: Consider all four cars together as one object. The horizontal force on this combined object is the force of the engine on the first car. The mass of all four cars together is $4m$.

EXECUTE:

$\sum F_x = ma_x$

$F = 4ma$

b) SET UP: Treat the last three cars together as one object. The horizontal force on this combined object is the force of the first car on the second car. The mass of these three cars together is $3m$.

EXECUTE:

$\sum F_x = ma_x$

$F = 3ma$

c) SET UP: Treat the last two cars together. Their total mass is $2m$.

EXECUTE:

$$\sum F_x = ma_x$$

$$F = 2ma$$

d) SET UP: By Newton's third law the force of the fourth car on the third car is equal and opposite to the force of the third car on the fourth. Free-body diagram for the fourth car:

EXECUTE:

$$\sum F_x = ma_x$$

$$F = ma$$

e) The forces would all be the same magnitude but would be in the opposite direction.

EVALUATE: All cars have the same acceleration. We could also look at the net force on each single car and verify that in each case it is equal to ma. for example, the second car has a force $3m$ to the right (exerted by the first car) and a force $2ma$ to the left (exerted by the third car).

4.45 IDENTIFY: Apply $\sum \vec{F} = m\vec{a}$ to the elevator to relate the forces on it to the acceleration.

a) SET UP:

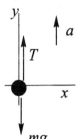

The net force is $T - mg$ (upward).

Take the $+y$-direction to be upward since that is the direction of the acceleration. The maximum upward acceleration is obtained from the maximum possible tension in the cables.

EXECUTE:

$\sum F_y = ma_y$ gives $T - mg = ma$

$$a = \frac{T - mg}{m} = \frac{28,000 \text{ N} - (2200 \text{ kg})(9.80 \text{ m/s}^2)}{2200 \text{ kg}} = 2.93 \text{ m/s}^2.$$

b) What changes is the weight mg of the elevator.

$$a = \frac{T - mg}{m} = \frac{28,000 \text{ N} - (2200 \text{ kg})(1.62 \text{ m/s}^2)}{2200 \text{ kg}} = 11.1 \text{ m/s}^2.$$

EVALUATE: The cables can give the elevator a greater acceleration on the moon since the downward force of gravity is less there and the same T then gives a greater net force.

4.49 IDENTIFY: Note that in this problem the mass of the rope is given, and that it is not negligible compared to the other masses. Apply $\sum \vec{F} = m\vec{a}$ to each object to relate the forces to the acceleration.

a) SET UP:

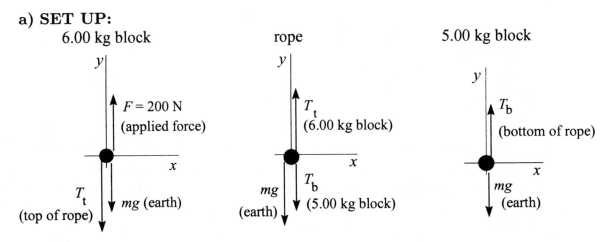

T_t is the tension at the top of the rope and T_b is the tension at the bottom of the rope.

EXECUTE:

b) Treat the rope and the two blocks together as a single object, with mass $m =$ 6.00 kg + 4.00 kg + 5.00 kg = 15.0 kg. Take $+y$ upward, since the acceleration is upward.

$$\sum F_y = ma_y$$
$$F - mg = ma$$
$$a = \frac{F - mg}{m}$$
$$a = \frac{200 \text{ N} - (15.0 \text{ kg})(9.80 \text{ m/s}^2)}{15.0 \text{ kg}} = 3.53 \text{ m/s}^2$$

c) Consider the forces on the top block ($m = 6.00$ kg), since the tension at the top of the rope (T_t) will be one of these forces.

$$\sum F_y = ma_y$$
$$F - mg - T_t = ma$$
$$T_t = F - m(g + a)$$
$$T = 200 \text{ N} - (6.00 \text{ kg})(9.80 \text{ m/s}^2 + 3.53 \text{ m/s}^2) = 120 \text{ N}$$

Alternatively, can consider the forces on the combined object rope plus bottom block ($m = 9.00$ kg):

$$\sum F_y = ma_y$$
$$T_t - mg = ma$$
$$T_t = m(g + a) = 9.00 \text{ kg}(9.80 \text{ m/s}^2 + 3.53 \text{ m/s}^2) = 120 \text{ N},$$
which checks

d) One way to do this is to consider the forces on the top half of the rope ($m = 2.00$ kg). Let T_m be the tension at the midpoint of the rope.

$$\sum F_y = ma_y$$
$$T_t - T_m - mg = ma$$
$$T_m = T_t - m(g + a) = 120 \text{ N} - 2.00 \text{ kg}(9.80 \text{ m/s}^2 + 3.53 \text{ m/s}^2) = 93.3 \text{ N}$$

To check this answer we can alternatively consider the forces on the bottom half of the rope plus the lower block taken together as a combined object ($m = 2.00$ kg + 5.00 kg = 7.00 kg):

$$\sum F_y = ma_y$$
$$T_m - mg = ma$$
$$T_m = m(g + a) = 7.00 \text{ kg}(9.80 \text{ m/s}^2 + 3.53 \text{ m/s}^2) = 93.3 \text{ N},$$
which checks

EVALUATE: The tension in the rope is not constant but increases from the bottom of the rope to the top. The tension at the top of the rope must accelerate the rope as well the 5.00-kg block. The tension at the top of the rope is less than F; there must be a net upward force on the 6.00-kg block.

CHAPTER 5
APPLICATIONS OF NEWTON'S LAWS

Exercises

5.3 **IDENTIFY:** Apply Newton's 1st law to the person. Each half of the rope exerts a force on him, directed along the rope and equal to the tension T in the rope.

SET UP:

a) Force diagram for the person:

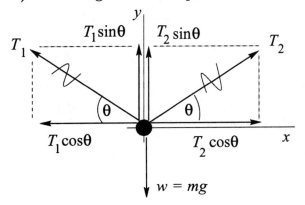

T_1 and T_2 are the tensions in each half of the rope.

EXECUTE: $\sum F_x = 0$

$T_2 \cos\theta - T_1 \cos\theta = 0$

This says that $T_1 = T_2 = T$ (The tension is the same on both sides of the person.)

$\sum F_y = 0$

$T_1 \sin\theta + T_2 \sin\theta - mg = 0$

But $T_1 = T_2 = T$, so $2T \sin\theta = mg$

$$T = \frac{mg}{2\sin\theta} = \frac{(90.0 \text{ kg})(9.80 \text{ m/s}^2)}{2\sin 10.0°} = 2540 \text{ N}$$

b) The relation $2T \sin\theta = mg$ still applies but now we are given that $T = 2.50 \times 10^4$ N (the breaking strength) and are asked to find θ.

$$\sin\theta = \frac{mg}{2T} = \frac{(90.0 \text{ kg})(9.80 \text{ m/s}^2)}{2(2.50 \times 10^4 \text{ N})} = 0.01764, \quad \theta = 1.01°.$$

EVALUATE: $T = mg/(2\sin\theta)$ says that $T = mg/2$ when $\theta = 90°$ (rope is vertical).

$T \to \infty$ when $\theta \to 0$ since the upward component of the tension becomes a smaller fraction of the tension.

5.5 **IDENTIFY:** Apply Newton's 1st law to the car. The forces are the same as in Example 5.4.

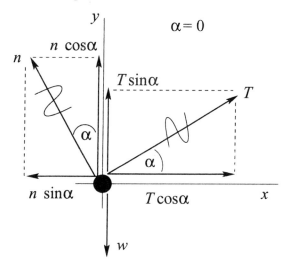

EXECUTE:

$\sum F_x = ma_x$
$T \cos \alpha - n \sin \alpha = 0$
$T \cos \alpha = n \sin \alpha$

$\sum F_y = ma_y$
$n \cos \alpha + T \sin \alpha - w = 0$
$n \cos \alpha + T \sin \alpha = w$

The first equation gives $n = T \left(\dfrac{\cos \alpha}{\sin \alpha} \right)$.

Use this in the second equation to eliminate n:

$\left(T \dfrac{\cos \alpha}{\sin \alpha} \right) \cos \alpha + T \sin \alpha = w$

Multiply this equation by $\sin \alpha$:

$T(\cos^2 \alpha + \sin^2 \alpha) = w \sin \alpha$

$T = w \sin \alpha$ (since $\cos^2 \alpha + \sin^2 \alpha = 1$).

Then $n = T \left(\dfrac{\cos \alpha}{\sin \alpha} \right) = w \sin \alpha \left(\dfrac{\cos \alpha}{\sin \alpha} \right) = w \cos \alpha$.

EVALUATE: These results are the same as obtained in Example 5.4. The choice of coordinate axes is up to us. Some choices may make the calculation easier, but the results are the same for any choice of axes.

5.7 **IDENTIFY:** Apply Newton's 1st law to the wrecking ball. Each cable exerts a force on the ball, directed along the cable.

SET UP: Force diagram for the wrecking ball:

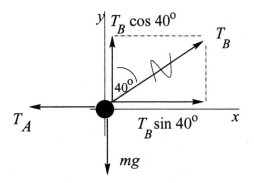

EXECUTE:

a) $\sum F_y = ma_y$

$T_B \cos 40° - mg = 0$

$$T_B = \frac{mg}{\cos 40°} = \frac{(4090 \text{ kg})(9.80 \text{ m/s}^2)}{\cos 40°} = 5.23 \times 10^4 \text{ N}$$

b) $\sum F_x = ma_x$

$T_B \sin 40° - T_A = 0$

$T_A = T_B \sin 40° = 3.36 \times 10^4 \text{ N}$

EVALUATE: If the angle 40° is replaced by 0° (cable B is vertical), then $T_B = mg$ and $T_A = 0$.

5.9   **IDENTIFY:** Apply Newton's 1st law to the car. The resistive force (target variable) is directed up the incline, opposite to the velocity of the car.

SET UP: Let $\vec{F}$ be the resistive force.

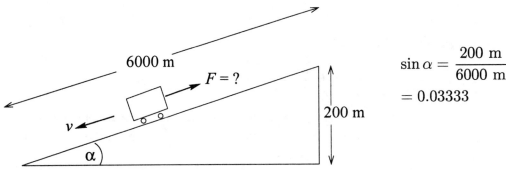

$$\sin \alpha = \frac{200 \text{ m}}{6000 \text{ m}}$$

$$= 0.03333$$

Free-body diagram for the car:

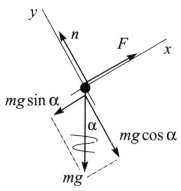

EXECUTE:

$\sum F_x = 0$ (constant velocity implies $\vec{a} = \mathbf{0}$).
$F - mg\sin\alpha = 0$
$F = mg\sin\alpha = (1600\ \text{kg})(9.80\ \text{m/s}^2)(0.03333)$
$F = 523\ \text{N}$

EVALUATE: $mg\sin\alpha$ is the component of gravity down the incline.

5.11 IDENTIFY: Apply Newton's 1st law to the hanging weight and to each knot. The tension force at each end of a string is the same.

a) Let the tensions in the three strings be T, T', and T''

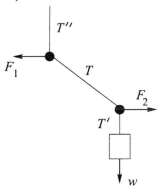

SET UP: Free-body diagram for the block

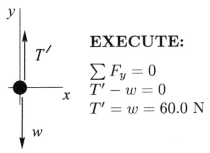

EXECUTE:

$\sum F_y = 0$
$T' - w = 0$
$T' = w = 60.0\ \text{N}$

SET UP: Free-body diagram for the lower knot

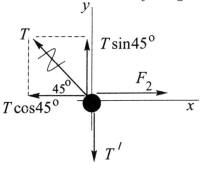

EXECUTE:

$\sum F_y = 0$
$T\sin 45° - T' = 0$

$$T = \frac{T'}{\sin 45°} = \frac{60.0\ \text{N}}{\sin 45°} = 84.9\ \text{N}$$

b) Apply $\sum F_x = 0$ to the force diagram for the lower knot:

$\sum F_x = 0$

$F_2 = T \cos 45° = (84.9 \text{ N}) \cos 45° = 60.0 \text{ N}$

SET UP: Free-body diagram for the upper knot:

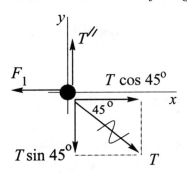

EXECUTE:

$\sum F_x = 0$

$\sum F_x = 0$
$T \cos 45° - F_1 = 0$
$F_1 = (84.9 \text{ N}) \cos 45°$
$F_1 = 60.0 \text{ N}$

Note that $F_1 = F_2$.

EVALUATE: Applying $\sum F_y = 0$ to the upper knot gives $T'' = T \sin 45° = 60.0 \text{ N} = w$. If we treat the whole system as a single object, the force disgram is

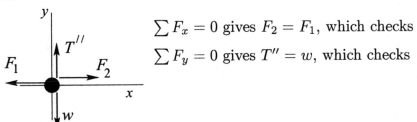

$\sum F_x = 0$ gives $F_2 = F_1$, which checks

$\sum F_y = 0$ gives $T'' = w$, which checks

5.15   IDENTIFY: Apply $\sum \vec{F} = m\vec{a}$ to the load of bricks and to the counterweight. The tension is the same at each end of the rope. The rope pulls up with the same force (T) on the bricks and on the counterweight. The counterweight accelerates downward and the bricks accelerate upward; these accelerations have the same magnitude.

a) SET UP: The free-body diagrams for the bricks and counterweight are

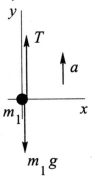

bricks   counterweight

b) **EXECUTE:** Apply $\sum F_y = ma_y$ to each object. The acceleration magnitude is the same for the two objects. For the bricks take $+y$ to be upward since $\vec{a}$ for the bricks is upward. For the counterweight take $+y$ to be downward since $\vec{a}$ is downward.

bricks: $\sum F_y = ma_y$

$T - m_1 g = m_1 a$

counterweight: $\sum F_y = ma_y$

$m_2 g - T = m_2 a$

Add these two equations to eliminate T:

$(m_2 - m_1)g = (m_1 + m_2)a$

$a = \left(\dfrac{m_2 - m_1}{m_1 + m_2} \right) g = \left(\dfrac{28.0 \text{ kg} - 15.0 \text{ kg}}{15.0 \text{ kg} + 28.0 \text{ kg}} \right) (9.80 \text{ m/s}^2) = 2.96 \text{ m/s}^2$

c) $T - m_1 g = m_1 a$ gives $T = m_1(a + g) = (15.0 \text{ kg})(2.96 \text{ m/s}^2 + 9.80 \text{ m/s}^2) = 191$ N

As a check, calculate T using the other equation:

$m_2 g - T = m_2 a$ gives $T = m_2(g - a) = 28.0 \text{ kg}(9.80 \text{ m/s}^2 - 2.96 \text{ m/s}^2) = 191$ N, which checks.

EVALUATE: The tension is 1.30 times the weight of the bricks; this causes the bricks to accelerate upward. The tension is 0.696 times the weight of the counterweight; this causes the counterweight to accelerate downward.

If $m_1 = m_2$, $a = 0$ and $T = m_1 g = m_2 g$. In this special case the objects don't move. If $m_1 = 0$, $a = g$ and $T = 0$; in this special case the counterweight is in free-fall. Our general result is correct in these two special cases.

5.17 **IDENTIFY:** Apply $\sum \vec{F} = m\vec{a}$ to each block. Since they are connected by the rope, the blocks have the same magnitude of acceleration. The target variables are the acceleration a of the blocks and the mass m of the hanging block. Since the mass of the rope is ignored, the tension has the same value at each end of the rope.

SET UP: A sketch of the situation is

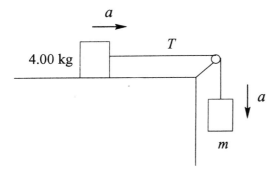

a) Free-body diagrams for each of the two blocks:

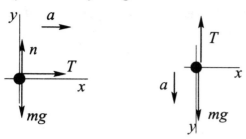

4.00 kg block   block of mass m

Note that for the hanging block we take the $+y$-direction to be downward since that is the direction of the acceleration of that block.

EXECUTE:

b) Apply $\sum F_x = ma_x$ to the 4.00 kg block:

$T = ma$

$$a = \frac{T}{m} = \frac{10.0 \text{ N}}{4.00 \text{ kg}} = 2.50 \text{ m/s}^2$$

Since the two blocks are connected by the rope, they have the same magnitude of acceleration.

c) Apply $\sum F_y = ma_y$ to the block of mass m:

$mg - T = ma$

$m(g - a) = T$

$$m = \frac{T}{g - a} = \frac{10.0 \text{ N}}{9.80 \text{ m/s}^2 - 2.50 \text{ m/s}^2} = 1.37 \text{ kg}$$

d) EVALUATE: The tension is less than the weight of the hanging block. The net force on the hanging block must be downward if that block is to accelerate downward. Note also that the tension force on the 4.00 kg block is less than the weight of that block. The tension and acceleration of this block are horizontal; the weight is vertical and doesn't enter into the $\sum F_x = ma_x$ equation for the block. Since the surface is frictionless, any T, no matter how small, produces a nonzero acceleration of the block. Note in fact that the mass of the hanging block is less than the mass of the other block, but the system still accelerates when released.

5.19 a) IDENTIFY: Apply $\sum \vec{F} = m\vec{a}$ to the student.

The reading on the scale equals the upward normal force exerted on the student. also, the student is at rest relative to the elevator, so the acceleration of the student equals that of the elevator.

SET UP: We can calculate the mass of the student from the weight:

$$m = \frac{w}{g} = \frac{550 \text{ N}}{9.80 \text{ m/s}^2} = 56.1 \text{ kg}$$

We know that $\vec{a}$ is downward since $w < N$. Thus take the $+y$-direction to be downward, in the direction of $\vec{a}$.

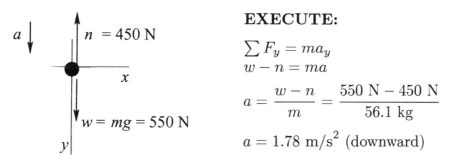

EXECUTE:

$$\sum F_y = ma_y$$
$$w - n = ma$$
$$a = \frac{w - n}{m} = \frac{550 \text{ N} - 450 \text{ N}}{56.1 \text{ kg}}$$
$$a = 1.78 \text{ m/s}^2 \text{ (downward)}$$

b) SET UP: Now $n > w$ so the student (and elevator) is accelerating upward. Take the $+y$-direction to be upward, in the direction of $\vec{a}$.

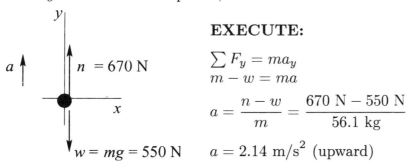

EXECUTE:

$$\sum F_y = ma_y$$
$$m - w = ma$$
$$a = \frac{n - w}{m} = \frac{670 \text{ N} - 550 \text{ N}}{56.1 \text{ kg}}$$
$$a = 2.14 \text{ m/s}^2 \text{ (upward)}$$

c) SET UP: Now $n = 0$. $n < w$ so the acceleration is downward.

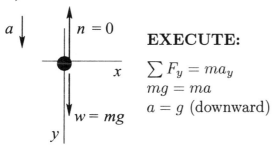

EXECUTE:

$$\sum F_y = ma_y$$
$$mg = ma$$
$$a = g \text{ (downward)}$$

The elevator and student are in free-fall; the elevator cable may have snapped.

EVALUATE: Note that in each case we take the $+y$ direction to be the direction of $\vec{a}$. The normal force exerted on the student by the scale is often called the apparent weight. If $a = 0$, $n = mg$.

5.23 a) IDENTIFY: Constant speed implies $a = 0$. Apply Newton's 1st law to the box. The friction force is directed opposite to the motion of the box.

SET UP: Consider the free-body diagram for the crate. Let $\vec{F}$ be the horizontal force applied by the worker. The friction is kinetic friction since the crate is sliding along the surface.

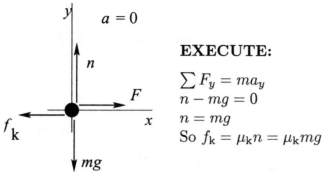

EXECUTE:

$$\sum F_y = ma_y$$
$$n - mg = 0$$
$$n = mg$$
So $f_k = \mu_k n = \mu_k mg$

$$\sum F_x = ma_x$$
$$F - f_k = 0$$
$$F = f_k = \mu_k mg = (0.20)(11.2 \text{ kg})(9.80 \text{ m/s}^2) = 22 \text{ N}$$

b) IDENTIFY: Now the only horizontal force on the crate is the kinetic friction force. Apply Newton's 2nd law to the box to calculate its acceleration. Once we have the acceleration, we can find the distance using a constant acceleration equation. The friction force is $f_k = \mu_k mg$, just as in part (a).

SET UP:

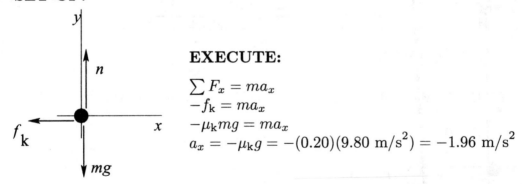

EXECUTE:

$$\sum F_x = ma_x$$
$$-f_k = ma_x$$
$$-\mu_k mg = ma_x$$
$$a_x = -\mu_k g = -(0.20)(9.80 \text{ m/s}^2) = -1.96 \text{ m/s}^2$$

Use the constant acceleration equations to find the distance the crate travels:

$$v_x = 0, \quad v_{0x} = 3.50 \text{ m/s}, \quad a_x = -1.96 \text{ m/s}^2, \quad x - x_0 = ?$$
$$v_x^2 = v_{0x}^2 + 2a_x(x - x_0)$$

$$x - x_0 = \frac{v_x^2 - v_{0x}^2}{2a_x} = \frac{0 - (3.50 \text{ m/s})^2}{2(-1.96 \text{ m/s}^2)} = 3.1 \text{ m}$$

EVALUATE: The normal force is the component of force exerted by a surface perpendicular to the surface. Its magnitude is determined by $\sum \vec{F} = m\vec{a}$. In this case n and mg are the only vertical forces and $a_y = 0$, so $n = mg$. Also note that f_k and n are proportional in magnitude but perpendicular in direction.

5.25 a) IDENTIFY: Apply $\sum \vec{F} = m\vec{a}$ to the motion of the box. The friction force is directed opposite to the motion of the box. Constant speed implies $a = 0$.

SET UP: Consider the free-body diagram for the box

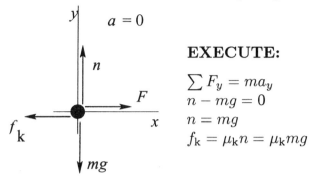

EXECUTE:

$$\sum F_y = ma_y$$
$$n - mg = 0$$
$$n = mg$$
$$f_k = \mu_k n = \mu_k mg$$

$$\sum F_x = ma_x$$
$$F - f_k = 0$$
$$F = f_k = \mu_k mg = (0.12)(6.00 \text{ kg})(9.80 \text{ m/s}^2) = 7.1 \text{ N}$$

b) SET UP:

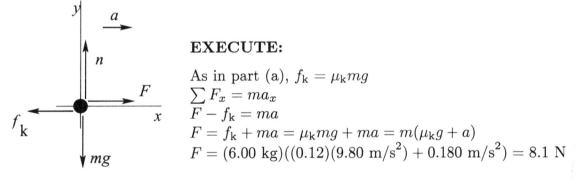

EXECUTE:

As in part (a), $f_k = \mu_k mg$
$$\sum F_x = ma_x$$
$$F - f_k = ma$$
$$F = f_k + ma = \mu_k mg + ma = m(\mu_k g + a)$$
$$F = (6.00 \text{ kg})((0.12)(9.80 \text{ m/s}^2) + 0.180 \text{ m/s}^2) = 8.1 \text{ N}$$

c) The normal force $n = mg$ is reduced. This in turn reduces the friction force, so the magnitude of the force $\vec{F}$ required to move the box is less.

part (a): $F = \mu_k mg = (0.12)(6.00 \text{ kg})(1.62 \text{ m/s}^2) = 1.2 \text{ N}$

part (b): $F = m(\mu_k g + a) = (6.00 \text{ kg})((0.12)(1.62 \text{ m/s}^2) + 0.180 \text{ m/s}^2) = 2.2 \text{ N}$

EVALUATE: The friction force has the same value in parts (a) and (b). In part (a), $F = f_k$ and $a = 0$. In part (b), $F > f_k$ and the box accelerates in the direction of $\vec{F}$.

5.27 IDENTIFY: Apply Newton's 2nd law to the safe. The target variable is the force $\vec{F}$ (magnitude and direction) needed to keep the box moving down the incline at constant speed. Constant speed implies $a = 0$.

The angle α of the incline is given by $\sin \alpha = \dfrac{2.00 \text{ m}}{20.0 \text{ m}}$, so $\alpha = 5.739°$.

a) SET UP: Consider the force diagram for the safe, with all the forces except for the applied force we are being asked to calculate. We must decide whether this force must be down the incline or up the incline, to make the total resultant force zero. Note that we know that the kinetic friction force f_k is directed up the incline, since the friction force opposes the motion. Use coordinates parallel and perpendicular to the incline.

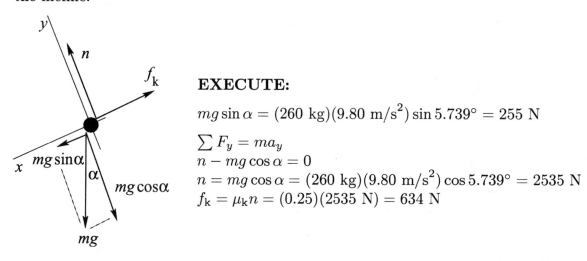

EXECUTE:

$mg \sin \alpha = (260 \text{ kg})(9.80 \text{ m/s}^2) \sin 5.739° = 255 \text{ N}$

$\sum F_y = ma_y$
$n - mg \cos \alpha = 0$
$n = mg \cos \alpha = (260 \text{ kg})(9.80 \text{ m/s}^2) \cos 5.739° = 2535 \text{ N}$
$f_k = \mu_k n = (0.25)(2535 \text{ N}) = 634 \text{ N}$

We see that $mg \sin \alpha < f_k$, so for the forces to balance ($a = 0$) more force directed down the incline is needed. The safe must be pulled down if it is to travel at constant speed.

b) SET UP: Now we can add the applied force $\vec{F}$ to the free-body diagram, since we have determined its direction.

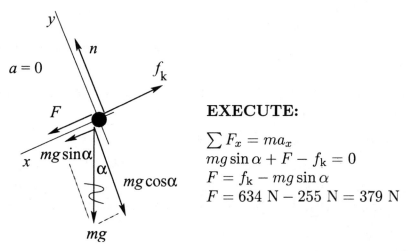

EXECUTE:

$\sum F_x = ma_x$
$mg \sin \alpha + F - f_k = 0$
$F = f_k - mg \sin \alpha$
$F = 634 \text{ N} - 255 \text{ N} = 379 \text{ N}$

EVALUATE: The normal force is not equal to the weight; $\sum \vec{F} = m\vec{a}$ says it is equal to $mg \cos \alpha$.

5.31 IDENTIFY: Apply $\sum \vec{F} = m\vec{a}$ to each crate. The rope exerts force T to the right on crate A and force T to the left on crate B. The target variables are the forces T and F. Constant v implies $a = 0$.

SET UP: Free-body diagram for A

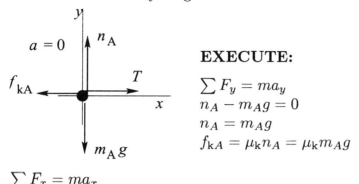

EXECUTE:

$$\sum F_y = ma_y$$
$$n_A - m_A g = 0$$
$$n_A = m_A g$$
$$f_{kA} = \mu_k n_A = \mu_k m_A g$$

$$\sum F_x = ma_x$$
$$T - f_{kA} = 0$$
$$T = \mu_k m_A g$$

SET UP: Free-body diagram for B

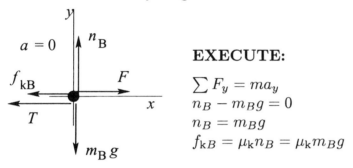

EXECUTE:

$$\sum F_y = ma_y$$
$$n_B - m_B g = 0$$
$$n_B = m_B g$$
$$f_{kB} = \mu_k n_B = \mu_k m_B g$$

$$\sum F_x = ma_x$$
$$F - T - f_{kB} = 0$$
$$F = T + \mu_k m_B g$$

Use the first equation to replace T in the second:

$$F = \mu_k m_A g + \mu_k m_B g.$$

a) $F = \mu_k(m_A + m_B)g$

b) $T = \mu_k m_A g$

EVALUATE: We can also consider both crates together as a single object of mass $(m_a + m_b)$. $\sum F_x = ma_x$ for this combined object gives $F = f_k = \mu_k(m_A + m_B)g$, in agreement with our answer in part (a).

5.35   IDENTIFY: Apply $\sum \vec{F} = m\vec{a}$ to each block. The target variables are the tension T in the cord and the acceleration a of the blocks. Then a can be used in a constant acceleration equation to find the speed of each block. The magnitude of the acceleration is the same for both blocks.

SET UP:

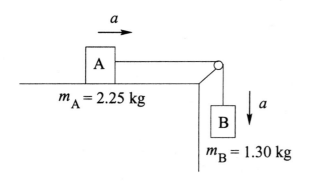

For each block take a positive coordinate direction to be the direction of the block's acceleration.

block on the table:

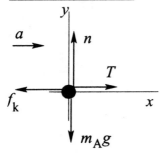

EXECUTE:

$$\sum F_y = ma_y$$
$$n - m_A g = 0$$
$$n = m_A g$$

$$f_k = \mu_k n = \mu_k m_A g$$

$$\sum F_x = ma_x$$
$$T - f_k = m_A a$$
$$T - \mu_k m_A g = m_A a$$

SET UP: hanging block:

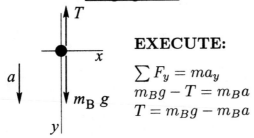

EXECUTE:

$$\sum F_y = ma_y$$
$$m_B g - T = m_B a$$
$$T = m_B g - m_B a$$

a) Use the second equation in the first

$$m_B g - m_B a - \mu_k m_A g = m_A a$$
$$(m_A + m_B)a = (m_B - \mu_k m_A)g$$
$$a = \frac{(m_B - \mu_k m_A)g}{m_A + m_B} = \frac{(1.30 \text{ kg} - (0.45)(2.25 \text{ kg}))(9.80 \text{ m/s}^2)}{2.25 \text{ kg} + 1.30 \text{ kg}} = 0.7937 \text{ m/s}^2$$

SET UP: Now use the constant acceleration equations to find the final speed. Note that the blocks have the same speeds.

$$x - x_0 = 0.0300 \text{ m}, \quad a_x = 0.7937 \text{ m/s}^2, \quad v_{0x} = 0, \quad v_x = ?$$
$$v_x^2 = v_{0x}^2 + 2a_x(x - x_0)$$

EXECUTE: $v_x = \sqrt{2a_x(x - x_0)} = \sqrt{2(0.7937 \text{ m/s}^2)(0.0300 \text{ m})} = 0.218 \text{ m/s} = 21.8 \text{ cm/s}$.

b) $T = m_B g - m_B a = m_B(g - a) = 1.30 \text{ kg}(9.80 \text{ m/s}^2 - 0.7937 \text{ m/s}^2) = 11.7 \text{ N}$

Or, to check,

$T - \mu_k m_A g = m_A a$

$T = m_A(a + \mu_k g) = 2.25 \text{ kg}(0.7937 \text{ m/s}^2 + (0.45)(9.80 \text{ m/s}^2)) = 11.7 \text{ N}$, which checks.

EVALUATE: The force T exerted by the cord has the same value for each block. $T < m_B g$ since the hanging block accelerates downward. Also, $f_k = \mu_k m_A g = 9.92 \text{ N}$. $T > f_k$ and the block on the table accelerates in the direction of T.

5.37 a) IDENTIFY: Apply $\sum \vec{F} = m\vec{a}$ to the crate. Constant v implies $a = 0$.

Crate moving says that the friction is kinetic friction. The target variable is the magnitude of the force applied by the woman.

SET UP: Free-body diagram for the crate

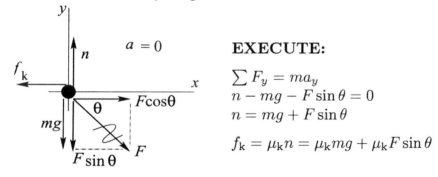

EXECUTE:

$\sum F_y = ma_y$
$n - mg - F\sin\theta = 0$
$n = mg + F\sin\theta$

$f_k = \mu_k n = \mu_k mg + \mu_k F\sin\theta$

$\sum F_x = ma_x$
$F\cos\theta - f_k = 0$
$F\cos\theta - \mu_k mg - \mu_k F\sin\theta = 0$
$F(\cos\theta - \mu_k \sin\theta) = \mu_k mg$

$F = \dfrac{\mu_k mg}{\cos\theta - \mu_k \sin\theta}$

b) IDENTIFY: and **SET UP:** "start the crate moving" means the same force diagram as in part (a), except that μ_k is replaced by μ_s.

Thus $F = \dfrac{\mu_s mg}{\cos\theta - \mu_s \sin\theta}$.

EXECUTE: $F \to \infty$ if $\cos\theta - \mu_s \sin\theta = 0$.

This gives $\mu_s = \dfrac{\cos\theta}{\sin\theta} = \dfrac{1}{\tan\theta}$.

EVALUATE: $\vec{F}$ has a downward component so $n > mg$. If $\theta = 0$ (woman pushes horizontally), $n = mg$ and $F = f_k = \mu_k mg$.

5.43 IDENTIFY: Apply $\sum \vec{F} = m\vec{a}$ to the stone. The stone moves in uniform circular motion so has acceleration $\vec{a}_{\text{rad}}$ in toward the center of the circular path. The target variable is the speed v, which is calculated from $a_{\text{rad}} = v^2/R$ once we have calculated a_{rad} from the forces.

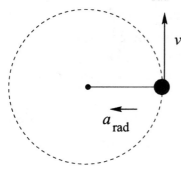

view from above

SET UP: Free-body diagram:

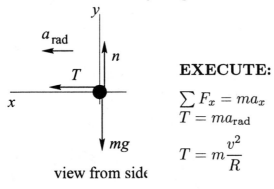

EXECUTE:

$$\sum F_x = ma_x$$
$$T = ma_{\text{rad}}$$
$$T = m\frac{v^2}{R}$$

view from side

The maximum v produces $T = 600$ N. Thus

$$v = \sqrt{\frac{RT}{m}} = \sqrt{\frac{(0.90 \text{ m})(600 \text{ N})}{0.80 \text{ kg}}} = 26.0 \text{ m/s}$$

EVALUATE: If v is larger, a_{rad} is larger and T would have to be larger.

5.45 IDENTIFY: Apply $\sum \vec{F} = m\vec{a}$ to the airplane. With the wings banked the airplane must be making a turn, with a net force toward the center of the circular arc of the turn. The target variable is the banking angle, the angle between the plane of the wings and the horizontal.

a) SET UP: Free-body diagram for the plane:

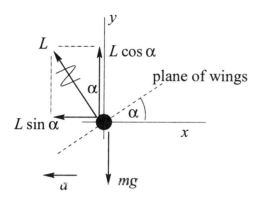

The lift force is perpendicular to the wings. The banking angle α is the angle between the plane of the wings and the horizontal.

For a constant-altitude turn the acceleration is horizontal, in toward the center of the circular path.

Use coordinates where x is horizontal and y is vertical. Then $a_y = 0$.

EXECUTE: $\sum F_y = ma_y$

$L\cos\alpha - w = 0$

Then $L = 3.8w$ gives $3.8w\cos\alpha = w$

$\cos\alpha = 1/3.8$ and $\alpha = 75°$.

EVALUATE: Larger banking angles require a lift force that is a larger multiple of the weight.

b) The answer to part (a) is determined from the vertical component of $\sum \vec{F} = m\vec{a}$ and the speed of the plane does not enter.

5.51 IDENTIFY: Apply $\sum \vec{F} = m\vec{a}$ to the motion of the pilot.
The pilot moves in a vertical circle. The apparent weight is the normal force exerted on him. At each point $\vec{a}_{\text{rad}}$ is directed toward the center of the circular path.

a) SET UP: "the pilot feels weightless" means that the vertical normal force n exerted on the pilot by the chair on which the pilot sits is zero. Force diagram for the pilot at the top of the path:

EXECUTE:

$\sum F_y = ma_y$

$mg = ma_{\text{rad}}$

$g = \dfrac{v^2}{R}$

Thus $v = \sqrt{gR} = \sqrt{(9.80 \text{ m/s}^2)(150 \text{ m})} = 38.34$ m/s

$v = (38.34 \text{ m/s}) \left(\dfrac{1 \text{ km}}{10^3 \text{ m}}\right) \left(\dfrac{3600 \text{ s}}{1 \text{ h}}\right) = 138$ km/h

b) SET UP: Force diagram for the pilot at the bottom of the path. Note that the vertical normal force exerted on the pilot by the chair on which the pilot sits is now upward.

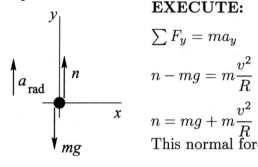

EXECUTE:

$$\sum F_y = ma_y$$

$$n - mg = m\frac{v^2}{R}$$

$$n = mg + m\frac{v^2}{R}$$

This normal force is the pilot's apparent weight.

$w = 700$ N, so $m = \dfrac{w}{g} = 71.43$ kg

$$v = (280 \text{ km/h})\left(\frac{1 \text{ h}}{3600 \text{ s}}\right)\left(\frac{10^3 \text{ m}}{1 \text{ km}}\right) = 77.78 \text{ m/s}$$

Thus $n = 700$ N $+ 71.43$ kg$\dfrac{(77.78 \text{ m/s})^2}{150 \text{ m}} = 3580$ N.

EVALUATE: In part (b), $n > mg$ since the acceleration is upward. The pilot feels he is much heavier than when at rest. The speed is not constant, but it is still true that $a_{\text{rad}} = v^2/R$ at each point of the motion.

5.53 **IDENTIFY:** Apply $\sum \vec{F} = m\vec{a}$ to the water. The water moves in a vertical circle. The target variable is the speed v; we will calculate a_{rad} and then get v from $a_{\text{rad}} = v^2/R$

SET UP: Consider the free-body diagram for the water when the pail is at the top of its circular path.

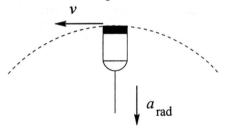

The radial acceleration is in toward the center of the circle so at this point is downward. n is the downward normal force exerted on the water by the bottom of the pail.

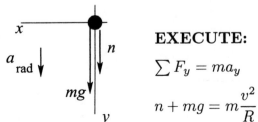

EXECUTE:

$$\sum F_y = ma_y$$

$$n + mg = m\frac{v^2}{R}$$

At the minimum speed the water is just ready to lose contact with the bottom of the pail, so at this speed, $n \to 0$. (Note that the force n cannot be upward.)

With $n \to 0$ the equation becomes $mg = m\dfrac{v^2}{R}$

$v = \sqrt{gR} = \sqrt{(9.80 \text{ m/s}^2)(0.600 \text{ m})} = 2.42 \text{ m/s}.$

EVALUATE: At the minumum speed $a_{\text{rad}} = g$. If v is less than this minimum speed, gravity pulls the water (and bucket) out of the circular path.

Problems

5.57 IDENTIFY: Apply $\sum \vec{F} = m\vec{a}$ to the refrigerator. The target variable is the applied force. No friction. Constant speed implies $a = 0$.

Since $a = 0$ it is equally convenient to use

(1) coordinate axes parallel and perpendicular to the incline;

(2) coordinate axes that are horizontal and vertical.

We will work the problem both ways.

SET UP: <u>coordinate axes parallel and perpendicular to the incline</u>

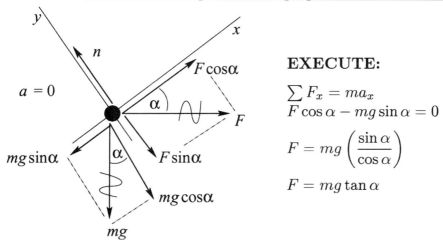

EXECUTE:

$\sum F_x = ma_x$
$F \cos \alpha - mg \sin \alpha = 0$

$F = mg \left(\dfrac{\sin \alpha}{\cos \alpha} \right)$

$F = mg \tan \alpha$

SET UP: <u>coordinates that are horizontal and vertical</u>

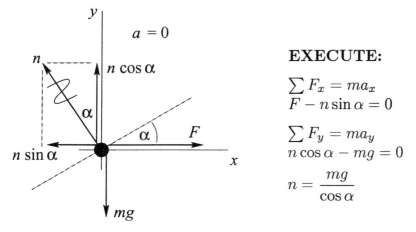

EXECUTE:

$\sum F_x = ma_x$
$F - n \sin \alpha = 0$

$\sum F_y = ma_y$
$n \cos \alpha - mg = 0$

$n = \dfrac{mg}{\cos \alpha}$

Combine these two equations to eliminate n:

$$F = n \sin \alpha = (mg / \cos \alpha) \sin \alpha$$

$F = mg \tan \alpha$, the same as before.

EVALUATE: If the force is applied parallel to the incline, $F = mg \sin \alpha$. $\tan \alpha > \sin \alpha$ so less force is required when the force is parallel to the incline. When the force is horizontal, only a component is directed up the incline. This problem illustrates that all choices of coordinate axes will yield the same results.

5.59 IDENTIFY: Apply $\sum \vec{F} = m\vec{a}$ to the combined rope plus block to find a. Then apply $\sum \vec{F} = m\vec{a}$ to a section of the rope of length x.

First note the limiting values of the tension.

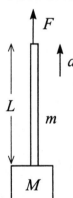

At the top of the rope $T = F$

At the bottom of the rope $T = M(g + a)$

SET UP: Consider the rope and block as one combined object, in order to calculate the acceleration:

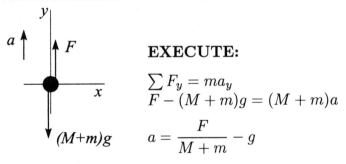

EXECUTE:

$$\sum F_y = ma_y$$
$$F - (M + m)g = (M + m)a$$

$$a = \frac{F}{M + m} - g$$

SET UP: Now consider the forces on a section of the rope that extends a distance $x < L$ below the top. The tension at the bottom of this section is $T(x)$ and the mass of this section is $m(x/L)$.

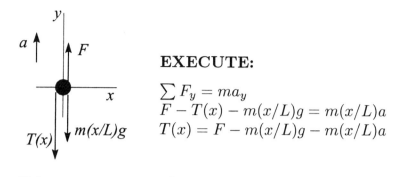

EXECUTE:

$\sum F_y = ma_y$
$F - T(x) - m(x/L)g = m(x/L)a$
$T(x) = F - m(x/L)g - m(x/L)a$

Using our expression for a and simplifying gives

$$T(x) = F\left(1 - \frac{mx}{L(M+m)}\right)$$

EVALUATE: Important to check this result for the limiting cases:

$x = 0$: The expression gives the correct value of $T = F$.

$x = L$: The expression gives $T = F(M/(M+m))$. This should equal $T = M(g+a)$, and when we use the expression for a we see that it does.

5.61 IDENTIFY: The system is in equilibirum. Apply Newton's 1st law to block A, to the hanging weight and to the knot where the cords meet. Target variables are the two forces.

a) SET UP: Free-body diagram for the hanging block:

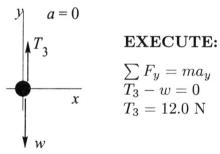

EXECUTE:

$\sum F_y = ma_y$
$T_3 - w = 0$
$T_3 = 12.0 \text{ N}$

SET UP: Free-body diagram for the knot:

EXECUTE:

$\sum F_y = ma_y$
$T_2 \sin 45.0° - T_3 = 0$

$$T_2 = \frac{T_3}{\sin 45.0°} = \frac{12.0 \text{ N}}{\sin 45.0°}$$

$T_2 = 17.0 \text{ N}$

$$\sum F_x = ma_x$$
$$T_2 \cos 45.0° - T_1 = 0$$
$$T_1 = T_2 \cos 45.0° = 12.0 \text{ N}$$

SET UP: Free-body diagram for block A:

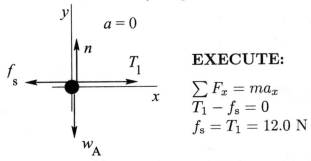

EXECUTE:

$$\sum F_x = ma_x$$
$$T_1 - f_s = 0$$
$$f_s = T_1 = 12.0 \text{ N}$$

EVALUATE: Also can apply $\sum F_y = ma_y$ to this block:

$$n - w_A = 0$$
$$n = w_A = 60.0 \text{ N}$$

Then $\mu_s n = (0.25)(60.0 \text{ N}) = 15.0 \text{ N}$; this is the maximum possible value for the static friction force.

We see that $f_s < \mu_s n$; for this value of w the static friction force can hold the blocks in place.

b) SET UP: We have all the same free-body diagrams and force equations as in part (a) but now the static friction force has its largest possbile value, $f_s = \mu_s n = 15.0$ N. Then $T_1 = f_s = 15.0$ N.

EXECUTE: From the equations for the forces on the knot

$$T_2 \cos 45.0° - T_1 = 0 \text{ implies } T_2 = T_1/\cos 45.0° = \frac{15.0 \text{ N}}{\cos 45.0°} = 21.2 \text{ N}$$

$$T_2 \sin 45.0° - T_3 = 0 \text{ implies } T_3 = T_2 \sin 45.0° = (21.2 \text{ N}) \sin 45.0° = 15.0 \text{ N}$$

And finally $T_3 - w = 0$ implies $w = T_3 = 15.0$ N.

EVALUATE: Compared to part (a), the friction is larger in part (b) by a factor of $(15.0/12.0)$ and w is larger by this same ratio.

5.63 IDENTIFY: Apply $\sum \vec{F} = m\vec{a}$ to the brush. Constant speed means $a = 0$. Target variables are two of the forces on the brush.

SET UP: Note that the normal force exerted by the wall is horizontal, since it is perpendicular to the wall. The kinetic friction force exerted by the wall is parallel to the wall and opposes the motion, so it is vertically downward.

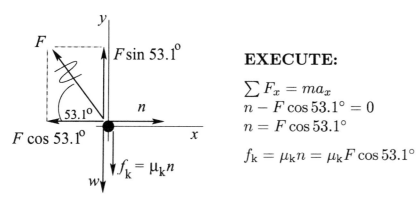

EXECUTE:

$$\sum F_x = ma_x$$
$$n - F \cos 53.1° = 0$$
$$n = F \cos 53.1°$$

$$f_k = \mu_k n = \mu_k F \cos 53.1°$$

$$\sum F_y = ma_y$$
$$F \sin 53.1° - w - f_k = 0$$
$$F \sin 53.1° - w - \mu_k F \cos 53.1° = 0$$
$$F(\sin 53.1° - \mu_k \cos 53.1°) = w$$

$$F = \frac{w}{\sin 53.1° - \mu_k \cos 53.1°}$$

a) $F = \dfrac{w}{\sin 53.1° - \mu_k \cos 53.1°} = \dfrac{12.0 \text{ N}}{\sin 53.1° - (0.15) \cos 53.1°} = 16.9 \text{ N}$

b) $n = F \cos 53.1° = (16.9 \text{ N}) \cos 53.1° = 10.1 \text{ N}$

EVALUATE: In the absence of friction $w = F \sin 53.1°$, which agrees with our expression.

5.73 IDENTIFY: Apply Newton's 1st law to the rope. Let m_1 be the mass of that part of the rope that is on the table, and let m_2 be the mass of that part of the rope that is hanging over the edge. ($m_1 + m_2 = m$, the total mass of the rope).

Since the mass of the rope is not being neglected, the tension in the rope varies along the length of the rope. Let T be the tension in the rope at that point that is at the edge of the table.

SET UP: Free-body diagram for the hanging section of the rope

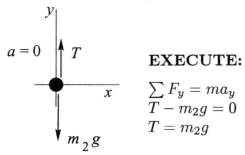

EXECUTE:

$$\sum F_y = ma_y$$
$$T - m_2 g = 0$$
$$T = m_2 g$$

SET UP: Free-body diagram for that part of the rope that is on the table:

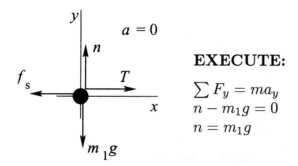

EXECUTE:

$$\sum F_y = ma_y$$
$$n - m_1 g = 0$$
$$n = m_1 g$$

When the maximum amount of rope hangs over the edge the static friction has its maximum value:

$$f_s = \mu_s n = \mu_s m_1 g$$

$$\sum F_x = ma_x$$
$$T - f_s = 0$$
$$T = \mu_s m_1 g$$

Use the first equation to replace T:

$$m_2 g = \mu_s m_1 g$$
$$m_2 = \mu_s m_1$$

The fraction that hangs over is $\dfrac{m_2}{m} = \dfrac{\mu_s m_1}{m_1 + \mu_s m_1} = \dfrac{\mu_s}{1 + \mu_s}$.

EVALUATE: As $\mu_s \to 0$, the fraction goes to zero and as $\mu_s \to \infty$, the fraction goes to unity.

5.75 IDENTIFY: First calculate the maximum acceleration that the static friction force can give to the case. Apply $\sum \vec{F} = m\vec{a}$ to the case.

a) SET UP: The static friction force is to the right in the sketch (northward) since it tries to make the case move with the truck. The maximum value it can have is $f_s = \mu_s N$.

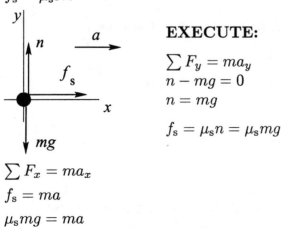

EXECUTE:

$$\sum F_y = ma_y$$
$$n - mg = 0$$
$$n = mg$$

$$f_s = \mu_s n = \mu_s mg$$

$$\sum F_x = ma_x$$
$$f_s = ma$$
$$\mu_s mg = ma$$

$$a = \mu_s g = (0.30)(9.80 \text{ m/s}^2) = 2.94 \text{ m/s}^2$$

The truck's acceleration is less than this so the case doesn't slip relative to the truck; the case's acceleration is $a = 2.20 \text{ m/s}^2$ (northward).

Then $f_s = ma = (30.0 \text{ kg})(2.20 \text{ m/s}^2) = 66 \text{ N}$, northward.

b) IDENTIFY: Now the acceleration of the truck is greater than the acceleration that static friction can give the case. Therefore, the case slips relative to the truck and the friction is kinetic friction. The friction force still tries to keep the case moving with the truck, so the acceleration of the case and the friction force are both southward.

SET UP:

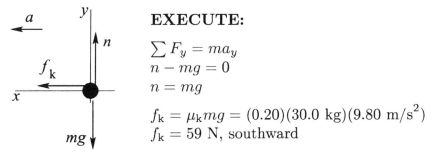

EXECUTE:

$$\sum F_y = ma_y$$
$$n - mg = 0$$
$$n = mg$$

$$f_k = \mu_k mg = (0.20)(30.0 \text{ kg})(9.80 \text{ m/s}^2)$$
$$f_k = 59 \text{ N, southward}$$

EVALUATE: $f_k = ma$ implies $a = \dfrac{f_k}{m} = \dfrac{59 \text{ N}}{30.0 \text{ kg}} = 2.0 \text{ m/s}^2$. The magnitude of the acceleration of the case is less than that of the truck and the case slides toward the front of the truck. In both parts (a) and (b) the friction is in the direction of the motion and accelerates the case. Friction opposes *relative* motion between two surfaces in contact.

5.81 Apply $\sum \vec{F} = m\vec{a}$ to each block. Forces between the blocks are related by Newton's 3rd law. The target variable is the force F.

Block B is pulled to the left at constant speed, so block A moves to the right at constant speed and $a = 0$ for each block.

SET UP: Free-body diagram for block A

n_{BA} is the normal force that B exerts on A.

$f_{BA} = \mu_k n_{BA}$ is the kinetic friction force that B exerts on A. Block A moves to the right relative to B, and f_{BA} opposes this motion, so f_{BA} is to the left.

Note also that F acts just on B, not on A.

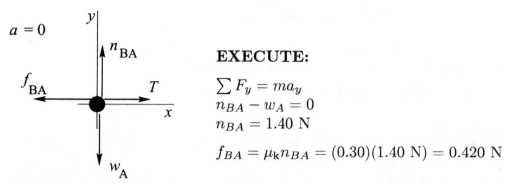

EXECUTE:

$\sum F_y = ma_y$
$n_{BA} - w_A = 0$
$n_{BA} = 1.40$ N

$f_{BA} = \mu_k n_{BA} = (0.30)(1.40 \text{ N}) = 0.420$ N

$\sum F_x = ma_x$
$T - f_{BA} = 0$
$T = f_{BA} = 0.420$ N

SET UP: Free-body diagram for block B

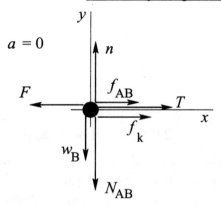

EXECUTE:

n_{AB} is the normal force that block A exerts on block B. By Newton's third law n_{AB} and n_{BA} are equal in magnitude and opposite in direction, so
$n_{AB} = 1.40$ N.

f_{AB} is the kinetic friction force that A exerts on B. Block B moves to the left relative to A and f_{AB} opposes this motion, so f_{AB} is to the right.

$f_{AB} = \mu_k n_{AB} = (0.30)(1.40 \text{ N}) = 0.420$ N.

n and f_k are the normal and friction force exerted by the floor on block B; $f_k = \mu_k n$. Note that block B moves to the left relative to the floor and f_k opposes this motion, so f_k is to the right.

$\sum F_y = ma_y$
$n - w_B - n_{AB} = 0$
$n = w_B + n_{AB} = 4.20 \text{ N} + 1.40 \text{ N} = 5.60$ N
Then $f_k = \mu_k n = (0.30)(5.60 \text{ N}) = 1.68$ N.

$\sum F_x = ma_x$

$$f_{AB} + T + f_k - F = 0$$
$$F = T + f_{AB} + f_k = 0.420 \text{ N} + 0.420 \text{ N} + 1.68 \text{ N} = 2.52 \text{ N}$$

EVALUATE: Note that f_{AB} and f_{BA} are a third law action-reaction pair, so they must be equal in magnitude and opposite in direction and this is indeed what our calculation gives.

5.83 Apply $\sum \vec{F} = m\vec{a}$ to each block. Parts (a) and (b) will be done together.

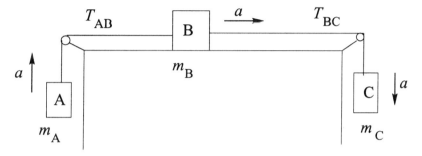

Note that each block has the same magnitude of acceleration, but in different directions. For each block let the direction of $\vec{a}$ be a positive coordinate direction.

SET UP: <u>Block A</u>

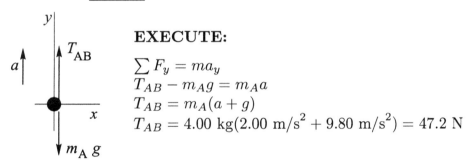

EXECUTE:

$$\sum F_y = ma_y$$
$$T_{AB} - m_A g = m_A a$$
$$T_{AB} = m_A(a + g)$$
$$T_{AB} = 4.00 \text{ kg}(2.00 \text{ m/s}^2 + 9.80 \text{ m/s}^2) = 47.2 \text{ N}$$

SET UP: <u>Block B</u>

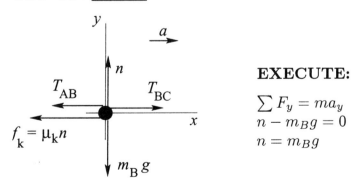

EXECUTE:

$$\sum F_y = ma_y$$
$$n - m_B g = 0$$
$$n = m_B g$$

$$f_k = \mu_k n = \mu_k m_B g = (0.25)(12.0 \text{ kg})(9.80 \text{ m/s}^2) = 29.4 \text{ N}$$

$$\sum F_x = ma_x$$

$$T_{BC} - T_{AB} - f_k = m_B a$$
$$T_{BC} = T_{AB} + f_k + m_B a = 47.2 \text{ N} + 29.4 \text{ N} + (12.0 \text{ kg})(2.00 \text{ m/s}^2)$$
$$T_{BC} = 100.6 \text{ N}$$

SET UP: Block C

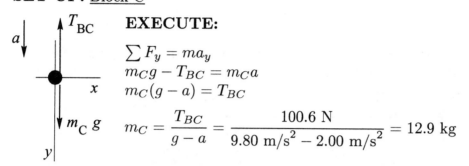

EXECUTE:

$$\sum F_y = ma_y$$
$$m_C g - T_{BC} = m_C a$$
$$m_C(g - a) = T_{BC}$$
$$m_C = \frac{T_{BC}}{g - a} = \frac{100.6 \text{ N}}{9.80 \text{ m/s}^2 - 2.00 \text{ m/s}^2} = 12.9 \text{ kg}$$

EVALUATE: If all three blocks are considered together as a single object and $\sum \vec{F} = m\vec{a}$ is applied to this combined object, $m_C g - m_A g - \mu_k m_B g = (m_A + m_B + m_C)a$. Using the values for μ_k, m_A and m_B given in the problem and the mass m_C we calculated, this equation gives $a = 2.00 \text{ m/s}^2$, which checks.

5.85   IDENTIFY: Let the tensions in the ropes be T_1 and T_2.

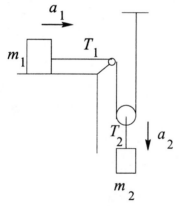

Consider the forces on each block. In each case take a positive coordinate direction in the direction of the acceleration of that block.

SET UP: Forces on m_1

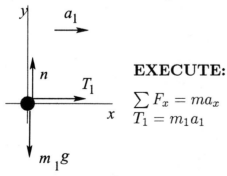

EXECUTE:

$$\sum F_x = ma_x$$
$$T_1 = m_1 a_1$$

SET UP: Forces on m_2

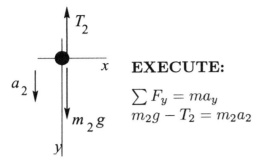

EXECUTE:

$$\sum F_y = ma_y$$
$$m_2g - T_2 = m_2a_2$$

This gives us two equations, but there are 4 unknowns (T_1, T_2, a_1, and a_2) so two more equations are required.

SET UP: Free-body diagram for the moveable pulley (mass m):

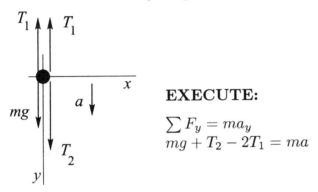

EXECUTE:

$$\sum F_y = ma_y$$
$$mg + T_2 - 2T_1 = ma$$

But our pulleys have negligible mass, so $mg = ma = 0$ and $T_2 = 2T_1$.

Combine these three equations to eliminate T_1 and T_2:

$m_2g - T_2 = m_2a_2$ gives $m_2g - 2T_1 = m_2a_2$.

And then with $T_1 = m_1a_1$ we have $m_2g - 2m_1a_1 = m_2a_2$.

SET UP: There are still two unknowns, a_1 and a_2. But the accelerations a_1 and a_2 are related. In any time interval, if m_1 moves to the right a distance d, then in the same time m_2 moves downward a distance $d/2$. One of the constant acceleration kinematic equations says $x - x_0 = v_0t + \frac{1}{2}at^2$, so if m_2 moves half the distance it must have half the acceleration of m_1: $a_2 = a_1/2$, or $a_1 = 2a_2$.

EXECUTE: This is the additional equation we need. Use it in the previous quation and get

$m_2g - 2m_1(2a_2) = m_2a_2$.

$a_2(4m_1 + m_2) = m_2g$

$$a_2 = \frac{m_2g}{4m_1 + m_2} \text{ and } a_1 = 2a_2 = \frac{2m_2g}{4m_1 + m_2}.$$

EVALUATE: If $m_2 \to 0$ or $m_1 \to \infty$, $a_1 = a_2 = 0$. If $m_2 >> m_1$, $a_2 = g$ and $a_1 = 2g$.

5.89 IDENTIFY: Apply $\sum \vec{F} = m\vec{a}$ to the block. The cart and the block have the same acceleration.

The normal force exerted by the cart on the block is perpendicular to the front of the cart, so is horizontal and to the right.

The friction force on the block is directed so as to hold the block up against the downward pull of gravity. We want to calculate the minimum a required, so take static friction to have its maximum value, $f_s = \mu_s n$.

SET UP: Free-body diagram for the block

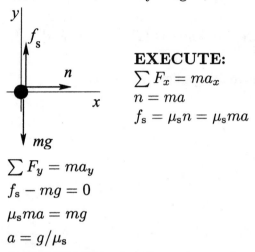

EXECUTE:
$$\sum F_x = ma_x$$
$$n = ma$$
$$f_s = \mu_s n = \mu_s ma$$

$$\sum F_y = ma_y$$
$$f_s - mg = 0$$
$$\mu_s ma = mg$$
$$a = g/\mu_s$$

EVALUATE: An observer on the cart sees the block pinned there, with no reason for a horizontal force on it because the block is at rest relative to the cart. Therefore, such an observer concludes that $n = 0$ and thus $f_s = 0$, and he doesn't understand what holds the block up against the downward force of gravity. The reason for this difficulty is that $\sum \vec{F} = m\vec{a}$ does not apply in a coordinate frame attached to the cart. This reference frame is accelerated, and hence not inertial.

The smaller μ_s is, the larger a must be to keep the block pinned against the front of the cart.

5.93 a) IDENTIFY: Apply $\sum \vec{F} = m\vec{a}$ to the two blocks taken as a single object.

SET UP: Force diagram for the two blocks taken as a single combined object:

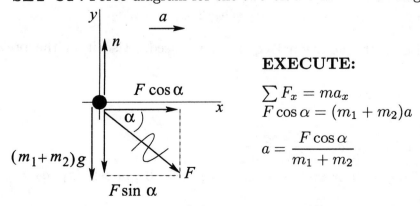

EXECUTE:
$$\sum F_x = ma_x$$
$$F \cos\alpha = (m_1 + m_2)a$$
$$a = \frac{F \cos\alpha}{m_1 + m_2}$$

$$\sum F_y = ma_y$$
$$n - F\sin\alpha - (m_1 + m_2)g = 0$$
$$n = F\sin\alpha + (m_1 + m_2)g$$

b) IDENTIFY: We now know the acceleration of each block if they move together. Now apply $\sum \vec{F} = m\vec{a}$ to each block.

SET UP: <u>bottom block</u>

n is the normal force applied by the surface, and was calculated in part (a). n' is the normal force applied by the top block. Note that $\vec{F}$ acts only on the top block so doesn't appear in this force diagram.

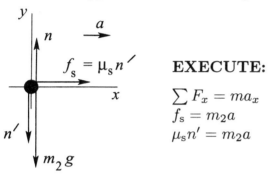

EXECUTE:

$$\sum F_x = ma_x$$
$$f_s = m_2 a$$
$$\mu_s n' = m_2 a$$

$$\sum F_y = ma_y$$
$$n - n' - m_2 g = 0$$
$$n' = n - m_2 g$$

In this equation use the value of n calculated in part (a);

$$n' = n - m_2 g = F\sin\alpha + m_1 g + m_2 g - m_2 g = F\sin\alpha + m_1 g.$$

Use this result, and also $a = F\cos\alpha/(m_1 + m_2)$ from part (a), in $\mu_s n' = m_2 a$:

$$\mu_s F\sin\alpha + \mu_s m_1 g = m_2 F\cos\alpha/(m_1 + m_2).$$

Solve for F:

$$(\mu_s \sin\alpha(m_1 + m_2) - m_2 \cos\alpha)F = -\mu_s m_1 g(m_1 + m_2)$$

$$F = \frac{\mu_s m_1(m_1 + m_2)g}{m_2 \cos\alpha - \mu_s \sin\alpha(m_1 + m_2)}, \text{ as was to be shown.}$$

IDENTIFY: An alternative approach is to apply $\sum \vec{F} = m\vec{a}$ to the top block.

SET UP:

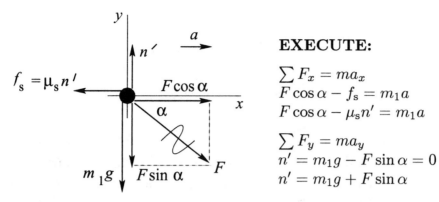

EXECUTE:

$$\sum F_x = ma_x$$
$$F \cos \alpha - f_s = m_1 a$$
$$F \cos \alpha - \mu_s n' = m_1 a$$

$$\sum F_y = ma_y$$
$$n' = m_1 g - F \sin \alpha = 0$$
$$n' = m_1 g + F \sin \alpha$$

Combine these two equations to eliminate n':

$$F \cos \alpha - \mu_s m_1 g - \mu_s F \sin \alpha = m_1 a$$

Use the value of a calculated in part (a):

$$F \cos \alpha - \mu_s m_1 g - \mu_s F \sin \alpha = \frac{m_1 F \cos \alpha}{m_1 + m_2}$$

Solve for F:

$$F((m_1 + m_2) \cos \alpha - \mu_s(m_1 + m_2) \sin \alpha - m_1 \cos \alpha) = \mu_s m_1 (m_1 + m_2) g$$

$$F = \frac{\mu_s m_1 (m_1 + m_2) g}{m_2 \cos \alpha - \mu_s \sin \alpha (m_1 + m_2)}, \text{ which is the same result.}$$

EVALUATE: When $\mu_s \to 0$, $F \to 0$ and $a = 0$. As the friction becomes less, the acceleration that can be given to the blocks becomes less.

5.95 IDENTIFY: Apply $\sum \vec{F} = m\vec{a}$ to the car. The car moves in the arc of a horizontal circle, so $\vec{a} = \vec{a}_{rad}$, directed toward the center of curvature of the roadway. The target variable is the speed of the car. a_{rad} will be calculated from the forces and then v will be calculated from $a_{rad} = v^2/R$.

a) To keep the car from sliding up the banking the static friction force is directed down the incline. At maximum speed the static friction force has its maximum value $f_s = \mu_s n$.

SET UP: Free-body diagram for the car

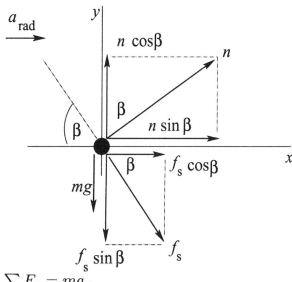

$\sum F_y = ma_y$

$n \cos \beta - f_\text{s} \sin \beta - mg = 0$

But $f_\text{s} = \mu_\text{s} n$, so

$n \cos \beta - \mu_\text{s} n \sin \beta - mg = 0$

$$n = \frac{mg}{\cos \beta - \mu_\text{s} \sin \beta}$$

$\sum F_x = ma_x$

$n \sin \beta + \mu_\text{s} n \cos \beta = ma_\text{rad}$

$n(\sin \beta + \mu_\text{s} \cos \beta) = ma$

Use the $\sum F_y$ equation to replace n:

$$\left(\frac{mg}{\cos \beta - \mu_\text{s} \sin \beta} \right) (\sin \beta + \mu_\text{s} \cos \beta) = ma_\text{rad}$$

$$a_\text{rad} = \left(\frac{\sin \beta + \mu_\text{s} \cos \beta}{\cos \beta - \mu_\text{s} \sin \beta} \right) g = \left(\frac{\sin 25° + (0.30) \cos 25°}{\cos 25° - (0.30) \sin 25°} \right) (9.80 \text{ m/s}^2) = 8.73 \text{ m/s}^2$$

$a_\text{rad} = v^2/R$ implies $v = \sqrt{a_\text{rad} R} = \sqrt{(8.73 \text{ m/s}^2)(50 \text{ m})} = 21 \text{ m/s}$.

b) IDENTIFY: To keep the car from sliding <u>down</u> the banking the static friction force is directed up the incline. At the minimum speed the static friction force has its maximum value $f_\text{s} = \mu_\text{s} n$.

SET UP: Free-body diagram for the car

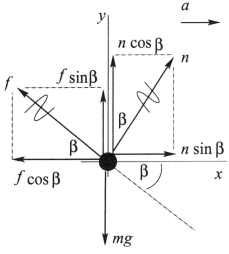

The free-body diagram is identical to that in part (a) except that now the components of f_s have opposite directions. The force equations are all the same except for the opposite sign for terms containing μ_s.

EXECUTE: $a_{\text{rad}} = \left(\dfrac{\sin\beta - \mu_s \cos\beta}{\cos\beta + \mu_s \sin\beta}\right) g = \left(\dfrac{\sin 25° - (0.30)\cos 25°}{\cos 25° + (0.30)\sin 25°}\right) (9.80 \text{ m/s}^2)$

$= 1.43 \text{ m/s}^2$

$v = \sqrt{a_{\text{rad}}R} = \sqrt{(1.43 \text{ m/s}^2)(50 \text{ m})} = 8.5 \text{ m/s}.$

EVALUATE: For v between these maximum and minimum values, the car is held on the road at a constant height by a static friction force that is less than $\mu_s n$. When $\mu_s \to 0$, $a_{\text{rad}} = g\tan\beta$. Our analysis agrees with the result of Example 5.24 in this special case.

5.109 a) IDENTIFY: Use the information given about Jena to find the time t for one revolution of the merry-go-round. Her acceleration is a_{rad}, directed in toward the axis. Let $\vec{F}_1$ be the horizontal force that keeps her from sliding off. Let her speed be v_1 and let R_1 be her distance from the axis. Apply $\sum \vec{F} = m\vec{a}$ to Jena, who moves in uniform circular motion.

SET UP:

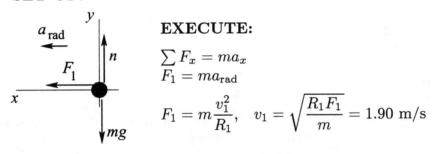

EXECUTE:

$\sum F_x = ma_x$

$F_1 = ma_{\text{rad}}$

$F_1 = m\dfrac{v_1^2}{R_1}, \quad v_1 = \sqrt{\dfrac{R_1 F_1}{m}} = 1.90 \text{ m/s}$

The time for one revolution is $t = \dfrac{2\pi R_1}{v_1} = 2\pi R_1 \sqrt{\dfrac{m}{R_1 F_1}}$.

Jackie goes around once in the same time but her speed (v_2) and the radius of her circular path (R_2) are different.

$v_2 = \dfrac{2\pi R_2}{t} = 2\pi R_2 \left(\dfrac{1}{2\pi R_1}\right)\sqrt{\dfrac{R_1 F_1}{m}} = \dfrac{R_2}{R_1}\sqrt{\dfrac{R_1 F_1}{m}}.$

IDENTIFY: Now apply $\sum \vec{F} = m\vec{a}$ to Jackie. She also moves in uniform circular motion.

SET UP: Free-body diagram for Jackie

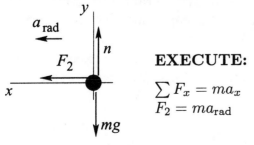

EXECUTE:

$\sum F_x = ma_x$

$F_2 = ma_{\text{rad}}$

$$F_2 = m\frac{v_2^2}{R_2} = \left(\frac{m}{R_2}\right)\left(\frac{R_2^2}{R_1^2}\right)\left(\frac{R_1F_1}{m}\right) = \left(\frac{R_2}{R_1}\right)F_1 = \left(\frac{3.60 \text{ m}}{1.80 \text{ m}}\right)(60.0 \text{ N})$$
$$= 120.0 \text{ N}$$

b) $F_2 = m\frac{v_2^2}{R_2}$, so $v_2 = \sqrt{\frac{F_2 R_2}{m}} = \sqrt{\frac{(120.0 \text{ N})(3.60 \text{ m})}{30.0 \text{ kg}}} = 3.79 \text{ m/s}$

EVALUATE: Both girls rotate together so have the same period T. By Eq.(5.16), a_{rad} is larger for Jackie so the force on her is larger. and Eq.(5.15) says $R_1/v_1 = R_2/v_2$ so $v_2 = v_1(R_2/R_1)$; this agrees with our result in (a).

5.111 IDENTIFY: Apply $\sum \vec{F} = m\vec{a}$ to the person. The person moves in a horizontal circle so his acceleration is $a_{\text{rad}} = v^2/R$, directed toward the center of the circle. The target variable is the coefficient of static friction between the person and the surface of the cylinder.

$$v = (0.60 \text{ rev/s})\left(\frac{2\pi R}{1 \text{ rev}}\right) = (0.60 \text{ rev/s})\left(\frac{2\pi(2.5 \text{ m})}{1 \text{ rev}}\right) = 9.425 \text{ m/s}$$

a) SET UP:

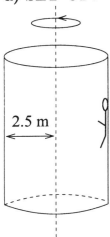

2.5 m

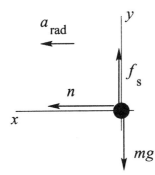

The person is held up against gravity by the static friction force exerted on him by the wall. The acceleration of the person is a_{rad}, directed in towards the axis of rotation.

b) EXECUTE: To calculate the minimum μ_s required, take f_s to have its maximum value, $f_s = \mu_s n$.
$$\sum F_y = ma_y$$

$$f_s - mg = 0$$

$$\mu_s n = mg$$

$$\sum F_x = ma_x$$

$$n = mv^2/R$$

Combine these two equations to eliminate n:

$$\mu_s mv^2/R = mg$$

$$\mu_s = \frac{Rg}{v^2} = \frac{(2.5 \text{ m})(9.80 \text{ m/s}^2)}{(9.425 \text{ m/s})^2} = 0.28$$

c) EVALUATE: No, the mass of the person divided out of the equation for μ_s.

Also, the smaller μ_s is, the larger v must be to keep the person from sliding down. For smaller μ_s the cylinder must rotate faster to make n large enough.

5.113 IDENTIFY: Apply $\sum \vec{F} = m\vec{a}$ to your friend. Your friend moves in the arc of a circle as the car turns.

a) Turn to the right.

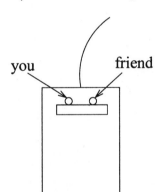

you   friend

As viewed in an inetial frame, in the absence of sufficient friction your friend doesn't make the turn completely and you move to the right toward your friend.

b) The maximum radius of the turn is the one that makes a_{rad} just equal to the maximum acceleration that static friction can give to your friend, and for this situation f_s has its maximum value $f_s = \mu_s n$.

SET UP: Free-body diagram for your friend, as viewed by someone standing behind the car:

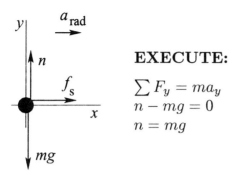

EXECUTE:

$$\sum F_y = ma_y$$
$$n - mg = 0$$
$$n = mg$$

$$\sum F_x = ma_x$$
$$f_s = ma_{rad}$$
$$\mu_s n = mv^2/R$$
$$\mu_s mg = mv^2/R$$

$$R = \frac{v^2}{\mu_s g} = \frac{(20 \text{ m/s}^2)^2}{(0.35)(9.80 \text{ m/s}^2)} = 120 \text{ m}$$

EVALUATE: The larger μ_s is, the smaller the radius R must be.

5.115 IDENTIFY: Apply $\sum \vec{F} = m\vec{a}$ to the circular motion of the bead. Also use Eq.(5.16) to relate a_{rad} to the period of rotation T.

SET UP:

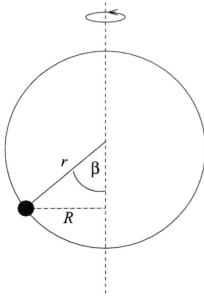

The bead moves in a circle of radius $R = r \sin \beta$.
The normal force exerted on the bead by the hoop is radially inward.

Free-body diagram for the bead:

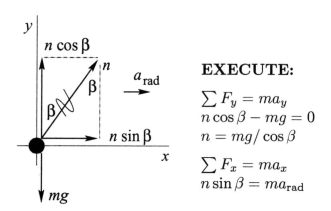

EXECUTE:

$$\sum F_y = ma_y$$
$$n \cos \beta - mg = 0$$
$$n = mg/\cos \beta$$

$$\sum F_x = ma_x$$
$$n \sin \beta = ma_{\text{rad}}$$

Combine these two equations to eliminate n:

$$\left(\frac{mg}{\cos \beta}\right) \sin \beta = ma_{\text{rad}}$$

$$\frac{\sin \beta}{\cos \beta} = \frac{a_{\text{rad}}}{g}$$

$a_{\text{rad}} = v^2/R$ and $v = 2\pi R/T$, so $a_{\text{rad}} = 4\pi^2 R/T^2$, where T is the time for one revolution.

$R = r \sin \beta$, so $a_{\text{rad}} = \dfrac{4\pi^2 r \sin \beta}{T^2}$

Use this in the above equation: $\dfrac{\sin \beta}{\cos \beta} = \dfrac{4\pi^2 r \sin \beta}{T^2 g}$

This equation is satisfied by $\sin \beta = 0$, so $\beta = 0$, or by

$\dfrac{1}{\cos \beta} = \dfrac{4\pi^2 r}{T^2 g}$, which gives $\cos \beta = \dfrac{T^2 g}{4\pi^2 r}$

a) 4.00 rev/s implies $T = (1/4.00)$ s $= 0.250$ s

Then $\cos \beta = \dfrac{(0.250 \text{ s})^2 (9.80 \text{ m/s}^2)}{4\pi^2 (0.100 \text{ m})}$ and $\beta = 81.1°$.

b) This would mean $\beta = 90°$. But $\cos 90° = 0$, so this requires $T \rightarrow 0$. So β approaches $90°$ as the hoop rotates very fast, but $\beta = 90°$ is not possible.

c) 1.00 rev/s implies $T = 1.00$ s

The $\cos \beta = \dfrac{T^2 g}{4\pi^2 r}$ equation then says $\cos \beta = \dfrac{(1.00 \text{ s})^2 (9.80 \text{ m/s}^2)}{4\pi^2 (0.100 \text{ m})} = 2.48$, which is not possible.

The only way to have the $\sum \vec{F} = m\vec{a}$ equations satisfied is for $\sin \beta = 0$. This means

$\beta = 0$; the bead sits at the bottom of the hoop.

EVALUATE: $\beta \to 90°$ as $T \to 0$ (hoop moves faster). The largest value T can have is given by $T^2 g/(4\pi^2 r) = 1$ so $T = 4\pi^2 r/g = 0.403$ s. This corresponds to a rotation rate of $(1/0.403)$ rev/s $= 2.48$ rev/s. For a rotation rate less than 2.48 rev/s, $\beta = 0$ is the only solution and the bead sits at the bottom of the hoop. Part (c) is an example of this.

CHAPTER 6
WORK AND KINETIC ENERGY

Exercises 1, 5, 7, 13, 19, 21, 25, 27, 29, 33, 35, 37, 39, 41, 47, 49, 53
Problems 57, 63, 65, 67, 69, 71, 73, 77, 83, 85, 89, 91, 93, 97, 99

Exercises

6.1 **IDENTIFY** and **SET UP:** The forces are constant so Eq.(6.2) can be used.

a) EXECUTE: $W_F = (F \cos \phi)s = (2.40 \text{ N})(\cos 0°)(1.50 \text{ m}) = +3.60 \text{ J}$

EVALUATE: The force and displacement vectors are in the same direction and the work done is positive.

b) EXECUTE: $W_F = (F \cos \phi)s = (0.600 \text{ N})(\cos 180°)(1.50 \text{ m}) = -0.900 \text{ J}$

EVALUATE: The force and displacement vectors are in opposite directions and the work done is negative.

c) EXECUTE: $W_{\text{tot}} = W_F + W_f = 3.60 \text{ J} - 0.900 \text{ J} = + 2.70 \text{ J}$

EVALUATE: The force you apply in the direction of the motion is larger than the opposing force of friction so the total work done is positive.

6.5 **IDENTIFY:** The forces are constant so Eq.(6.2) can be used to calculate the work. Constant speed implies $a = 0$. We must use $\sum \vec{F} = m\vec{a}$ applied to the crate to find the forces acting on it.

a) SET UP: Free-body diagram for the crate:

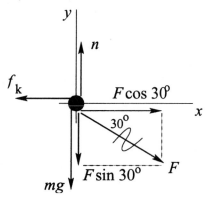

EXECUTE:
$\sum F_y = ma_y$
$n - mg - F \sin 30° = 0$
$n = mg + F \sin 30°$

$f_k = \mu_k n = \mu_k mg + F \mu_k \sin 30°$

$\sum F_x = ma_x$
$F \cos 30° - f_k = 0$
$F \cos 30° - \mu_k mg - \mu_k \sin 30° F = 0$

$$F = \frac{\mu_k mg}{\cos 30° - \mu_k \sin 30°} = \frac{0.25(30.0 \text{ kg})(9.80 \text{ m/s}^2)}{\cos 30° - (0.25) \sin 30°} = 99.2 \text{ N}$$

b) $W_F = (F \cos \phi)s = (99.2 \text{ N})(\cos 30°)(4.5 \text{ m}) = 387 \text{ J}$

($F \cos 30°$ is the horizontal component of $\vec{F}$; the work done by $\vec{F}$ is the displacement times the component of $\vec{F}$ in the direction of the displacement.)

c) We have an expression for f_k from part (a):

$f_k = \mu_k(mg + F \sin 30°) = (0.250)[(30.0 \text{ kg})(9.80 \text{ m/s}^2) + (99.2 \text{ N})(\sin 30°)] =$ 85.9 N

$\phi = 180°$ since f_k is opposite to the displacement.

Thus $W_f = (f_k \cos \phi)s = (85.9 \text{ N})(\cos 180°)(4.5 \text{ m}) = -387 \text{ J}$

d) The normal force is perpendicular to the displacement so $\phi = 90°$ and $W_n = 0$. The gravity force (the weight) is perpendicular to the displacement so $\phi = 90°$ and $W_w = 0$.

e) $W_{\text{tot}} = W_F + W_f + W_n + W_w = +387 \text{ J} + (-387 \text{ J}) = 0$

EVALUATE: Forces with a component in the direction of the displacement do positive work, forces opposite to the displacement do negative work and forces perpendicular to the displacement do zero work. The total work, obtained as the sum of the work done by each force, equals the work done by the net force. In this problem, $F_{\text{net}} = 0$ since $a = 0$ and $W_{\text{tot}} = 0$, which agrees with the sum calculated in part (e).

6.7 **IDENTIFY** and **SET UP:** $W_F = (F \cos \phi)s$, since the forces are constant. We can calculate the totalwork by summing the work done by each force.

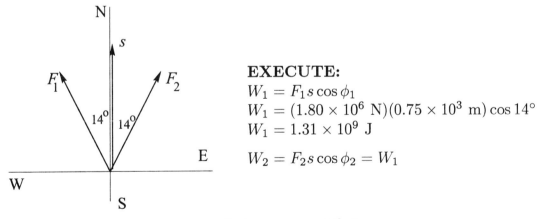

EXECUTE:
$W_1 = F_1 s \cos \phi_1$
$W_1 = (1.80 \times 10^6 \text{ N})(0.75 \times 10^3 \text{ m}) \cos 14°$
$W_1 = 1.31 \times 10^9 \text{ J}$

$W_2 = F_2 s \cos \phi_2 = W_1$

$W_{\text{tot}} = W_1 + W_2 = 2(1.31 \times 10^9 \text{ J}) = 2.62 \times 10^9 \text{ J}$

EVALUATE: Only the component $F \cos \phi$ of force in the direction of the displacement does work. These components are in the direction of $\vec{s}$ so the forces do positive work.

6.13 **IDENTIFY** and **SET UP:** Apply Eq.(6.6) to the box. Let point 1 be at the bottom of the incline and let point 2 be at the skier. Work is done by gravity and by friction. Solve for K_1 and from that obtain the required initial speed.

EXECUTE:

$W_{tot} = K_2 - K_1$

$K_1 = \frac{1}{2}mv_0^2$, $K_2 = 0$

Work is done by gravity and friction, so $W_{tot} = W_{mg} + W_f$.

$W_{mg} = -mg(y_2 - y_1) = -mgh$

$W_f = -fs$. The normal force is $n = mg\cos\alpha$ and $s = h/\sin\alpha$, where s is the distance the box travels along the incline.

$W_f = -(\mu_k mg\cos\alpha)(h/\sin\alpha) = -\mu_k mgh/\tan\alpha$

Substituting these expressions into the work-energy theorem gives

$-mgh - \mu_k mgh/\tan\alpha = -\frac{1}{2}mv_0^2$.

Solving for v_0 then gives $v_0 = \sqrt{2gh(1 + \mu_k/\tan\alpha)}$.

EVALUATE: The result is independent of the mass of the box. As $\alpha \to 90°$, $h = s$ and $v_0 = \sqrt{2gh}$, the same as throwing the box straight up into the air. For $\alpha = 90°$ the normal force is zero so there is no friction.

6.19   **IDENTIFY** and **SET UP:** Apply Eq.(6.6). The relation between the speeds v_1 and v_2 tell us the relation between K_1 and K_2.

EXECUTE:

a) $W = K_2 - K_1$

$K_1 = \frac{1}{2}mv_1^2$, $K_2 = \frac{1}{2}mv_2^2$

$v_2 = \frac{1}{4}v_1$ gives that $K_2 = \frac{1}{2}m(\frac{1}{4}v_1)^2 = \frac{1}{16}(\frac{1}{2}mv_1^2) = \frac{1}{16}K_1$

$W = K_2 - K_1 = \frac{1}{16}K_1 - K_1 = -\frac{15}{16}K_1$

b) EVALUATE: K depends only on the magnitude of $\vec{v}$ not on its direction, so the answer for W in part (a) does <u>not</u> depend on the final direction of the electron's motion. The electron slows down, so its kinetic energy decreases and the total work done on it is negative.

6.21   **IDENTIFY** and **SET UP:** Use Eq.(6.6) to calcualte the work done by the foot on the ball. Then use Eq.(6.2) to find the distance over which this force acts.

EXECUTE:

$W_{tot} = K_2 - K_1$

$K_1 = \frac{1}{2}mv_1^2 = \frac{1}{2}(0.420 \text{ kg})(2.00 \text{ m/s})^2 = 0.84 \text{ J}$

$K_2 = \frac{1}{2}mv_2^2 = \frac{1}{2}(0.420 \text{ kg})(6.00 \text{ m/s})^2 = 7.56 \text{ J}$

$W_{tot} = K_2 - K_1 = 7.56 \text{ J} - 0.84 \text{ J} = 6.72 \text{ J}$

The 40.0 N force is the only force doing work on the ball, so it must do 6.72 J of work.

$W_F = (F \cos \phi)s$ gives that $s = \dfrac{W}{F \cos \phi} = \dfrac{6.72 \text{ J}}{(40.0 \text{ N})(\cos 0)} = 0.168$ m

EVALUATE: The force is in the direction of the motion so positive work is done and this is consistent with an increase in kinetic energy.

6.25 a) IDENTIFY and **SET UP:** Use Eq.(6.2) to find the work done by the positive force. Then use Eq.(6.6) to find the final kinetic energy, and then $K_2 = \frac{1}{2}mv_2^2$ gives the final speed.

EXECUTE: $W_{\text{tot}} - K_2 - K_1$, so $K_2 = W_{\text{tot}} + K_1$

$K_1 = \frac{1}{2}mv_1^2 = \frac{1}{2}(7.00 \text{ kg})(4.00 \text{ m/s})^2 = 56.0$ J

The only force that does work on the wagon is the 10.0 N force. This force is in the direction of the displacement so $\phi = 0°$ and the force does positive work:

$W_F = (F \cos \phi)s = (10.0 \text{ N})(\cos 0)(3.0 \text{ m}) = 30.0$ J

Then $K_2 = W_{\text{tot}} + K_1 = 30.0 \text{ J} + 56.0 \text{ J} = 86.0$ J.

$K_2 = \frac{1}{2}mv_2^2; \quad v_2 = \sqrt{\dfrac{2K_2}{m}} = \sqrt{\dfrac{2(86.0 \text{ J})}{7.00 \text{ kg}}} = 4.96$ m/s

b) IDENTIFY: Apply $\sum \vec{F} = m\vec{a}$ to the wagon to calculate a. Then use a constant acceleration equation to calculate the final speed.

SET UP:

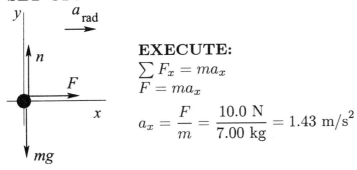

EXECUTE:
$\sum F_x = ma_x$
$F = ma_x$

$a_x = \dfrac{F}{m} = \dfrac{10.0 \text{ N}}{7.00 \text{ kg}} = 1.43 \text{ m/s}^2$

$v_{2x}^2 = v_{1x}^2 + 2a_x(x - x_0)$

$v_{2x} = \sqrt{v_{1x}^2 + 2a_x(x - x_0)} = \sqrt{(4.00 \text{ m/s})^2 + 2(1.43 \text{ m/s}^2)(3.0 \text{ m})} = 4.96$ m/s

EVALUATE: This agrees with the result calculated in part (a).

The force in the direction of the motion does positive work and the kinetic energy and speed increase. In part (b),the equivalent statement is that the force produces an acceleration in the direction of the velocity and this causes the magnitude of the velocity to increase.

6.27 IDENTIFY: Apply Eq.(6.6) to the motion of the car.
SET UP:

a) $W_{\text{tot}} = K_2 - K_1$

$f_k = \mu_k n = \mu_k mg$

f_k is the only force that does work, so $W_{\text{tot}} = W_f$

f_k is opposite to the displacement so $\phi = 180°$ and

$W_f = f(\cos\phi)s = \mu_k mg(\cos 180°)s = -\mu_k mgs$

Let point 1 be when $v = v_0$ and point 2 be where the car has stopped.

EXECUTE: $K_2 = 0$ (car stops)

$K_1 = \frac{1}{2}mv_0^2$

$W_{\text{tot}} = K_2 - K_1$, so $-\mu_k mgs = -\frac{1}{2}mv_0^2$ and $s = \dfrac{v_0^2}{2\mu_k g}$

b) $s = \dfrac{v_0^2}{2\mu_k g}$ says that $\dfrac{v_0^2}{s} = 2\mu_k g = $ constant, so $\dfrac{s_1}{v_{01}^2} = \dfrac{s_2}{v_{02}^2}$

$s_2 = s_1 \left(\dfrac{v_{02}}{v_{01}}\right)^2 = 91.2 \text{ m} \left(\dfrac{60.0 \text{ km/h}}{80.0 \text{ km/h}}\right)^2 = 51.3 \text{ m}$

EVALUATE: The car stops when the friction force has done enough negative work to take away all of the initial kinetic energy of the car. The stopping distance is proportional to the initial kinetic energy, and hence is proportional to v_0^2.

6.29 IDENTIFY and **SET UP:** Use Eq.(6.8) to calculate k for the spring. Then Eq.(6.10), with $x_1 = 0$, can be used to calculate the work done to stretch or compress the spring an amount x_2.

EXECUTE: Use the information given to calculate the force constant of the spring.

$F_x = kx$ gives $k = \dfrac{F_x}{x} = \dfrac{160 \text{ N}}{0.050 \text{ s}} = 3200 \text{ N/m}$

a) $F_x = kx = (3200 \text{ N/m})(0.015 \text{ m}) = 48 \text{ N}$

$F_x = kx = (3200 \text{ N/m})(-0.020 \text{ m}) = -64 \text{ N}$ (magnitude 64 N)

b) $W = \frac{1}{2}kx^2 = \frac{1}{2}(3200 \text{ N/m})(0.015 \text{ m})^2 = 0.36 \text{ J}$

$W = \frac{1}{2}kx^2 = \frac{1}{2}(3200 \text{ N/m})(-0.020 \text{ m})^2 \doteq 0.64 \text{ J}$

Note that in each case the work done is positive.

EVALUATE: The force is not constant during the displacement so Eq.(6.2) <u>cannot</u> be used. A force in the $+x$ direction is required to stretch the spring and a force in the opposite direction to compress it. The force F_x is in the same direction as the displacement, so positive work is done in both cases.

6.33 IDENTIFY: Apply Eq.(6.6) to the box.

SET UP: Let point 1 be just before the box reaches the end of the spring and let point 2 be where the spring has maximum compression and the box has momentarily come to rest.

EXECUTE: $W_{\text{tot}} = K_2 - K_1$

$K_1 = \frac{1}{2}mv_0^2$, $K_2 = 0$

Work is done by the spring force. $W_{\text{tot}} = -\frac{1}{2}kx_2^2$, where x_2 is the amount the spring is compressed.

$-\frac{1}{2}kx_2^2 = -\frac{1}{2}mv_0^2$ and $x_2 = v_0\sqrt{m/k} = (3.0 \text{ m/s})\sqrt{(6.0 \text{ kg})/(7500 \text{ N/m})} = 8.5$ cm

EVALUATE: The compression of the spring increases when either v_0 or m increases and decreases when k increases (stiffer spring).

6.35 IDENTIFY: Apply $\sum \vec{F} = m\vec{a}$ to calculate the μ_s required for the static friction force to equal the spring force.

SET UP: a) Free-body diagram for the glider:

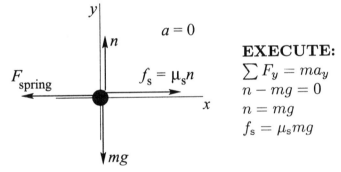

EXECUTE:

$\sum F_y = ma_y$

$n - mg = 0$

$n = mg$

$f_s = \mu_s mg$

$\sum F_x = ma_x$

$f_s - F_{\text{spring}} = 0$

$\mu_s mg - kd = 0$

$\mu_s = \dfrac{kd}{mg} = \dfrac{(20.0 \text{ N/m})(0.086 \text{ m})}{(0.100 \text{ kg})(9.80 \text{ m/s}^2)} = 1.76$

b) IDENTIFY and **SET UP:** Apply $\sum \vec{F} = m\vec{a}$ to find the maximum amount the spring can be compressed and still have the spring force balanced by friction. Then use $W_{\text{tot}} = K_2 - K_1$ to find the initial speed that results in this compression of the spring when the glider stops.

EXECUTE:

$\mu_s mg = kd$

$d = \dfrac{\mu_s mg}{k} = \dfrac{(0.60)(0.100 \text{ kg})(9.80 \text{ m/s}^2)}{20.0 \text{ N/m}} = 0.0294$ m

Now apply the work-energy theorem to the motion of the glider:

$W_{\text{tot}} = K_2 - K_1$

$K_1 = \frac{1}{2}mv_1^2$, $K_2 = 0$ (instantaneously stops)

$W_{\text{tot}} = W_{\text{spring}} + W_{\text{fric}} = -\frac{1}{2}kd^2 - \mu_k mgd$ (as in Example 6.6)

$W_{\text{tot}} = -\frac{1}{2}(20.0 \text{ N/m})(0.0294 \text{ m})^2 - 0.47(0.100 \text{ kg})(9.80 \text{ m/s}^2)(0.0294 \text{ m}) = -0.02218$ J

Then $W_{\text{tot}} = K_2 - K_1$ gives -0.02218 J $= -\frac{1}{2}mv_1^2$.

$$v_1 = \sqrt{\frac{2(0.02218 \text{ J})}{0.100 \text{ kg}}} = 0.67 \text{ m/s}$$

EVALUATE: In Example 6.6 an initial speed of 1.50 m/s compresses the spring 0.086 m and in part (a) of this problem we found that the glider doesn't stay at rest. In part (b) we found that a smaller displacement of 0.0294 m when the glider stops is required if it is to stay at rest. And we calculate a smaller initial speed (0.67 m/s) to produce this smaller displacement.

6.37 **IDENTIFY** and **SET UP:** The magnitude of the work done by F_x equals the area under the F_x versus x curve. The work is positive when F_x and the displacement are in the same direction; it is negative when they are in opposite directions.

EXECUTE:

a) F_x is positive and the displacement Δx is positive, so $W > 0$.

$W = \frac{1}{2}(2.0 \text{ N})(2.0 \text{ m}) + (2.0 \text{ N})(1.0 \text{ m}) = +4.0$ J

b) During this displacement $F_x = 0$, so $W = 0$.

c) F_x is negative, Δx is positive, so $W < 0$.

$W = -\frac{1}{2}(1.0 \text{ N})(2.0 \text{ m}) = -1.0$ J

d) The work is the sum of the answers to parts (a), (b), and (c), so

$W = 4.0 \text{ J} + 0 - 1.0 \text{ J} = +3.0$ J

e) The work done for $x = 7.0$ m to $x = 3.0$ m is $+1.0$ J. This work is positive since the displacement and the force are both in the $-x$-direction. The magnitude of the work done for $x = 3.0$ m to $x = 2.0$ m is 2.0 J, the area under F_x versus x. This work is negative since the displacement is in the $-x$-direction and the force is in the $+x$-direction. Thus $W = +1.0 \text{ J} - 2.0 \text{ J} = -1.0$ J

EVALUATE: The work done when the car moves from $x = 2.0$ m to $x = 0$ is $-\frac{1}{2}(2.0 \text{ N})(2.0 \text{ m}) = -2.0$ J. Adding this to the work for $x = 7.0$ m to $x = 2.0$ m gives a total of $W = -3.0$ J for $x = 7.0$ m to $x = 0$. The work for $x = 7.0$ m to $x = 0$ is the negative of the work for $x = 0$ to $x = 7.0$ m.

6.39 **IDENTIFY** and **SET UP:** Apply Eq.(6.6). Let point 1 be where the sled is released and point 2 be at $x = 0$ for part (a) and at $x = -0.200$ m for part (b). Use Eq.(6.10)

for the work done by the spring and calculate K_2. Then $K_2 = \frac{1}{2}mv_2^2$ gives v_2.

EXECUTE:

a) $W_{\text{tot}} = K_2 - K_1$ so $K_2 = K_1 + W_{\text{tot}}$

$K_1 = 0$ (released with no initial velocity), $K_2 = \frac{1}{2}mv_2^2$

The only force doing work is the spring force. Eq.(6-10) gives the work done <u>on</u> the spring to move its end from x_1 to x_2. The force the spring exerts on an object attached to it is $F = -kx$, so the work the spring does is

$W_{\text{spr}} = -(\frac{1}{2}kx_2^2 - \frac{1}{2}kx_1^2) = \frac{1}{2}kx_1^2 - \frac{1}{2}kx_2^2$. Here $x_1 = -0.375$ m and $x_2 = 0$.

Thus $W_{\text{spr}} = \frac{1}{2}(4000 \text{ N/m})(-0.375 \text{ m})^2 - 0 = 281$ J.

$K_2 = K_1 + W_{\text{tot}} = 0 + 281 \text{ J} = 281$ J

Then $K_2 = \frac{1}{2}mv_2^2$ implies $v_2 = \sqrt{\dfrac{2K_2}{m}} = \sqrt{\dfrac{2(281 \text{ J})}{70.0 \text{ kg}}} = 2.83$ m/s.

b) $K_2 = K_1 + W_{\text{tot}}$

$K_1 = 0$

$W_{\text{tot}} = W_{\text{spr}} = \frac{1}{2}kx_1^2 - \frac{1}{2}kx_2^2$. Now $x_2 = -0.200$ m, so

$W_{\text{spr}} = \frac{1}{2}(4000 \text{ N/m})(-0.375 \text{ m})^2 - \frac{1}{2}(4000 \text{ N/m})(-0.200 \text{ m})^2 = 281 \text{ J} - 80 \text{ J} = 201$ J

Thus $K_2 = 0 + 201 \text{ J} = 201$ J and $K_2 = \frac{1}{2}mv_2^2$ gives

$v_2 = \sqrt{\dfrac{2K_2}{m}} = \sqrt{\dfrac{2(201 \text{ J})}{70.0 \text{ kg}}} = 2.40$ m/s.

EVALUATE: The spring does positive work and the sled gains speed as it returns to $x = 0$. More work is done during the larger displacement in part (a), so the speed there is larger than in part (b).

6.41 IDENTIFY and **SET UP:** Apply Eq.(6.6) to the glider.

Work is done by the spring and by gravity. Take point 1 to be where the glider is released. In part (a) point 2 is where the glider has traveled 1.80 m and $K_2 = 0$. In part (b) point 2 is where the glider has traveled 0.80 m.

EXECUTE:

a) $W_{\text{tot}} = K_2 - K_1 = 0$. Solve for x_1, the amount the spring is initially compressed.

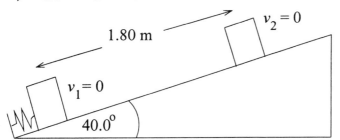

$W_{\text{tot}} = W_{\text{spr}} + W_w = 0$
So $W_{\text{spr}} = -W_w$
(The spring does positive work on the glider since the spring force is directed up the incline, the same as the direction of the displacement.

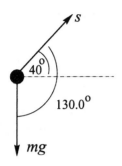

$$W_w = (w \cos \phi)s = (mg \cos 130.0°)s$$
$$W_w = (0.0900 \text{ kg})(9.80 \text{ m/s}^2)(\cos 130.0°)(1.80 \text{ m}) = -1.020 \text{ J}$$
(The component of w parallel to the incline is directed down the incline, opposite to the displacement, so gravity does negative work.)

$$W_{\text{spr}} = -W_w = +1.020 \text{ J}$$

$$W_{\text{spr}} = \tfrac{1}{2}kx_1^2 \text{ so } x_1 = \sqrt{\frac{2W_{\text{spr}}}{k}} = \sqrt{\frac{2(1.020 \text{ J})}{640 \text{ N/m}}} = 0.0565 \text{ m}$$

b) The spring was compressed only 0.0565 m so at this point in the motion the glider is no longer in contact with the spring.

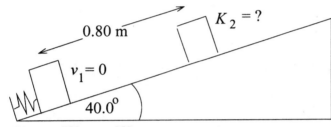

$$W_{\text{tot}} = K_2 - K_1$$
$$K_2 = K_1 + W_{\text{tot}}$$
$$K_1 = 0$$

$$W_{\text{tot}} = W_{\text{spr}} + W_w$$

From part (a), $W_{\text{spr}} = 1.020 \text{ J}$ and

$$W_w = (mg \cos 130.0°)s = (0.0900 \text{ kg})(9.80 \text{ m/s}^2)(\cos 130.0°)(0.80 \text{ m}) = -0.454 \text{ J}$$

Then $K_2 = W_{\text{spr}} + W_w = +1.020 \text{ J} - 0.454 \text{ J} = +0.57 \text{ J}$.

EVALUATE: The kinetic energy in part (b) is positive, as it must be. In part (a), $x_2 = 0$ since the spring force is no longer applied past this point. In computing the work done by gravity we use the full 1.80 m the glider moves.

6.47 **IDENTIFY** and **SET UP:** Calculate the power used to make the plane climb against gravity. Consider the vertical motion since gravity is vertical.

EXECUTE: The rate at which work is being done against gravity is $P = Fv = mgv = (700 \text{ kg})(9.80 \text{ m/s}^2)(2.5 \text{ m/s}) = 17.15 \text{ kW}$.

This is the part of the engine power that is being used to make the airplane climb. The fraction this is of the total is

17.15 kw/75 kW = 0.23

EVALUATE: The power we calculate for making the airplane climb is considerably less than the power output of the engine.

6.49 **IDENTIFY** and **SET UP:** Use Eq.(6.15) to relate the power provided and the amount of work done against gravity in 16.0 s. The work done against gravity depends on the total weight which depends on the number of passengers.

EXECUTE:

Find the total mass that can be lifted:

$$P_{av} = \frac{\Delta W}{\Delta t} = \frac{mgh}{t}, \text{ so } m = \frac{P_{av}t}{gh}$$

$$P_{av} = (40 \text{ hp}) \left(\frac{746 \text{ W}}{1 \text{ hp}} \right) = 2.984 \times 10^4 \text{ W}$$

$$m = \frac{P_{av}t}{gh} = \frac{(2.984 \times 10^4 \text{ W})(16.0 \text{ s})}{(9.80 \text{ m/s}^2)(20.0 \text{ m})} = 2.436 \times 10^3 \text{ kg}$$

This is the total mass of elevator plus passengers. The mass of the passengers is $2.436 \times 10^3 \text{ kg} - 600 \text{ kg} = 1.836 \times 10^3 \text{ kg}$.

The number of passengers is $\dfrac{1.836 \times 10^3 \text{ kg}}{65.0 \text{ kg}} = 28.2.$

28 passengers can ride.

EVALUATE: Typical elevator capacities are about half this, in order to have a margin of safety.

6.53 IDENTIFY and **SET UP:** Use Eq.(6.18). Use Newton's 2nd law to relate F to a and a constant acceleration equation to relate v to a and t.

EXECUTE:

a) $F = ma$. $v = v_0 + at$ and $v_0 = 0$, so $v = at$.
Since the motion is in one direction we have replaced the comonents of $\vec{v}$, $\vec{a}$, and $\vec{F}$ by their magnitudes.
The instantaneous power is $P = Fv = (ma)(at) = ma^2t$.

b) P is proportional to a^2.
Triple a says that increase P by a factor of 9.

c) $\dfrac{P}{t} = ma^2 = $ constant, so $\dfrac{P_1}{t_1} = \dfrac{P_2}{t_2}$

$$P_2 = P_1 \left(\frac{t_2}{t_1} \right) = 36 \text{ W} \left(\frac{15.0 \text{ s}}{5.0 \text{ s}} \right) = 108 \text{ W}.$$

EVALUATE: The instantaneous power increases linearly with time even though the force is constant because the speed increases linearly with time.

Problems

6.57 IDENTIFY and **SET UP:** Since the forces are constant, Eq.(6.2) can be used to calculate the work done by each force.

Forces on the suitcase:

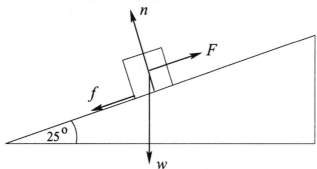

In part (f), Eq.(6.6) is used to relate the total work to the initial and final kinetic energy.

EXECUTE:

a) $W_F = (F \cos \phi)s$

Both $\vec{F}$ and $\vec{s}$ are parallel to the incline and in the same direction, so $\phi = 0°$ and $W_F = Fs = (140 \text{ N})(3.80 \text{ m}) = 532 \text{ J}$

b)

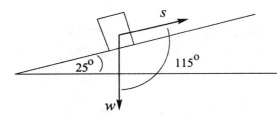

$W_w = (w \cos \phi)s$
$\phi = 115°$, so
$W_w = (196 \text{ N})(\cos 115°)(3.80 \text{ m})$
$W_w = -315 \text{ J}$

Alternatively, the component of w parallel to the incline is $w \sin 25°$. This component is down the incline so its angle with $\vec{s}$ is $\phi = 180°$.

$W_{w \sin 25°} = (196 \text{ N} \sin 25°)(\cos 180°)(3.80 \text{ m}) = -315 \text{ J}$.

The other component of w, $w \cos 25°$, is perpendicular to $\vec{s}$ and hence does no work.

Thus $W_w = W_{w \sin 25°} = -315 \text{ J}$, which agrees with the above.

c) The normal force is perpendicular to the displacement ($\phi = 90°$), so $W_n = 0$.

d) $n = w \cos 25°$ so $f_k = \mu_k n = \mu_k w \cos 25° = (0.30)(196 \text{ N}) \cos 25° = 53.3 \text{ N}$
$W_f = (f_k \cos \phi)s = (53.3 \text{ N})(\cos 180°)(3.80 \text{ m}) = -202 \text{ J}$

e) $W_{\text{tot}} = W_F + W_w + W_N + W_f = +532 \text{ J} - 315 \text{ J} + 0 - 202 \text{ J} = 15 \text{ J}$

f) $W_{\text{tot}} = K_2 - K_1$, $K_1 = 0$, so $K_2 = W_{\text{tot}}$

$\frac{1}{2}mv_2^2 = W_{\text{tot}}$ so $v_2 = \sqrt{\dfrac{2W_{\text{tot}}}{m}} = \sqrt{\dfrac{2(15 \text{ J})}{20.0 \text{ kg}}} = 1.2 \text{ m/s}$

EVALUATE: The total work done is positive and the kinetic energy of the suitcase increases as it moves up the incline.

6.63 IDENTIFY and **SET UP:** Use Eq.(6.6) to relate the work to the kinetic energy, and hence the distance to the speed. Use $\sum \vec{F} = m\vec{a}$ and the force diagram for the package to calculate the normal force and from this the friction force. Work is done by friction and by gravity.

Point 1 is at the top of the ramp and point 2 is where the package stops.

EXECUTE:

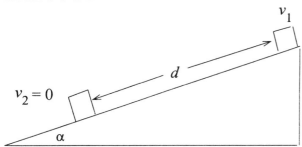

Apply $W_{\text{tot}} = K_2 - K_1$ to the motion of the package from point 1 to point 2.

$K_2 = 0$
$K_1 = \frac{1}{2}mv_1^2$

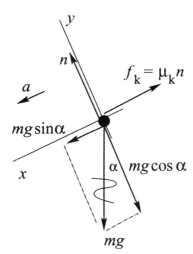

$\sum F_y = ma_y$
$n - mg\cos\alpha = 0$
$n = mg\cos\alpha$

$f_k = \mu_k n = \mu_k mg\cos\alpha$

$W_{\text{tot}} = W_n + W_f + W_{\text{mg}}$

$W_n = 0, \quad W_f = -\mu_k mg\cos\alpha\, d, \quad W_{mg} = mg\sin\alpha\, d$

Then $W_{\text{tot}} = K_2 - K_1$ gives $mg(\sin\alpha - \mu_k\cos\alpha)d = -\frac{1}{2}mv_1^2$.

$$d = \frac{-v_1^2}{2g(\sin\alpha - \mu_k\cos\alpha)} = \frac{-(2.20 \text{ m/s})^2}{2(9.80 \text{ m/s})(\sin 12.0° - (0.310)\cos 12.0°)} = 2.59 \text{ m}$$

EVALUATE: $\mu_k\cos\alpha > \sin\alpha$ so $f_k > mg\sin\alpha$ and $|W_f| > W_{mg}$. The negative work done by friction is greater in magnitude than the positive work done by gravity, so the package loses kinetic energy as it moves down the incline.

6.65 IDENTIFY Apply Eq.(6.6) to the motion of the asteriod.

SET UP: Let point 1 be at a great distance and let point 2 be at the surface of the earth. Assume $K_1 = 0$. From the information given about the gravitational force its magnitude as a function of distance r from the center of the earth must be $F = mg(R_E/r)^2$. This force is directed in the $-\hat{r}$ direction since it is a "pull". F is not constant so Eq.(6.7) must be used to calculate the work it does.

EXECUTE:

$$W = -\int_1^2 F\,ds = -\int_\infty^{R_E} \left(\frac{mgR_E^2}{r^2}\right) dr = -mgR_E^2(-(1/r)|_\infty^{R_E}) = mgR_E$$

$W_{\text{tot}} = K_2 - K_1,\ K_1 = 0$

This gives $K_2 = mgR_E = 1.25 \times 10^{12}$ J

$K_2 = \frac{1}{2}mv_2^2$ so $v_2 = \sqrt{2K_2/m} = 11{,}000$ m/s

EVALUATE: Note that $v_2 = \sqrt{2gR_E}$, the impact speed is independent of the mass of the asteroid.

6.67 IDENTIFY and **SET UP:** The force is not constant so Eq.(6.2) cannot be used. Instead, use Eq.(6.7).

EXECUTE:

$F = \alpha x^3$, with $\alpha = 4.00$ N/m^3

a) For $x = 1.00$ m, $F = (4.00\text{ N/m}^3)(1.00\text{ m})^3 = 4.00$ N.

Force directed toward the origin means that $\vec{F}$ is in the $-x$-direction; $\vec{F} = -(4.00\text{ N})\hat{i}$.

b) For $x = 2.00$ m, $F = (4.00\text{ N/m}^3)(2.00\text{ m})^3 = 32.0$ N; $\vec{F} = -(32.0\text{ N})\hat{i}$.

c) $W = \int_{x_1}^{x_2} F_x\,dx$

$F_x = -\alpha x^3$, negative since $\vec{F}$ is toward the origin.

$W = \int_{x_1}^{x_2}(-\alpha x^3)\,dx = -(\alpha/4)(x^4|_{x_1}^{x_2}) = -(\alpha/4)(x_2^4 - x_1^4) =$

$W = -\frac{1}{4}(4.00\text{ N/m}^3)((2.00\text{ m})^4 - (1.00\text{ m})^4) = -15.0$ J

EVALUATE: The work done is negative since the object moves away from the origin and the force is in the opposite direction, toward the origin.

6.69 IDENTIFY and **SET UP:** Use $\sum \vec{F} = m\vec{a}$ to find the tension force T. The block moves in uniform circular motion and $\vec{a} = \vec{a}_{\text{rad}}$.

a) Free-body diagram for the block:

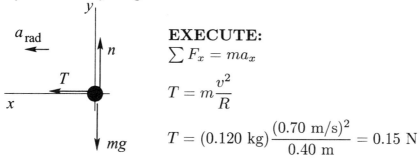

EXECUTE:

$$\sum F_x = ma_x$$

$$T = m\frac{v^2}{R}$$

$$T = (0.120 \text{ kg})\frac{(0.70 \text{ m/s})^2}{0.40 \text{ m}} = 0.15 \text{ N}$$

b) $T = m\dfrac{v^2}{R} = (0.120 \text{ kg})\dfrac{(2.80 \text{ m/s})^2}{0.10 \text{ m}} = 9.4 \text{ N}$

c) SET UP: The tension changes as the distance of the block from the hole changes. We could use $W = \int_{x_1}^{x_2} F_x \, dx$ to calculate the work. But a much simpler approach is to use $W_{\text{tot}} = K_2 - K_1$.

EXECUTE: The only force doing work on the block is the tension in the cord, so $W_{\text{tot}} = W_T$.

$K_1 = \frac{1}{2}mv_1^2 = \frac{1}{2}(0.120 \text{ kg})(0.70 \text{ m/s})^2 = 0.0294 \text{ J}$
$K_2 = \frac{1}{2}mv_2^2 = \frac{1}{2}(0.120 \text{ kg})(2.80 \text{ m/s})^2 = 0.470 \text{ J}$

$W_{\text{tot}} = K_2 - K_1 = 0.470 \text{ J} - 0.029 \text{ J} = 0.44 \text{ J}$

This is the amount of work done by the person who pulled the cord.

EVALUATE: The block moves inward, in the direction of the tension, so T does positive work and the kinetic energy increases.

6.71 IDENTIFY and **SET UP:** Use $v_x = dx/dt$ and $a_x = dv/dt$. Use $\sum \vec{F} = m\vec{a}$ to calculate $\vec{F}$ from $\vec{a}$.

EXECUTE:

a) $x(t) = \alpha t^2 + \beta t^3 \quad v_x(t) = \dfrac{dx}{dt} = 2\alpha t + 3\beta t^2$

$t = 4.00 \text{ s}: v_x = 2(0.200 \text{ m/s}^2)(4.00 \text{ s}) + 3(0.0200 \text{ m/s}^3)(4.00 \text{ s})^2 = 2.56 \text{ m/s}.$

b) $a_x(t) = \dfrac{dv_x}{dt} = 2\alpha + 6\beta t$

$F_x = ma_x = m(2\alpha + 6\beta t)$

$t = 4.00 \text{ s}: F_x = 6.00 \text{ kg}(2(0.200 \text{ m/s}^2) + 6(0.0200 \text{ m/s}^3)(4.00 \text{ s})) = 5.28 \text{ N}$

c) IDENTIFY and **SET UP:** Use Eq.(6.6) to calculate the work.

EXECUTE: $W_{\text{tot}} = K_2 - K_1$

At $t_1 = 0$, $v_1 = 0$ so $K_1 = 0$.

$W_{\text{tot}} = W_F$

$K_2 = \frac{1}{2}mv_2^2 = \frac{1}{2}(6.00 \text{ kg})(2.56 \text{ m/s})^2 = 19.7 \text{ J}$

Then $W_{\text{tot}} = K_2 - K_1$ gives that $W_F = 19.7 \text{ J}$

EVALUATE: v increases with t so the kinetic energy increases and the work done is positive. We can also calculate W_F directly from Eq.(6.7), by writing dx as $v_x \, dt$ and performing the integral.

6.73 IDENTIFY and **SET UP:** Use Eq.(6.6). You do positive work and gravity does negative work. Let point 1 be at the base of the bridge and point 2 be at the top of the bridge.

EXECUTE:

a) $W_{\text{tot}} = K_2 - K_1$

$K_1 = \frac{1}{2}mv_1^2 = \frac{1}{2}(80.0 \text{ kg})(5.00 \text{ m/s})^2 = 1000 \text{ J}$

$K_2 = \frac{1}{2}mv_2^2 = \frac{1}{2}(80.0 \text{ kg})(1.50 \text{ m/s})^2 = 90 \text{ J}$

$W_{\text{tot}} = 90 \text{ J} - 1000 \text{ J} = -910 \text{ J}$

b) Neglecting friction, work is done by you (with the force you apply to the pedals) and by gravity: $W_{\text{tot}} = W_{\text{you}} + W_{\text{gravity}}$.

The gravity force is $w = mg = (80.0 \text{ kg})(9.80 \text{ m/s}^2) = 784 \text{ N}$, downward. The displacement is 5.20 m, upward. Thus $\phi = 180°$ and

$W_{\text{gravity}} = (F \cos \phi)s = (784 \text{ N})(5.20 \text{ m}) \cos 180° = -4077 \text{ J}$

Then $W_{\text{tot}} = W_{\text{you}} + W_{\text{gravity}}$ gives

$W_{\text{you}} = W_{\text{tot}} - W_{\text{gravity}} = -910 \text{ J} - (-4077 \text{ J}) = +3170 \text{ J}$

EVALUATE: The total work done is negative and you lose kinetic energy.

6.77 IDENTIFY and **SET UP:** Use Eq.(6.6). Work is done by the spring and by gravity. Let point 1 be where the textbook is released and point 2 be where it stops sliding. $x_2 = 0$ since at point 2 the spring is neither stretched nor compressed.

EXECUTE:

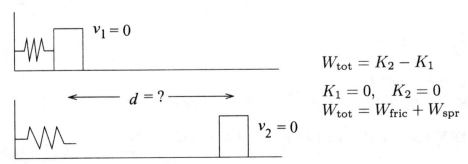

$W_{\text{tot}} = K_2 - K_1$

$K_1 = 0, \quad K_2 = 0$

$W_{\text{tot}} = W_{\text{fric}} + W_{\text{spr}}$

$W_{\text{spr}} = \frac{1}{2}kx_1^2$, where $x_1 = 0.250$ m (Spring force is in direction of motion of block so it does positive work.)

$W_{\text{fric}} = -\mu_{\text{k}}mgd$

Then $W_{\text{tot}} = K_2 - K_1$ gives $\frac{1}{2}kx_1^2 - \mu_{\text{k}}mgd = 0$

$$d = \frac{kx_1^2}{2\mu_{\text{k}}mg} = \frac{(250\ \text{N/m})(0.250\ \text{m})^2}{2(0.30)(2.50\ \text{kg})(9.80\ \text{m/s}^2)} = 1.1\ \text{m, measured from the point where}$$
the block was released.

EVALUATE: The positive work done by the spring equals the magnitude of the negative work done by friction. The total work done during the motion between points 1 and 2 is zero and the textbook starts and ends with zero kinetic energy.

6.83 **IDENTIFY** and **SET UP:** Apply $W_{\text{tot}} = K_2 - K_1$ to the system consisting of both blocks. Since they are connected by the cord, both blocks have the same speed at every point in the motion. Also, when the 6.00-kg block has moved downward 1.50 m, the 8.00-kg block has moved 1.50 m to the right. The target variable, μ_{k}, will be a factor in the work done by friction.

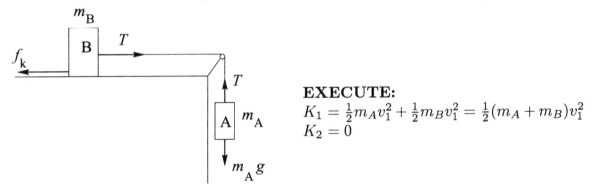

EXECUTE:

$K_1 = \frac{1}{2}m_A v_1^2 + \frac{1}{2}m_B v_1^2 = \frac{1}{2}(m_A + m_B)v_1^2$

$K_2 = 0$

The tension T in the rope does positive work on block B and the same magnitude of negative work on block A, so T does no net work on the system.

Gravity does work $W_{mg} = m_A gd$ on block A, where $d = 2.00$ m. (Block B moves horizontally, so no work is done on it by gravity.)

Friction does work $W_{\text{fric}} = -\mu_{\text{k}}m_B gd$ on block B.

Thus $W_{\text{tot}} = W_{mg} + W_{\text{fric}} = m_A gd - \mu_{\text{k}}m_B gd$.

Then $W_{\text{tot}} = K_2 - K_1$ gives $m_A gd - \mu_{\text{k}}m_B gd = -\frac{1}{2}(m_A + m_B)v_1^2$ and

$$\mu_{\text{k}} = \frac{m_A}{m_B} + \frac{\frac{1}{2}(m_A + m_B)v_1^2}{m_B gd} = \frac{6.00\ \text{kg}}{8.00\ \text{kg}} + \frac{(6.00\ \text{kg} + 8.00\ \text{kg})(0.900\ \text{m/s})^2}{2(8.00\ \text{kg})(9.80\ \text{m/s}^2)(2.00\ \text{m})} = 0.786$$

EVALUATE: The weight of block A does positive work and the friction force on block B does negative work, so the net work is positive and the kinetic energy of the blocks increases as block A descends. Note that K_2 includes the kinetic energy of both blocks. We could have applied the work-energy theorem to block A alone, but then W_{tot} includes the work done on block A by the tension force.

6.85 IDENTIFY: Apply Eq.(6.6) to the skater.

SET UP: Let point 1 be just before she reaches the rough patch and let point 2 be where she exits from the patch. Work is done by friction. We don't know the skater's mass so can't calculate either friction or the initial kinetic energy. Leave her mass m as a variable and expect that it will divide out of the final equation.

EXECUTE:

$f_{\text{k}} = 0.25mg$ so $W_f = W_{\text{tot}} = -(0.25mg)s$, where s is the length of the rough patch.

$W_{\text{tot}} = K_2 - K_1$

$K_1 = \frac{1}{2}mv_0^2$, $K_2 = \frac{1}{2}mv_2^2 = \frac{1}{2}m(0.45v_0^2) = 0.2025(\frac{1}{2}mv_0^2)$

The work-energy relation gives $-(0.25mg)s = (0.2025 - 1)\frac{1}{2}mv_0^2$

The mass divides out, and solving gives $s = 1.5$ m.

EVALUATE: Friction does negative work and this reduces her kinetic energy.

6.89 IDENTIFY and SET UP: Energy is $P_{\text{av}}t$. The total energy expended in one day is the sum of the energy expended in each type of activity.

EXECUTE: 1 day $= 8.64 \times 10^4$ s

Let t_{walk} be the time she spends walking and t_{other} be the time she spends in other activities; $t_{\text{other}} = 8.64 \times 10^4$ s $- t_{\text{walk}}$.

The energy expended in each activity is the power output times the time, so

$E = Pt = (280\text{ W})t_{\text{walk}} + (100\text{ W})t_{\text{other}} = 1.1 \times 10^7$ J

$(280\text{ W})t_{\text{walk}} + (100\text{ W})(8.64 \times 10^4\text{ s} - t_{\text{walk}}) = 1.1 \times 10^7$ J

$(180\text{ W})t_{\text{walk}} = 2.36 \times 10^6$ J

$t_{\text{walk}} = 1.31 \times 10^4$ s $= 218$ min $= 3.6$ h.

EVALUATE: Her average power for one day is $(1.1 \times 10^7$ J$)/([24][3600\text{ s}]) = 127$ W. This is much closer to her 100 W rate than to her 280 W rate, so most of her day is spent at the 100 W rate.

6.91 IDENTIFY and SET UP: Use Eq.(6.15). The work done on the water by gravity is mgh, where $h = 170$ m. Solve for the mass m of water for 1.00 s and then calculate the volume of water that has this mass.

EXECUTE: The power output is $P_{\text{av}} = 2000$ MW $= 2.00 \times 10^9$ W.

$P_{\text{av}} = \dfrac{\Delta W}{\Delta t}$ and 92% of the work done on the water by gravity is converted to electrical power output, so in 1.00 s the amount of work done on the water by gravity is

$$W = \frac{P_{\text{av}}\Delta t}{0.92} = \frac{(2.00 \times 10^9\text{ W})(1.00\text{ s})}{0.92} = 2.174 \times 10^9 \text{ J}$$

$W = mgh$, so the mass of water flowing over the dam in 1.00 s must be

$$m = \frac{W}{gh} = \frac{2.174 \times 10^9 \text{ J}}{(9.80 \text{ m/s}^2)(170 \text{ m})} = 1.30 \times 10^6 \text{ kg}$$

$$\text{density} = \frac{m}{V} \text{ so } V = \frac{m}{\text{density}} = \frac{1.30 \times 10^6 \text{ kg}}{1.00 \times 10^3 \text{ kg/m}^3} = 1.30 \times 10^3 \text{ m}^3.$$

EVALUATE: The dam is 1270 m long, so this volume corresponds to about a m^3 flowing over each 1 m length of the dam, a reasonable amount.

6.93 **IDENTIFY** and **SET UP:** For part (a) calculate m from the volume of blood pumped by the heart in one day. For part (b) use W calculated in part (a) in Eq.(6.15).

EXECUTE:

a) $W = mgh$, as in Example 6.10.

We need the mass of blood lifted; we are given the volume

$$V = (7500 \text{ L}) \left(\frac{1 \times 10^{-3} \text{ m}^3}{1 \text{ L}} \right) = 7.50 \text{ m}^3.$$

$m = \text{density X volume} = (1.05 \times 10^3 \text{ kg/m}^3)(7.50 \text{ m}^3) = 7.875 \times 10^3 \text{ kg}$

Then $W = mgh = (7.875 \times 10^3 \text{ kg})(9.80 \text{ m/s}^2)(1.63 \text{ m}) = 1.26 \times 10^5 \text{ J}$.

b) $P_{av} = \dfrac{\Delta W}{\Delta t} = \dfrac{1.26 \times 10^5 \text{ J}}{(24 \text{ h})(3600 \text{ s/h})} = 1.46 \text{ W}.$

EVALUATE: Compared to light bulbs or common electrical devices, the power output of the heart is rather small.

6.97 **IDENTIFY** and **SET UP:** Use Eq.(6.18) to relate the forces to the power required. The air resistance force is $F_{air} = \frac{1}{2}CA\rho Av^2$, where C is the drag coefficient.

EXECUTE:

a) $P = F_{tot}v$, with $F_{tot} = F_{roll} + F_{air}$

$F_{air} = \frac{1}{2}CA\rho v^2 = \frac{1}{2}(1.0)(0.463 \text{ m}^3)(1.2 \text{ kg/m}^3)(12.0 \text{ m/s})^2 = 40.0 \text{ N}$

$F_{roll} = \mu_r n = \mu_r w = (0.0045)(490 \text{ N} + 118 \text{ N}) = 2.74 \text{ N}$

$P = (F_{roll} + F_{air})v = (2.74 \text{ N} + 40.0 \text{ N})(12.0 \text{ s}) = 513 \text{ W}$

b) $F_{air} = \frac{1}{2}CA\rho v^2 = \frac{1}{2}(0.88)(0.366 \text{ m}^3)(1.2 \text{ kg/m}^3)(12.0 \text{ m/s})^2 = 27.8 \text{ N}$

$F_{roll} = \mu_r n = \mu_r w = (0.0030)(490 \text{ N} + 88 \text{ N}) = 1.73 \text{ N}$

$P = (F_{roll} + F_{air})v = (1.73 \text{ N} + 27.8 \text{ N})(12.0 \text{ s}) = 354 \text{ W}$

c) $F_{\mathrm{air}} = \frac{1}{2}CA\rho v^2 = \frac{1}{2}(0.88)(0.366\ \mathrm{m}^3)(1.2\ \mathrm{kg/m}^3)(6.0\ \mathrm{m/s})^2 = 6.96\ \mathrm{N}$

$F_{\mathrm{roll}} = \mu_r n = 1.73\ \mathrm{N}$ (unchanged)

$P = (F_{\mathrm{roll}} + F_{\mathrm{air}})v = (1.73\ \mathrm{N} + 6.96\ \mathrm{N})(6.0\ \mathrm{s}) = 52.1\ \mathrm{W}$

EVALUATE: Since F_{air} is proportional to v^2 and $P = Fv$, reducing the speed greatly reduces the power required.

6.99 IDENTIFY and **SET UP:** Use Eq.(6.18) to relate F and P.

In part (a), F is the retarding force. In parts (b) and (c), F includes gravity.

EXECUTE:

a) $P = Fv$, so $F = P/v$.

$$P = (8.00\ \mathrm{hp})\left(\frac{746\ \mathrm{W}}{1\ \mathrm{hp}}\right) = 5968\ \mathrm{W}$$

$$v = (60.0\ \mathrm{km/h})\left(\frac{1000\ \mathrm{m}}{1\ \mathrm{km}}\right)\left(\frac{1\ \mathrm{h}}{3600\ \mathrm{s}}\right) = 16.67\ \mathrm{m/s}$$

$$F = \frac{P}{v} = \frac{5968\ \mathrm{W}}{16.67\ \mathrm{m/s}} = 358\ \mathrm{N}.$$

b) The power required is the 8.00 hp of part (a) plus the power P_g required to lift the car against gravity.

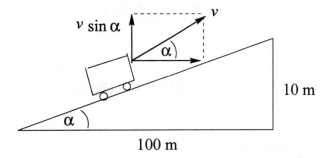

$$\tan\alpha = \frac{10\ \mathrm{m}}{100\ \mathrm{m}} = 0.10$$

$$\alpha = 5.71°$$

The vertical component of the velocity of the car is $v\sin\alpha = (16.67\ \mathrm{m/s})\sin 5.71° = 1.658\ \mathrm{m/s}$.

Then $P_g = F(v\sin\alpha) = mgv\sin\alpha = (1800\ \mathrm{kg})(9.80\ \mathrm{m/s}^2)(1.658\ \mathrm{m/s}) = 2.92 \times 10^4\ \mathrm{W}$

$$P_g = 2.92 \times 10^4\ \mathrm{W}\left(\frac{1\ \mathrm{hp}}{746\ \mathrm{W}}\right) = 39.1\ \mathrm{hp}$$

The total power required is $8.00\ \mathrm{hp} + 39.1\ \mathrm{hp} = 47.1\ \mathrm{hp}$.

c) The power required from the engine is <u>reduced</u> by the rate at which gravity does positive work.

The road incline angle α is given by $\tan\alpha = 0.0100$, so $\alpha = 0.5729°$.

$P_g = mg(v\sin\alpha) = (1800\ \mathrm{kg})(9.80\ \mathrm{m/s}^2)(16.67\ \mathrm{m/s})\sin 0.5729° = 2.94 \times 10^3\ \mathrm{W} = 3.94\ \mathrm{hp}$.

The power required from the engine is then 8.00 hp - 3.94 hp = 4.06 hp.

d) No power is needed from the engine if gravity does work at the rate of
$P_g = 8.00$ hp $= 5968$ W

$$P_g = mgv \sin \alpha, \text{ so } \sin \alpha = \frac{P_g}{mgv} = \frac{5968 \text{ W}}{(1800 \text{ kg})(9.80 \text{ m/s}^2)(16.67 \text{ m/s})} = 0.02030$$

$\alpha = 1.163°$ and $\tan \alpha = 0.0203$, a 2.03% grade.

EVALUATE: More power is required when the car goes uphill and less when it goes downhill. In part (d), at this angle the component of gravity down the incline is $mg \sin \alpha = 358$ N and this force cancels the retarding force and no force from the engine is required. The retarding force depends on the speed so it is the same in parts (a), (b), and (c).

CHAPTER 7
POTENTIAL ENERGY AND ENERGY CONSERVATION

Exercises 5, 9, 11, 13, 17, 19, 21, 23, 27, 31, 33, 35, 37
Problems 43, 47, 49, 51 53, 55, 59, 61, 63, 65, 67, 73, 75, 81, 83

Exercises

7.5 **IDENTIFY** and **SET UP:** Use energy methods.

a) $K_1 + U_1 + W_{other} = K_2 + U_2$. Solve for K_2 and then use $K_2 = \frac{1}{2}mv_2^2$ to obtain v_2.

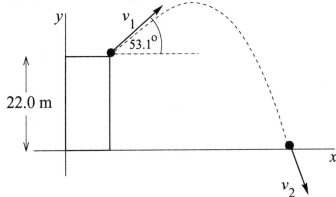

$W_{other} = 0$ (The only force on the ball while it is in the air is gravity.)
$K_1 = \frac{1}{2}mv_1^2$; $K_2 = \frac{1}{2}mv_2^2$
$U_1 = mgy_1$, $y = 22.0$ m
$U_2 = mgy_2 = 0$, since $y_2 = 0$ for our choice of coordinates.

EXECUTE: $\frac{1}{2}mv_1^2 + mgy_1 = \frac{1}{2}mv_2^2$

$v_2 = \sqrt{v_1^2 + 2gy_1} = \sqrt{(12.0 \text{ m/s})^2 + 2(9.80 \text{ m/s}^2)(22.0 \text{ m})} = 24.0 \text{ m/s}$

EVALUATE: The projection angle of 53.1° doesn't enter into the calculation. The kinetic energy depends only on the magnitude of the velocity; it is independent of the direction of the velocity.

b) Nothing changes in the calculation. The expression derived in part (a) for v_2 is independent of the angle, so $v_2 = 24.0$ m/s, the same as in part (a).

c) The ball travels a shorter distance in part (b), so in that case air resistance will have less effect.

7.9 **IDENTIFY:** Use energy methods.

SET UP: The forces on the object are gravity, the normal force n and friction. The normal force is at all points in the motion perpendicular to the displacement, so it does no work. Hence $W_{other} = W_f$, the work done by friction.

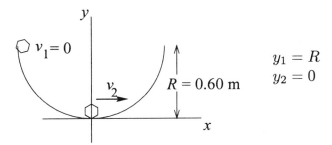

EXECUTE: $K_1 + U_1 + W_{\text{other}} = K_2 + U_2$

$K_1 = \frac{1}{2}mv_1^2 = 0.$ $K_2 = \frac{1}{2}mv_2^2$

$U_1 = mgy_1 = (0.20 \text{ kg})(9.80 \text{ m/s}^2)(0.50 \text{ m}) = 0.980 \text{ J},$ $U_2 = mgy_2 = 0$

$W_{\text{other}} = W_f = -0.22 \text{ J}$

Thus $0 + U_1 + W_f = K_2 + 0$

$0.980 \text{ J} - 0.22 \text{ J} = \frac{1}{2}(0.20 \text{ kg})v_2^2$

$v_2 = \sqrt{\dfrac{2(0.760 \text{ J})}{0.20 \text{ kg}}} = 2.8 \text{ m/s}$

EVALUATE: In the absence of friction v_2 is the same as it the object was dropped from a height of 0.5 m; in this case $v_2 = \sqrt{2gy_1} = 3.1 \text{ m/s}$. Friction does negative work and decreases v_2.

7.11   IDENTIFY: Apply Eq.(7.7) to the motion of the car.

SET UP: Take $y = 0$ at point A. Let point 1 be A and point 2 be B.

$K_1 + U_1 + W_{\text{other}} = K_2 + U_2$

EXECUTE: $U_1 = 0$, $U_2 = mg(2R) = 28,224 \text{ J}$, $W_{\text{other}} = W_f$

$K_1 = \frac{1}{2}mv_1^2 = 37,500 \text{ J}$, $K_2 = \frac{1}{2}mv_2^2 = 3840 \text{ J}$

The work-energy relation then gives $W_f = K_2 + U_2 - K_1 = -5400 \text{ J}$.

EVALUATE: Friction does negative work. The final mechanical energy ($K_2 + U_2 = 32,064 \text{ J}$) is less than the initial mechanical energy ($K_1 + U_1 = 37,500 \text{ J}$) because of the energy removed by friction work.

7.13

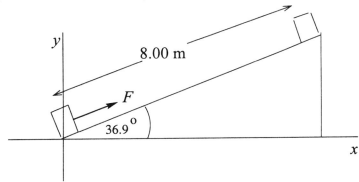

a) IDENTIFY and **SET UP:** $\vec{F}$ is constant so Eq.(6.2) can be used.

EXECUTE: $W_F = (F\cos\phi)s = (110 \text{ N})(\cos 0°)(8.00 \text{ m}) = 880 \text{ J}$

EVALUATE: $\vec{F}$ is in the direction of the displacement and does positive work.

b) IDENTIFY and **SET UP:** Calculate W using Eq.(6.2) but first must calculate the friction force. Use the free-body diagram for the oven to calculate the normal force n; then the friction force can be calculated from $f_k = \mu_k n$. For this calculation use coordinates parallel and perpendicular to the incline.

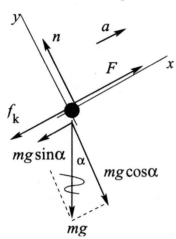

EXECUTE:

$\sum F_y = ma_y$

$n - mg\cos 36.9° = 0$

$n = mg\cos 36.9°$

$f_k = \mu_k n = \mu_k mg\cos 36.9°$

$f_k = (0.25)(10.0 \text{ kg})(9.80 \text{ m/s}^2)\cos 36.9° = 19.6 \text{ N}$

$W_f = (f_k\cos\phi)s = (19.6 \text{ N})(\cos 180°)(8.00 \text{ m}) = -157 \text{ J}$

EVALUATE: Friction does negative work.

c) IDENTIFY and **SET UP:** $U = mgy$; take $y = 0$ at the bottom of the ramp.

EXECUTE:

$\Delta U = U_2 - U_1 = mg(y_2 - y_1) = (10.0 \text{ kg})(9.80 \text{ m/s}^2)(4.80 \text{ m} - 0) = 470 \text{ J}$

EVALUATE: The object moves upward and U increases.

d) IDENTIFY and **SET UP:** Use Eq.(7.7). Solve for ΔK.

EXECUTE: $K_1 + U_1 + W_{\text{other}} = K_2 + U_2$

$\Delta K = K_2 - K_1 = U_1 - U_2 + W_{\text{other}}$

$\Delta K = W_{\text{other}} - \Delta U$

$W_{\text{other}} = W_F + W_f = 880 \text{ J} - 157 \text{ J} = 723 \text{ J}$

$\Delta U = 470 \text{ J}$

Thus $\Delta K = 723 \text{ J} - 470 \text{ J} = 253 \text{ J}$.

EVALUATE: W_{other} is positive. Some of W_{other} goes to increasing U and the rest goes to increasing K.

e) IDENTIFY: Apply $\sum \vec{F} = m\vec{a}$ to the oven. Solve for $\vec{a}$ and then use a constant acceleration equation to calculate v_2.

SET UP: We can use the free-body diagram that is in part (b):

$$\sum F_x = ma_x$$

$$F - f_k - mg\sin 36.9° = ma$$

EXECUTE: $a = \dfrac{F - f_k - mg\sin 36.9°}{m} =$

$$\dfrac{110\text{ N} - 19.6\text{ N} - (10.0\text{ kg})(9.80\text{ m/s}^2)\sin 36.9°}{10.0\text{ kg}} = 3.16\text{ m/s}^2$$

SET UP: $v_{1x} = 0,\quad a_x = 3.16\text{ m/s}^2,\quad x - x_0 = 8.00\text{ m},\quad v_{2x} = ?$

$$v_{2x}^2 = v_{1x}^2 + 2a_x(x - x_0)$$

EXECUTE: $v_{2x} = \sqrt{2a_x(x - x_0)} = \sqrt{2(3.16\text{ m/s}^2)(8.00\text{ m})} = 7.11\text{ m/s}^2$

Then $\Delta K = K_2 - K_1 = \frac{1}{2}mv_2^2 = \frac{1}{2}(10.0\text{ kg})(7.11\text{ m/s})^2 = 253$ J.

EVALUATE: This agrees with the result calculated in part (d) using energy methods.

7.17 IDENTIFY and **SET UP:** Use energy methods. There are changes in both elastic and gravitational potential energy; elastic: $U = \frac{1}{2}kx^2$, gravitational: $U = mgy$.

EXECUTE:

a) $U = \frac{1}{2}kx^2$ so $x = \sqrt{\dfrac{2U}{k}} = \sqrt{\dfrac{2(3.20\text{ J})}{1600\text{ N/m}}} = 0.0632$ m $= 6.32$ cm

b)

$K_1 + U_1 + W_{\text{other}} = K_2 + U_2$

$W_{\text{other}} = 0$ (only work is that done by gravity and spring force)

$K_1 = 0, K_2 = 0$

$y = 0$ at final position of book

$U_1 = mg(h + d),\quad U_2 = \frac{1}{2}kd^2$

$$0 + mg(h + d) + 0 = \tfrac{1}{2}kd^2$$

The original gravitational potential energy of the system is converted into potential energy of the compressed spring.

$$\tfrac{1}{2}kd^2 - mgd - mgh = 0$$

$$d = \frac{1}{k}\left(mg \pm \sqrt{(mg)^2 + 4(\tfrac{1}{2}k)(mgh)}\right)$$

d must be positive, so $d = \dfrac{1}{k}(mg + \sqrt{(mg)^2 + 2kmgh})$

$d = \dfrac{1}{1600 \text{ N/m}}((1.20 \text{ kg})(9.80 \text{ m/s}^2)+$

$\overline{\sqrt{((1.20 \text{ kg})(9.80 \text{ m/s}^2)^2 + 2(1600 \text{ N/m})(1.20 \text{ kg})(9.80 \text{ m/s}^2)(0.80 \text{ m}))}}$

$d = 0.0074 \text{ m} + 0.1087 \text{ m} = 0.12 \text{ m} = 12 \text{ cm}$

EVALUATE: It was important to recognize that the total displacement was $h + d$; gravity continues to do work as the book moves against the spring. Also note that with the spring compressed 0.12 m it exerts an upward force (23.0 N) greater than the weight of the book (11.8 N). The book will be accelerated upward from this position.

7.19 IDENTIFY: Use energy methods. There are changes in both elastic and gravitational potential energy.

SET UP: $K_1 + U_1 + W_{\text{other}} = K_2 + U_2$

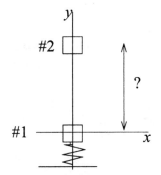

The spring force and gravity are the only forces doing work on the cheese, so $W_{\text{other}} = 0$ and $U = U_{\text{grav}} + U_{\text{el}}$.

EXECUTE: Cheese released from rest implies $K_1 = 0$.

At the maximum height $v_2 = 0$ so $K_2 = 0$.

$U_1 = U_{1,\text{el}} + U_{1,\text{grav}}$

$y_1 = 0$ implies $U_{1,\text{grav}} = 0$

$U_{1,\text{el}} = \tfrac{1}{2}kx_1^2 = \tfrac{1}{2}(1800 \text{ N/m})(0.15 \text{ m})^2 = 20.25 \text{ J}$

(Here x_1 refers to the amount the spring is stretched or compressed when the cheese is at poisition 1; it is not the x-coordinate of the cheese in the coordinate system shown in the sketch.)

$U_2 = U_{2,\text{el}} + U_{2,\text{grav}}$

$U_{2,\text{grav}} = mgy_2$, where y_2 is the height we are solving for

$U_{2,\text{el}} = 0$ since now the spring is no longer compressed

Putting all this into $K_1 + U_1 + W_{\text{other}} = K_2 + U_2$ gives $U_{1,\text{el}} = U_{2,\text{grav}}$

$y_2 = \dfrac{20.25 \text{ J}}{mg} = \dfrac{20.25 \text{ J}}{(1.20 \text{ kg})(9.80 \text{ m/s}^2)} = 1.72 \text{ m}$

EVALUATE: The description in terms of energy is very simple; the elastic potential energy originally stored in the spring is converted into gravitational potential energy of the system.

7.21 **IDENTIFY** and **SET UP:** Use energy methods. The elastic potential energy changes. In part (a) solve for K_2 and from this obtain v_2. In part (b) solve for U_1 and from this obtain x_1.

a) $K_1 + U_1 + W_{other} = K_2 + U_2$

point 1: the glider is at its initial position, where $x_1 = 0.100$ m and $v_1 = 0$

point 2: the glider is at $x = 0$

EXECUTE:

$K_1 = 0$ (released from rest), $K_2 = \frac{1}{2}mv_2^2$

$U_1 = \frac{1}{2}kx_1^2$, $U_2 = 0$, $W_{other} = 0$ (only the spring force does work)

Thus $\frac{1}{2}kx_1^2 = \frac{1}{2}mv_2^2$. (The initial potential energy of the stretched spring is converted entirely into kinetic energy of the glider.)

$$v_2 = x_1\sqrt{\frac{k}{m}} = (0.100 \text{ m})\sqrt{\frac{5.00 \text{ N/m}}{0.200 \text{ kg}}} = 0.500 \text{ m/s}$$

b) The maximum speed occurs at $x = 0$, so the same equation applies.

$\frac{1}{2}kx_1^2 = \frac{1}{2}mv_2^2$

$$x_1 = v_2\sqrt{\frac{m}{k}} = 2.50 \text{ m/s}\sqrt{\frac{0.200 \text{ kg}}{5.00 \text{ N/m}}} = 0.500 \text{ m}$$

EVALUATE: Elastic potential energy is converted into kinetic energy. A larger x_1 gives a larger v_2.

7.23 **a) IDENTIFY** and **SET UP:** Use energy methods. Both elastic and gravitational potential energy changes. Work is done by friction.

Choose point 1 as in Example 7.10 and let that be the origin, so $y_1 = 0$. Let point 2 be 1.00 m below point 1, so $y_2 = -1.00$ m.

EXECUTE:

$K_1 + U_1 + W_{other} = K_2 + U_2$

$K_1 = \frac{1}{2}mv_1^2 = \frac{1}{2}(2000 \text{ kg})(25 \text{ m/s})^2 = 625,000 \text{ J}$, $U_1 = 0$

$W_{other} = -f \mid y_2 \mid = -(17.000 \text{ N})(1.00 \text{ m}) = -17,000 \text{ J}$

$K_2 = \frac{1}{2}mv_2^2$

$U_2 = U_{2,grav} + U_{2,el} = mgy_2 + \frac{1}{2}ky_2^2$

$U_2 = (2000 \text{ kg})(9.80 \text{ m/s}^2)(-1.00 \text{ m}) + \frac{1}{2}(1.41 \times 10^5 \text{ N/m})(1.00 \text{ m})^2$

$U_2 = -19,600 \text{ J} + 70,500 \text{ J} = +50,900 \text{ J}$

Thus $625,000 \text{ J} - 17,000 \text{ J} = \frac{1}{2}mv_2^2 + 50,900 \text{ J}$

$\frac{1}{2}mv_2^2 = 557,100 \text{ J}$

$$v_2 = \sqrt{\frac{2(557,100 \text{ J})}{2000 \text{ kg}}} = 23.6 \text{ m/s}$$

EVALUATE: The elevator stops after descending 3.00 m. After descending 1.00 m it is still moving but has slowed down.

b) IDENTIFY: Apply $\sum \vec{F} = m\vec{a}$ to the elevator. We know the forces and can solve for $\vec{a}$.

SET UP: Free-body diagram for the elevator:

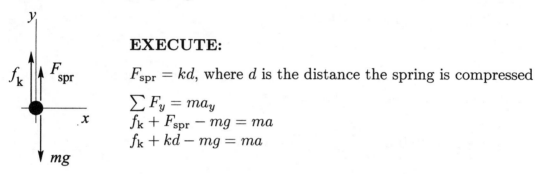

EXECUTE:

$F_{\text{spr}} = kd$, where d is the distance the spring is compressed

$\sum F_y = ma_y$

$f_k + F_{\text{spr}} - mg = ma$

$f_k + kd - mg = ma$

$$a = \frac{f_k + kd - mg}{m} = \frac{17,000 \text{ N} + (1.41 \times 10^5 \text{ N/m})(1.00 \text{ m}) - (2000 \text{ kg})(9.80 \text{ m/s}^2)}{2000 \text{ kg}}$$

$= 69.2 \text{ m/s}^2$

We calculate that a is positive, so the acceleration is upward.

EVALUATE: The velocity is downward and the acceleration is upward, so the elevator is slowing down at this point. Note that $a = 7.1g$; this is unacceptably high for an elevator.

7.27 IDENTIFY and **SET UP:** The force is not constant so we must use Eq.(6.14) to calculate W. The properties of work done by a conservative force are described in Section 7.3.

$W = \int_1^2 \vec{F} \cdot d\vec{l}, \quad \vec{F} = -\alpha x^2 \hat{i}$

EXECUTE:

a) $d\vec{l} = dy\hat{j}$ (x is constant; the displacement is in the $+y$-direction)

$\vec{F} \cdot d\vec{l} = 0$ (since $\hat{i} \cdot \hat{j} = 0$) and thus $W = 0$.

b) $d\vec{l} = dx\hat{i}$

$\vec{F} \cdot d\vec{l} = (-\alpha x^2) \cdot (dx\hat{i}) = -\alpha x^2 \, dx$

$W = \int_{x_1}^{x_2}(-\alpha x^2)\,dx = -\frac{1}{3}\alpha x^3\big|_{x_1}^{x_2} = -\frac{1}{3}\alpha(x_2^3 - x_1^3) = -\frac{12\ \text{N/m}^2}{3}((0.30\ \text{m})^3 - (0.10\ \text{m})^3) = -0.10\ \text{J}$

c) $d\vec{l} = dx\hat{i}$ as in part (b), but now $x_1 = 0.30$ m and $x_2 = 0.10$ m
$W = -\frac{1}{3}\alpha(x_2^3 - x_1^3) = +0.10\ \text{J}$

d) **EVALUATE:** The total work for the displacement along the x-axis from 0.10 m to 0.30 m and then back to 0.10 m is the sum of the results of parts (b) and (c), which is zero. The total work is zero when the starting and ending points are the same, so the force is conservative.

EXECUTE: $W_{x_1 \to x_2} = -\frac{1}{3}\alpha(x_2^3 - x_1^3) = \frac{1}{3}\alpha x_1^3 - \frac{1}{3}\alpha x_2^3$

The definition of the potential energy function is $W_{x_1 \to x_2} = U_1 - U_2$. Comparison of the two expressions for W gives $U = \frac{1}{3}\alpha x^3$. This does correspond to $U = 0$ when $x = 0$.

EVALUATE: In part (a) the work done is zero because the force and displacement are perpendicular. In part (b) the force is directed opposite to the displacement and the work done is negative. In part (c) the force and displacement are in the same direction and the work done is positive.

7.31 IDENTIFY and **SET UP:** The friction force is constant during each displacement and Eq.(6.2) can be used to calculate work, but the direction of the friction force can be different for different displacements.

$f = \mu_k mg = (0.25)(1.5\ \text{kg})(9.80\ \text{m/s}^2) = 3.675\ \text{N}$; direction of $\vec{f}$ is opposite to the motion.

EXECUTE:

a)

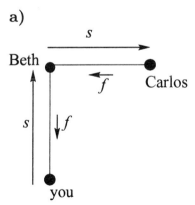

For the motion from you to Beth the friction force is directed opposite to the displacement $\vec{s}$ and $W_1 = -fs = -(3.675\ \text{N})(8.0\ \text{m}) = -29.4\ \text{J}$.

For the motion from Beth to Carlos the friction force is again directed opposite to the displacement and $W_2 = -29.4\ \text{J}$.

$W_{\text{tot}} = W_1 + W_2 = -29.4\ \text{J} - 29.4\ \text{J} = -59\ \text{J}$

b)

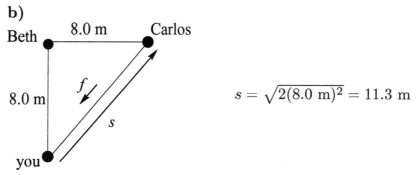

$$s = \sqrt{2(8.0 \text{ m})^2} = 11.3 \text{ m}$$

$\vec{f}$ is opposite to $\vec{s}$, so $W = -fs = -(3.675 \text{ N})(11.3 \text{ m}) = -42 \text{ J}$

c)

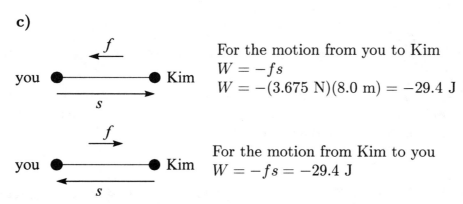

For the motion from you to Kim
$W = -fs$
$W = -(3.675 \text{ N})(8.0 \text{ m}) = -29.4 \text{ J}$

For the motion from Kim to you
$W = -fs = -29.4 \text{ J}$

The total work for the round trip is $-29.4 \text{ J} -29.4 \text{ J} = -59 \text{ J}$.

d) EVALUATE: Parts (a) and (b) show that for two different paths between you and Carlos, the work done by friction is different. Part (c) shows that when the starting and ending points are the same, the total work is not zero. Both these results show that the friction force is nonconservative.

7.33   IDENTIFY and **SET UP:** Use Eq.(7.17) to calculate the force from $U(x)$. Use coordinates where the origin is at one atom. The other atom then has coordinate x.

EXECUTE:
$$F_x = -\frac{dU}{dx} = -\frac{d}{dx}\left(-\frac{C_6}{x^6}\right) = +C_6\frac{d}{dx}\left(\frac{1}{x^6}\right) = -\frac{6C_6}{x^7}$$
The minus sign mean that F_x is directed in the $-x$-direction, toward the origin.
The force has magnitude $6C_6/x^7$ and is attractive.

EVALUATE: U depends only on x so $\vec{F}$ is along the x-axis; it has no y or z components.

7.35   IDENTIFY and **SET UP:** Use Eq.(7.17) to calculate the force from U.
EXECUTE:

$$U(x,y) = k(x^2 + y^2) + k'xy \quad F_x = -\frac{\partial U}{\partial x} = -2kx - k'y$$

$$F_y = -\frac{\partial U}{\partial y} = -2ky - k'x$$

$$\vec{F} = -(2kx + k'y)\hat{i} - (2ky + k'x)\hat{j}$$

EVALUATE: U depends on both x and y so has both x and y components.

7.37 IDENTIFY and **SET UP:** Use Eq.(7.17) to calculate the force from U. At equilibrium $F = 0$.

a) EXECUTE:

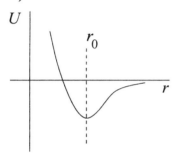
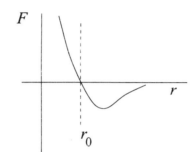

$$U = \frac{a}{r^{12}} - \frac{b}{r^6}$$

$$F = -\frac{dU}{dr} = +\frac{12a}{r^{13}} - \frac{6b}{r^7}$$

b) At equilibrium $F = 0$, so $\dfrac{dU}{dr} = 0$

$F = 0$ implies $\dfrac{+12a}{r^{13}} - \dfrac{6b}{r^7} = 0$

$6br^6 = 12a$; solution is the equilibrium distance $r_0 = (2a/b)^{1/6}$

U is a minimum at this r; the equilibrium is stable.

c) At $r = (2a/b)^{1/6}$, $U = a/r^{12} - b/r^6 = a(b/2a)^2 - b(b/2a) = -b^2/4a$.
At $r \to \infty$, $U = 0$. The energy that must be added is $-\Delta U = b^2/4a$.

d) $r_0 = (2a/b)^{1/6} = 1.13 \times 10^{-10}$ m gives that
$2a/b = 2.082 \times 10^{-60}$ m^6 and $b/4a = 2.402 \times 10^{59}$ m^{-6}
$b^2/4a = b(b/4a) = 1.54 \times 10^{-18}$ J
$b(2.402 \times 10^{59}$ m$^{-6}) = 1.54 \times 10^{-18}$ J and $b = 6.41 \times 10^{-78}$ J·m^6.

Then $2a/b = 2.082 \times 10^{-60}$ m^6 gives $a = (b/2)(2.082 \times 10^{-60}$ m$^6) = $
$\frac{1}{2}(6.41 \times 10^{-78}$ J·m$^6)(2.082 \times 10^{-60}$ m$^6) = 6.67 \times 10^{-138}$ J·m^{12}

EVALUATE: As the graphs in part (a) show, $F(r)$ is the slope of $U(r)$ at each r. $U(r)$ has a minimum where $F = 0$.

Problems

7.43 IDENTIFY: Use the work-energy theorem, Eq(7.7). The target variable μ_k will be a factor in the work done by friction.

SET UP: Let point 1 be where the block is released and let point 2 be where the block stops.

$$K_1 + U_1 + W_{other} = K_2 + U_2$$

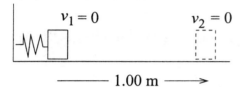

Work is done on the block by the spring and by friction, so $W_{other} = W_f$ and $U = U_{el}$.

EXECUTE: $K_1 = K_2 = 0$

$U_1 = U_{1,el} = \frac{1}{2}kx_1^2 = \frac{1}{2}(100 \text{ N/m})(0.200)^2 = 2.00 \text{ J}$

$U_2 = U_{2,el} = 0$, since after the block leaves the spring has given up all its stored energy

$W_{other} = W_f = (f_k \cos\phi)s = \mu_k mg(\cos\phi)s = -\mu_k mgs$, since $\phi = 180°$ (The friction force is directed opposite to the displacement and does negative work.)

Putting all this into $K_1 + U_1 + W_{other} = K_2 + U_2$ gives

$U_{1,el} + W_f = 0$

$\mu_k mgs = U_{1,el}$

$$\mu_k = \frac{U_{1,el}}{mgs} = \frac{2.00 \text{ J}}{(0.50 \text{ kg})(9.80 \text{ m/s}^2)(1.00 \text{ m})} = 0.41.$$

EVALUATE: $U_{1,el} + W_f = 0$ says that the potential energy originally stored in the spring is taken out of the system by the negative work done by friction.

7.47 a) IDENTIFY: Use work-energy relation to find the kinetic energy of the wood as it enters the rough bottom.

SET UP: Let point 1 be where the piece of wood is released and point 2 be just before it enters the rough bottom. Let $y = 0$ be at point 2.

EXECUTE: $U_1 = K_2$ gives $K_2 = mgy_1 = 78.4 \text{ J}$.

IDENTIFY: Now apply work-energy relation to the motion along the rough bottom.

SET UP: Let point 1 be where it enters the rough bottom and point 2 be where it stops.

$$K_1 + U_1 + W_{other} = K_2 + U_2$$

EXECUTE: $W_{other} = W_f = -\mu_k mgs$, $K_2 = U_1 = U_2 = 0$; $K_1 = 78.4 \text{ J}$

$78.4 \text{ J} - \mu_k mgs = 0$; solving for s gives $s = 20.0 \text{ m}$.

The wood stops after traveling 20.0 m along the rough bottom.

b) Friction does -78.4 J of work.

EVALUATE: The piece of wood stops before it makes one trip across the rough bottom. The final mechanical energy is zero. The negative friction work takes away all the mechanical energy initially in the system.

7.49 a) IDENTIFY: Apply Eq.(7.7) to the motion of the stone.

SET UP: $K_1 + U_1 + W_{\text{other}} = K_2 + U_2$

Let point 1 be point A and point 2 be point B. Take $y = 0$ at point B.

EXECUTE: $mgy_1 + \frac{1}{2}mv_1^2 = \frac{1}{2}mv_2^2$, with $h = 20.0$ m and $v_1 = 10.0$ m/s

$v_2 = \sqrt{v_1^2 + 2gh} = 22.2$ m/s

EVALUATE: The loss of gravitational potential energy equals the gain of kinetic energy.

b) IDENTIFY: Apply Eq.(7.15) to the motion of the stone from point B to where it comes to rest against the spring.

SET UP: Use $K_1 + U_1 + W_{\text{other}} = K_2 + U_2$, with point 1 at B and point 2 where the spring has its maximum compression x.

EXECUTE: $U_1 = U_2 = K_2 = 0$; $K_1 = \frac{1}{2}mv_1^2$ with $v_1 = 22.2$ m/s

$W_{\text{other}} = W_f + W_{\text{el}} = -\mu_{\text{k}}mgs - \frac{1}{2}kx^2$, with $s = 100$ m $+ x$

The work-energy relation gives $K_1 + W_{\text{other}} = 0$.

$\frac{1}{2}mv_1^2 - \mu_{\text{k}}mgs - \frac{1}{2}kx^2 = 0$

Putting in the numerical values gives $x^2 + 29.4x - 750 = 0$. The positive root to this equation is $x = 16.4$ m.

EVALUATE: Part of the initial mechanical (kinetic) energy is removed by friction work and the rest goes into the potential energy stored in the spring.

c) IDENTIFY and **SET UP:** Consider the forces.

EXECUTE: When the spring is compressed $x = 16.4$ m the force it exerts on the stone is $F_{\text{el}} = kx = 32.8$ N. The maximum possible static friction force is

$\max f_{\text{s}} = \mu_{\text{s}}mg = (0.80)(15.0 \text{ kg})(9.80 \text{ m/s}^2) = 118$ N.

EVALUATE: The spring force is less than the maximum possible static friction force so the stone remains at rest.

7.51 IDENTIFY: Apply $K_1 + U_1 + W_{\text{other}} = K_2 + U_2$ to the motion of the person.

SET UP: Point 1 is where he steps off the platform and point 2 is where he is stopped by the cord. Let $y = 0$ at point 2. $y_1 = 41.0$ m. $W_{\text{other}} = -\frac{1}{2}kx^2$, where $x = 11.0$ m is the amount the cord is stretched at point 2. The cord does negative work.

EXECUTE: $K_1 = K_2 = U_2 = 0$, so $mgy_1 - \frac{1}{2}kx^2 = 0$ and $k = 631$ N/m.

Now apply $F = kx$ to the test pulls:

$F = kx$ so $x = F/k = 0.602$ m.

EVALUATE: All his initial gravitational potential energy is taken away by the negative work done by the force exerted by the cord, and this amount of energy is stored as elastic potential energy in the stretched cord.

7.53 IDENTIFY: Use the work-energy theorem, Eq.(7.7). Solve for K_2 and then for v_2.

SET UP: Let point 1 be at his initial position against the compressed spring and let point 2 be at the end of the barrel. Use $F = kx$ to find the amount the spring is initially compressed by the 4400 N force.

$K_1 + U_1 + W_{\text{other}} = K_2 + U_2$

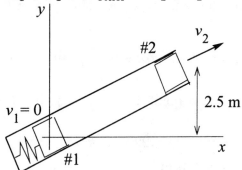

Take $y = 0$ at his initial position.

EXECUTE:

$K_1 = 0, \quad K_2 = \frac{1}{2}mv_2^2$

$W_{\text{other}} = W_{\text{fric}} = -fs$

$W_{\text{other}} = -(40 \text{ N})(4.0\text{m}) = -160$ J

$U_{1,\text{grav}} = 0, \quad U_{1,\text{el}} = \frac{1}{2}kd^2$, where d is the distance the spring is initially compressed.

$F = kd$ so $d = \dfrac{F}{k} = \dfrac{4400 \text{ N}}{1100 \text{ N/m}} = 4.00$ m

and $U_{1,\text{el}} = \frac{1}{2}(1100 \text{ N/m})(4.00 \text{ m})^2 = 8800$ J

$U_{2,\text{grav}} = mgy_2 = (60 \text{ kg})(9.80 \text{ m/s}^2)(2.5 \text{ m}) = 1470 \text{ J}, U_{2,\text{el}} = 0$

Then $K_1 + U_1 + W_{\text{other}} = K_2 + U_2$ gives

$8800 \text{ J} - 160 \text{ J} = \frac{1}{2}mv_2^2 + 1470$ J

$\frac{1}{2}mv_2^2 = 7170$ J and $v_2 = \sqrt{\dfrac{2(7170 \text{ J})}{60 \text{ kg}}} = 15.5$ m/s

EVALUATE: Some of the potential energy stored in the compressed spring is taken away by the work done by friction. The rest goes partly into gravitational potential energy and partly into kinetic energy.

7.55 IDENTIFY: Apply Eq.(7.7) to the system consisting of the two buckets. If we ignore the inertia of the pulley we ignore the kinetic energy it has.

SET UP: $K_1 + U_1 + W_{\text{other}} = K_2 + U_2$

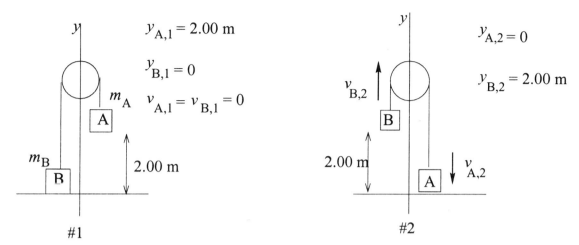

#1   #2

The tension force does positive work on the 4.0 kg bucket and an equal amount of negative work on the 12.0 kg bucket, so the net work done by the tension is zero.

Work is done on the system only by gravity, so $W_{\text{other}} = 0$ and $U = U_{\text{grav}}$

EXECUTE: $K_1 = 0$

$K_2 = \frac{1}{2}m_A v_{A,2}^2 + \frac{1}{2}m_B v_{B,2}^2$ But since the two buckets are connected by a rope they move together and have the same speed: $v_{A,2} = v_{B,2} = v_2$.

Thus $K_2 = \frac{1}{2}(m_A + m_B)v_2^2 = (8.00 \text{ kg})v_2^2$.

$U_1 = m_A g y_{A,1} = (12.0 \text{ kg})(9.80 \text{ m/s}^2)(2.00 \text{ m}) = 235.2 \text{ J}$.

$U_2 = m_B g y_{B,2} = (4.0 \text{ kg})(9.80 \text{ m/s}^2)(2.00 \text{ m}) = 78.4 \text{ J}$.

Putting all this into $K_1 + U_1 + W_{\text{other}} = K_2 + U_2$ gives

$U_1 = K_2 + U_2$

$235.2 \text{ J} = (8.00 \text{ kg})v_2^2 + 78.4 \text{ J}$

$$v_2 = \sqrt{\frac{235.2 \text{ J} - 78.4 \text{ J}}{8.00 \text{ kg}}} = 4.4 \text{ m/s}$$

EVALUATE: The gravitational potential energy decreases and the kinetic energy increases by the same amount. We could apply Eq.(7.7) to one bucket, but then we would have to include in W_{other} the work done on the bucket by the tension T.

7.59 a) IDENTIFY and **SET UP:** Apply Eq.(7.7) to the motion of the potato.
Let point 1 be where the potato is released and point 2 be at the lowest point in its motion.

$K_1 + U_1 + W_{\text{other}} = K_2 + U_2$

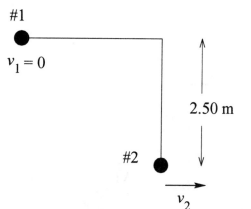

$y_1 = 2.50$ m
$y_2 = 0$

The tension in the string is at all points in the motion perpendicular to the displacement, so $W_T = 0$

The only force that does work on the potato is gravity, so $W_{\text{other}} = 0$.

EXECUTE: $K_1 = 0$, $K_2 = \frac{1}{2}mv_2^2$,
$U_1 = mgy_1$, $U_2 = 0$

Thus $U_1 = K_2$.
$$mgy_1 = \frac{1}{2}mv_2^2$$

$$v_2 = \sqrt{2gy_1} = \sqrt{2(9.80 \text{ m/s}^2)(2.50 \text{ m})} = 7.00 \text{ m/s}$$

EVALUATE: v_2 is the same as if the potato fell through 2.50 m.

b) IDENTIFY: Apply $\sum \vec{F} = m\vec{a}$ to the potato. The potato moves in an arc of a circle so its acceleration is $\vec{a}_{\text{rad}}$, where $a_{\text{rad}} = v^2/R$ and is directed toward the center of the circle. Solve for one of the forces, the tension T in the string.

SET UP: Free-body diagram for the potato as it swings through its lowest point

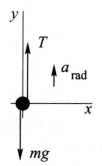

The acceleration $\vec{a}_{\text{rad}}$ is directed in toward the center of the circular path, so at this point it is upward.

EXECUTE: $\sum F_y = ma_y$
$$T - mg = ma_{\text{rad}}$$

$T = m(g + a_{\text{rad}}) = m(g + \frac{v_2^2}{R})$, where the radius R for the circular motion is the length L of the string.

It is instructive to use the algebraic expression for v_2 from part (a) rather than just putting in the numerical value:
$v_2 = \sqrt{2gy_1} = \sqrt{2gL}$, so $v_2^2 = 2gL$

Then $T = m(g + \frac{v_2^2}{L}) = m(g + \frac{2gL}{L}) = 3mg$; the tension at this point is three times the weight of the potato.

$T = 3mg = 3(0.100 \text{ kg})(9.80 \text{ m/s}^2) = 2.94 \text{ N}$

EVALUATE: The tension is greater than the weight; the acceleration is upward so the net force must be upward.

7.61 IDENTIFY and **SET UP:** There are two situations to compare: stepping off a platform and sliding down a pole. Apply the work-energy theorem to each.

a) EXECUTE: Speed at ground if steps off platform at height h:

$K_1 + U_1 + W_{other} = K_2 + U_2$

$mgh = \frac{1}{2}mv_2^2$, so $v_2^2 = 2gh$

Motion from top to bottom of pole: (take $y = 0$ at bottom)

$K_1 + U_1 + W_{other} = K_2 + U_2$

$mgd - fd = \frac{1}{2}mv_2^2$

Use $v_2^2 = 2gh$ and get $mgd - fd = mgh$

$fd = mg(d - h)$

$f = mg(d - h)/d = mg(1 - h/d)$

EVALUATE: For $h = d$ this gives $f = 0$ as it should (friction has no effect).

For $h = 0$, $v_2 = 0$ (no motion). The equation for f gives $f = mg$ in this special case. When $f = mg$ the forces on him cancel and he doesn't accelerate down the pole, which agrees with $v_2 = 0$.

b) EXECUTE: $f = mg(1 - h/d) = (75 \text{ kg})(9.80 \text{ m/s}^2)(1 - 1.0 \text{ m}/2.5 \text{ m}) = 441 \text{ N}$.

c) Take $y = 0$ at bottom of pole, so $y_1 = d$ and $y_2 = y$.

$K_1 + U_1 + W_{other} = K_2 + U_2$

$0 + mgd - f(d - y) = \frac{1}{2}mv^2 + mgy$

$\frac{1}{2}mv^2 = mg(d - y) - f(d - y)$

Using $f = mg(1 - h/d)$ gives $\frac{1}{2}mv^2 = mg(d - y) - mg(1 - h/d)(d - y)$

$\frac{1}{2}mv^2 = mg(h/d)(d - y)$ and $v = \sqrt{2gh(1 - y/d)}$

EVALUATE: This gives the correct results for $y = 0$ and for $y = d$.

7.63 IDENTIFY and **SET UP:** First apply $\sum \vec{F} = m\vec{a}$ to the skier. Find the angle α where the normal force becomes zero, in terms of the speed v_2 at this point. Then apply the work-energy theorem to the motion of the skier to obtain another equation that relates v_2 and α. Solve these two equations for α.

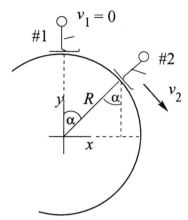

Let point 2 be where the skier loses contact with the snowball.

Loses contact implies $n \to 0$.

$y_1 = R, \; y_2 = R\cos\alpha$

First, analyze the forces on the skier when she is at point 2. For this use coordinates that are in the tangential and radial directions. The skier moves in an arc of a circle, so her acceleration is $a_{\text{rad}} = v^2/R$, directed in towards the center of the snowball.

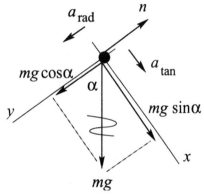

EXECUTE:

$\sum F_y = ma_y$
$mg\cos\alpha - n = mv_2^2/R$
But $n = 0$ so $mg\cos\alpha = mv_2^2/R$
$v_2^2 = Rg\cos\alpha$

Now use conservation of energy to get another equation relating v_2 to α:

$K_1 + U_1 + W_{\text{other}} = K_2 + U_2$

The only force that does work on the skier is gravity, so $W_{\text{other}} = 0$.

$K_1 = 0, \quad K_2 = \frac{1}{2}mv_2^2$
$U_1 = mgy_1 = mgR \qquad U_2 = mgy_2 = mgR\cos\alpha$
Then $mgR = \frac{1}{2}mv_2^2 + mgR\cos\alpha$
$v_2^2 = 2gR(1 - \cos\alpha)$

Combine this with the $\sum F_y = ma_y$ equation:

$Rg\cos\alpha = 2gR(1 - \cos\alpha)$

$\cos\alpha = 2 - 2\cos\alpha$

$3\cos\alpha = 2$ so $\cos\alpha = 2/3$ and $\alpha = 48.2°$

EVALUATE: She speeds up and her a_{rad} increases as she loses gravitational potential energy. She loses contact when she is going so fast that the radially inward component of her weight isn't large enough to keep her in the circular path. Note that α where she loses contact does not depend on her mass or on the radius of the snowball.

7.65 IDENTIFY and **SET UP:**

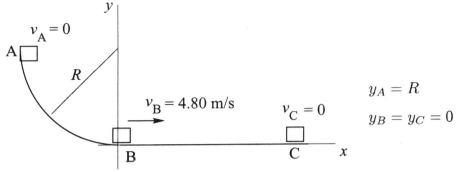

$y_A = R$

$y_B = y_C = 0$

a) Apply conservation of energy to the motion from B to C:

$K_B + U_B + W_{\text{other}} = K_C + U_C$

EXECUTE: The only force that does work on the package during this part of the motion is friction, so

$W_{\text{other}} = W_f = f_k(\cos\phi)s = \mu_k mg(\cos 180°)s = -\mu_k mgs$

$K_B = \frac{1}{2}mv_B^2, \quad K_C = 0$

$U_B = 0, \quad U_C = 0$

Thus $K_B + W_f = 0$

$\frac{1}{2}mv_B^2 - \mu_k mgs = 0$

$\mu_k = \dfrac{v_B^2}{2gs} = \dfrac{(4.80 \text{ m/s})^2}{2(9.80 \text{ m/s}^2)(3.00 \text{ m})} = 0.392$

EVALUATE: The negative friction work takes away all the kinetic energy.

b) IDENTIFY and **SET UP:** Apply conservation of energy to the motion from A to B:

$K_A + U_A + W_{\text{other}} = K_B + U_B$

EXECUTE: Work is done by gravity and by friction, so $W_{\text{other}} = W_f$.

$K_A = 0, \quad K_B = \frac{1}{2}mv_B^2 = \frac{1}{2}(0.200 \text{ kg})(4.80 \text{ m/s})^2 = 2.304 \text{ J}$

$U_A = mgy_A = mgR = (0.200 \text{ kg})(9.80 \text{ m/s}^2)(1.60 \text{ m}) = 3.136 \text{ J}, \quad U_B = 0$

Thus $U_A + W_f = K_B$

$W_f = K_B - U_A = 2.304 \text{ J} - 3.136 \text{ J} = -0.83 \text{ J}$

EVALUATE: W_f is negative as expected; the friction force does negative work since it is directed opposite to the displacement.

7.67 $F_x = -\alpha x - \beta x^2$, $\alpha = 60.0 \text{ N/m}$ and $\beta = 18.0 \text{ N/m}^2$

a) IDENTIFY: Use Eq.(6.7) to calculate W and then use $W = -\Delta U$ to identify the potential energy function $U(x)$.

SET UP: $W_{F_x} = U_1 - U_2 = \int_{x_1}^{x_2} F_x(x)\, dx$

Let $x_1 = 0$ and $U_1 = 0$. Let x_2 be some arbitrary point x, so $U_2 = U(x)$.
EXECUTE:

$U(x) = -\int_0^x F_x(x)\,dx = -\int_0^x (-\alpha x - \beta x^2)\,dx = \int_0^x (\alpha x + \beta x^2)\,dx = \frac{1}{2}\alpha x^2 + \frac{1}{3}\beta x^3$.

EVALUATE: If $\beta = 0$, the spring does obey Hooke's law, with $k = \alpha$, and our result reduces to $\frac{1}{2}kx^2$.

b) IDENTIFY: Apply Eq.(7.15) to the motion of the object.
SET UP:

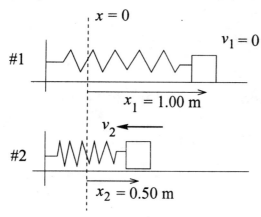

$K_1 + U_1 + W_{\text{other}} = K_2 + U_2$

The only force that does work on the object is the spring force, so $W_{\text{other}} = 0$.

EXECUTE: $K_1 = 0$, $\quad K_2 = \frac{1}{2}mv_2^2$

$U_1 = U(x_1) = \frac{1}{2}\alpha x_1^2 + \frac{1}{3}\beta x_1^3 = \frac{1}{2}(60.0 \text{ N/m})(1.00 \text{ m})^2 + \frac{1}{3}(18.0 \text{ N/m}^2)(1.00 \text{ m})^3$
$= 36.0 \text{ J}$

$U_2 = U(x_2) = \frac{1}{2}\alpha x_2^2 + \frac{1}{3}\beta x_2^3 = \frac{1}{2}(60.0 \text{ N/m})(0.500 \text{ m})^2 + \frac{1}{3}(18.0 \text{ N/m}^2)(0.500 \text{ m})^3$
$= 8.25 \text{ J}$

Thus $36.0 \text{ J} = \frac{1}{2}mv_2^2 + 8.25 \text{ J}$

$v_2 = \sqrt{\dfrac{2(36.0 \text{ J} - 8.25 \text{ J})}{0.900 \text{ kg}}} = 7.85 \text{ m/s}$

EVALUATE: The elastic potential energy stored in the spring decreases and the kinetic energy of the object increases.

7.73 IDENTIFY: Apply Eq.(7.15) to the motion of the block.
SET UP:

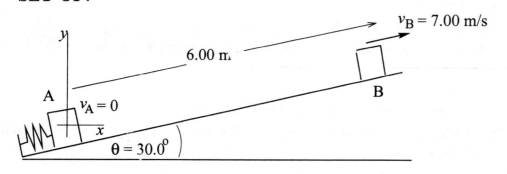

The normal force is $n = mg \cos \theta$, so $f_k = \mu_k n = \mu_k mg \cos \theta$.

$y_A = 0; \quad y_B = (6.00 \text{ m}) \sin 30.0° = 3.00 \text{ m}$

$K_A + U_A + W_{\text{other}} = K_B + U_B$

EXECUTE:

Work is done by gravity, by the spring force, and by friction, so $W_{\text{other}} = W_f$ and $U = U_{\text{el}} + U_{\text{grav}}$

$K_A = 0, \quad K_B = \frac{1}{2}mv_B^2 = \frac{1}{2}(1.50 \text{ kg})(7.00 \text{ m/s})^2 = 36.75 \text{ J}$

$U_A = U_{\text{el},A} + U_{\text{grav},A} = U_{\text{el},A}$, since $U_{\text{grav},A} = 0$

$U_B = U_{\text{el},B} + U_{\text{grav},B} = 0 + mgy_B = (1.50 \text{ kg})(9.80 \text{ m/s}^2)(3.00 \text{ m}) = 44.1 \text{ J}$

$W_{\text{other}} = W_f = (f_k \cos \phi)s = \mu_k mg \cos \theta (\cos 180°)s = -\mu_k mg \cos \theta\, s$

$W_{\text{other}} = -(0.50)(1.50 \text{ kg})(9.80 \text{ m/s}^2)(\cos 30.0°)(6.00 \text{ m}) = -38.19 \text{ J}$

Thus $U_{\text{el},A} - 38.19 \text{ J} = 36.75 \text{ J} + 44.10 \text{ J}$

$U_{\text{el},A} = 38.19 \text{ J} + 36.75 \text{ J} + 44.10 \text{ J} = 119 \text{ J}$

EVALUATE: U_{el} must always be repulsive. Part of the energy initially stored in the spring was taken away by friction work; the rest went partly into kinetic energy and partly into an increase in gravitational potential energy.

7.75 a) IDENTIFY and **SET UP:** Apply $K_A + U_A + W_{\text{other}} = K_B + U_B$ to the motion from A to B.

EXECUTE:

$K_A = 0, \quad K_B = \frac{1}{2}mv_B^2$

$U_A = 0, \quad U_B = U_{\text{el},B} = \frac{1}{2}kx_B^2$, where $x_B = 0.25 \text{ m}$

$W_{\text{other}} = W_F = Fx_B$

Thus $Fx_B = \frac{1}{2}mv_B^2 + \frac{1}{2}kx_B^2$. (The work done by F goes partly to the potential energy of the stretched spring and partly to the kinetic energy of the block.)

$Fx_B = (20.0 \text{ N})(0.25 \text{ m}) = 5.0 \text{ J}$ and $\frac{1}{2}kx_B^2 = \frac{1}{2}(40.0 \text{ N/m})(0.25 \text{ m})^2 = 1.25 \text{ J}$

Thus $5.0 \text{ J} = \frac{1}{2}mv_B^2 + 1.25 \text{ J}$ and $v_B = \sqrt{\dfrac{2(3.75 \text{ J})}{0.500 \text{ kg}}} = 3.87 \text{ m/s}$

b) IDENTIFY: Apply Eq.(7.15) to the motion of the block. Let point C be where the block is closest to the wall. When the block is at point C the spring is compressed an amount $| x_C |$, so the block is $0.60 \text{ m}- | x_C |$ from the wall, and the distance between B and C is $x_B+ | x_C |$.

SET UP:

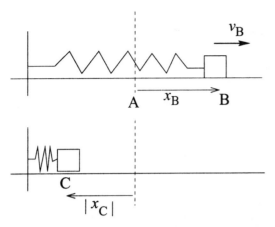

$$K_B + U_B + W_{\text{other}} = K_C + U_C$$

EXECUTE:

$W_{\text{other}} = 0$

$K_B = \frac{1}{2}mv_B^2 = 5.0 \text{ J} - 1.25 \text{ J} = 3.75 \text{ J}$
(from part (a))

$U_B = \frac{1}{2}kx_B^2 = 1.25 \text{ J}$

$K_C = 0$ (instantaneously at rest at
point closest to wall)

$U_C = \frac{1}{2}k \mid x_C \mid^2$

Thus $3.75 \text{ J} + 1.25 \text{ J} = \frac{1}{2}k \mid x_C \mid^2$

$$\mid x_C \mid = \sqrt{\frac{2(5.0 \text{ J})}{40.0 \text{ N/m}}} = 0.50 \text{ m}$$

The distance of the block from the wall is $0.60 \text{ m} - 0.50 \text{ m} = 0.10 \text{ m}$.

EVALUATE: The work $(20.0 \text{ N})(0.25 \text{ m}) = 5.0 \text{ J}$ done by F puts 5.0 J of mechancial energy into the system. No mechanical energy is taken away by friction, so the total energy at points B and C is 5.0 J.

7.81 **IDENTIFY** and **SET UP:** The potential energy of a horizontal layer of thickness dy, area A, and height y is $dU = (dm)gy$. Let ρ be the density of water.

EXECUTE: $dm = \rho \, dV = \rho A \, dy$, so $dU = \rho Agy \, dy$.

The total potential energy U is

$U = \int_0^h dU = \rho Ag \int_0^h y \, dy = \frac{1}{2}\rho Agh^2$.

$A = 3.0 \times 10^6 \text{ m}^2$ and $h = 150 \text{ m}$, so $U = 3.3 \times 10^{14} \text{ J} = 9.2 \times 10^7 \text{ kWh}$.

EVALUATE: The volume is Ah and the mass of water is $\rho V = \rho Ah$. The average depth is $h_{\text{av}} = h/2$, so $U = mgh_{\text{av}}$.

7.83 $\vec{F} = -\alpha xy^2\hat{j}$, $\alpha = 2.50 \text{ N/m}^3$

IDENTIFY: $\vec{F}$ is not constant so use Eq.(6.14) to calculate W. $\vec{F}$ must be evaluated along the path.

a) SET UP:

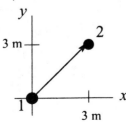

$d\vec{l} = dx\hat{i} + dy\hat{j}$
$\vec{F} \cdot d\vec{l} = -\alpha xy^2 \, dy$
On the path, $x = y$ so $\vec{F} \cdot d\vec{l} = -\alpha y^3 \, dy$

EXECUTE: $W = \int_1^2 \vec{F} \cdot d\vec{l} = \int_{y_1}^{y_2} (-\alpha y^3)\, dy = -(\alpha/4)(y^4|_{y_1}^{y_2}) = -(\alpha/4)(y_2^4 - y_1^4)$

$y_1 = 0$, $y_2 = 3.00$ m, so $W = -\frac{1}{4}(2.50 \text{ N/m}^3)(3.00 \text{ m})^4 = -50.6$ J

b) SET UP:

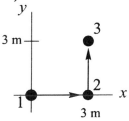

For the displacement from point 1 to point 2, $d\vec{l} = dx\hat{i}$, so $\vec{F} \cdot d\vec{l} = 0$ and $W = 0$. (The force is perpendicular to the displacement at each point along the path, so $W = 0$.)

For the displacement from point 2 to point 3, $d\vec{l} = dy\hat{j}$, so $\vec{F} \cdot d\vec{l} = -\alpha x y^2\, dy$.

On this path, $x = 3.00$ m, so

$\vec{F} \cdot d\vec{l} = -(2.50 \text{ N/m}^3)(3.00 \text{ m})y^2\, dy = -(7.50 \text{ N/m}^2)y^2\, dy$.

EXECUTE: $W = \int_2^3 \vec{F} \cdot d\vec{l} = -(7.50 \text{ N/m}^2) \int_{y_2}^{y_3} y^2\, dy = -(7.50 \text{ N/m}^2)\frac{1}{3}(y_3^3 - y_2^3)$

$W = -(7.50 \text{ N/m}^2)(\frac{1}{3})(3.00 \text{ m})^3 = -67.5$ J

c) EVALUATE: For these two paths between the same starting and ending points the work done is different, so the force is nonconservative.

CHAPTER 8
MOMENTUM, IMPULSE, AND COLLISIONS

Exercises

8.5 **IDENTIFY:** The momentum of the system is the vector sum of the momenta of the two objects.

SET UP: $P_y = p_{1y} + p_{2y}$

EXECUTE:

$p_{1y} = mv_{1y} = (0.145 \text{ kg})(1.30 \text{ m/s}) = 0.1885 \text{ kg} \cdot \text{m/s}$

$p_{2y} = mv_{2y} = (0.057 \text{ kg})(-7.80 \text{ m/s}) = -0.4446 \text{ kg} \cdot \text{m/s}$

$P_y = p_{1y} + p_{2y} = 0.1885 \text{ kg} \cdot \text{m/s} - 0.4446 \text{ kg} \cdot \text{m/s} = -0.256 \text{ kg} \cdot \text{m/s}$

(minus sign means that is in the $-y$-direction)

EVALUATE: The momenta are in opposite directions so subtract their magnitudes to find the total momentum; must add the momenta as vectors.

8.9 **IDENTFIY:** Use Eq.(8.9). We know the initial momentum and the impulse so can solve for the final momentum and then the final velocity.

SET UP: Take the x-axis to be toward the right, so $v_{1x} = +3.00 \text{ m/s}$. Use Eq.(8.5) to calculate the impulse, since the force is constant.

EXECUTE:

a) $J_x = p_{2x} - p_{1x}$

$J_x = F_x(t_2 - t_1) = (+25.0\text{N})(0.050 \text{ s}) = +1.25 \text{ kg} \cdot \text{m/s}$

Thus $p_{2x} = J_x + p_{1x} = +1.25 \text{ kg} \cdot \text{m/s} + (0.160 \text{ kg})(+3.00 \text{ m/s}) = +1.73 \text{ kg} \cdot \text{m/s}$

$$v_{2x} = \frac{p_{2x}}{m} = \frac{1.73 \text{ kg} \cdot \text{m/s}}{0.160 \text{ kg}} = +10.8 \text{ m/s} \text{ (to the right)}$$

b) $J_x = F_x(t_2 - t_1) = (-12.0 \text{ N})(0.050 \text{ s}) = -0.600 \text{ kg} \cdot \text{m/s}$ (negative since force is to left)

$p_{2x} = J_x + p_{1x} = -0.600 \text{ kg} \cdot \text{m/s} + (0.160 \text{ kg})(+3.00 \text{ m/s}) = -0.120 \text{ kg} \cdot \text{m/s}$

$$v_{2x} = \frac{p_{2x}}{m} = \frac{-0.120 \text{ kg} \cdot \text{m/s}}{0.160 \text{ kg}} = -0.75 \text{ m/s} \text{ (to the left)}$$

EVALUATE: In part (a) the impulse and initial momentum are in the same direction and v_x increases. In part (b) the impulse and initial momentum are in opposite directions and the velocity decreases.

8.11 IDENTIFY: Calculate the impulse of each force and then use Eq.(8.6) to find the final momentum. We must use the total impulse, the vector sum of the impulse due to each force.

SET UP: $\vec{F}$ is not constant so we must do the integral in Eq.(8.7). Gravity is constant and for its impulse we can use Eq.(8.5). Assume that $+y$ is upward. Express the vectors $\vec{J}$, $\vec{p}_1$, and $\vec{p}_2$ in terms of unit vectors.

EXECUTE:

a) $\vec{J} = \int_{t_1}^{t_2} \vec{F}\, dt = \int_{t_1}^{t_2} [(1.60 \times 10^7 \text{ N/s})t - (6.00 \times 10^9 \text{ N/s}^2)t^2]\hat{i}\, dt$, where $t_1 = 0$ and $t_2 = 2.50 \times 10^{-3}$ s.

$\vec{J} = [(1.60 \times 10^7 \text{ N/s})(\frac{1}{2}t_2^2) - (6.00 \times 10^9 \text{ N/s}^2)(\frac{1}{3}t_2^3)]\hat{i}$

$\vec{J} = [\frac{1}{2}(1.60 \times 10^7 \text{ N/s})(2.50 \times 10^{-3} \text{ s})^2 - \frac{1}{3}(6.00 \times 10^9 \text{ N/s}^2)(2.50 \times 10^{-3} \text{ s})^3]\hat{i} =$
$(18.8 \text{ kg} \cdot \text{m/s})\hat{i}$

b) $\vec{w} = -(mg)\hat{j}$ (constant)

$\vec{J} = \vec{w}(t_2 - t_1) = -mgt_2\hat{j} = -(0.145 \text{ kg})(9.80 \text{ m/s}^2)(2.50 \times 10^{-3} \text{ s})\hat{j}$

$\vec{J} = -(0.00355 \text{ kg} \cdot \text{m/s})\hat{j}$

c) $\vec{F}_{\text{av}} = \dfrac{\vec{J}}{(t_2 - t_1)} = \dfrac{(18.8 \text{ kg} \cdot \text{m/s})\vec{i}}{2.50 \times 10^{-3} \text{ s}} = (7520 \text{ N})\hat{i}$

d) $\vec{J} = \vec{p}_2 - \vec{p}_1$, so $\vec{p}_2 = \vec{p}_1 + \vec{J}$

$\vec{J} = (18.8 \text{ kg} \cdot \text{m/s})\hat{i} - (0.00335 \text{ kg} \cdot \text{m/s})\hat{j}$

$\vec{p}_1 = m\vec{v}_1 = (0.145 \text{ kg})[(-40.0 \text{ m/s})\hat{i} - (5.0 \text{ m/s})\hat{j}] = (-5.80 \text{ kg} \cdot \text{m/s})\hat{i} -$
$(0.725 \text{ kg} \cdot \text{m/s})\hat{j}$

Then $\vec{p}_2 = (-5.80 \text{ kg·m/s})\hat{i} - (0.725 \text{ kg·m/s})\hat{j} + (18.8 \text{ kg·m/s})\hat{i} - (0.00335 \text{ kg·m/s})\hat{j}$

$\vec{p}_2 = (13.0 \text{ kg} \cdot \text{m/s})\hat{i} - (0.73 \text{ kg} \cdot \text{m/s})\hat{j}$

$\vec{v}_2 = \vec{p}_2/m = (89.7 \text{ m/s})\hat{i} - (5.0 \text{ m/s})\hat{j}$

EVALUATE: The bat exerts a large impulse in the $+x$-direction and v_x has a large positive change: the ball's velocity changes from being in the $-x$-direction to the $+x$-direction. J_y due to gravity is small and has little effect on v_y.

8.13 IDENTIFY and **SET UP:** Evaluate the integral in Eq.(8.9) to calculate J_x. Then solve for the final momentum and final velocity. Take the x-axis to be toward the right, so $F_x = +(A + Bt^2)$.

EXECUTE:

a) $J_x = \int_{t_1}^{t_2} F_x\, dt = \int_0^{t_2}(A + Bt^2)\, dt = (At + \frac{1}{3}Bt^3)\, \big|_0^{t_2} = At_2 + \frac{1}{3}Bt_2^3$.

The impulse has magnitude $J = At_2 + \frac{1}{3}Bt_2^3$ and is directed to the right.

b) $J_x = p_{2x} - p_{1x}$ so $p_{2x} = J_x + p_{1x}$

Initially at rest implies $p_{1x} = 0$.

$v_{2x} = p_{2x}/m = J_x/m = (A/m)t_2 + (B/3m)t_2^3$

Her speed is $v = (A/m)t_2 + (B/3m)t_2^3$ and she is moving to the right.

EVALUATE: The force in the $+x$ direction produces an impulse in the $+x$ direction and this increases v_x.

8.17 a) IDENTIFY: External horizontal forces during the collsion can be neglected so P_x is conserved.

SET UP: Let Gretzky be object A and the defender be object B. Let the x-direction be the direction in which Gretzky is moving initially.

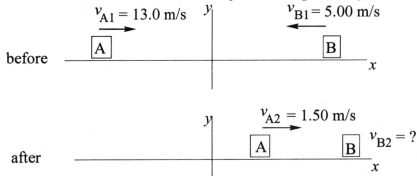

EXECUTE:

P_x constant implies $m_A v_{A1x} + m_B v_{B1x} = m_A v_{A2x} + m_B v_{B2x}$

If we multiply through by g we get an equation that uses the object's weight rather than its mass:

$w_A v_{A1x} + w_B v_{B1x} = w_A v_{A2x} + w_B v_{B2x}$

The components can be positive or negative depending on the directions of the velocities relative to our coordinate system.

Putting in the numbers:

$(756 \text{ N})(+13.0 \text{ m/s}) + (900 \text{ N})(-5.00 \text{ m/s}) = (756 \text{ N})(+1.50 \text{ m/s}) + (900 \text{ N})v_{B2x}$

$9828 \text{ N} \cdot \text{m/s} - 4500 \text{ N} \cdot \text{m/s} = 1134 \text{ N} \cdot \text{m/s} + (900 \text{ N})v_{B2x}$

$$v_{B2x} = \frac{9828 \text{ N} \cdot \text{m/s} - 4500 \text{ N} \cdot \text{m/s} - 1134 \text{ N} \cdot \text{m/s}}{900 \text{ N}} = 4.66 \text{ m/s}$$

After the collision the defender is moving at 4.66 m/s in the same direction as Gretzky.

b) IDENTIFY and **SET UP:** Use $K = \frac{1}{2}mv^2$ to calculate the kinetic energy for each object before the collsion; add to get the total kinetic energy before the collision

(K_1). Repeat for after the collision (K_2). The change in K is $\Delta K = K_2 - K_1$.

EXECUTE:

$K_1 = \frac{1}{2}m_A v_{A1}^2 + \frac{1}{2}m_B v_{B1}^2$

$K_1 = \frac{1}{2}\left(\dfrac{756 \text{ N}}{9.80 \text{ m/s}^2}\right)(13.0 \text{ m/s})^2 + \frac{1}{2}\left(\dfrac{900 \text{ N}}{9.80 \text{ m/s}^2}\right)(5.00 \text{ m/s})^2$

$K_1 = 6519 \text{ J} + 1148 \text{ J} = 7667 \text{ J}$

$K_2 = \frac{1}{2}m_A v_{A2}^2 + \frac{1}{2}m_b v_{B2}^2$

$K_2 = \frac{1}{2}\left(\dfrac{756 \text{ N}}{9.80 \text{ m/s}^2}\right)(1.50 \text{ m/s})^2 + \frac{1}{2}\left(\dfrac{900 \text{ N}}{9.80 \text{ m/s}^2}\right)(4.66 \text{ m/s})^2$

$K_2 = 87 \text{ J} + 997 \text{ J} = 1084 \text{ J}$

$\Delta K = K_2 - K_1 = 1084 \text{ J} - 7667 \text{ J} = -6580 \text{ J}$

The kinetic energy decreases by 6580 J.

EVALUATE: The kinetic energy of an object is always positive, and that it does not depend on the direction of the object's velocity.

8.19 IDENTIFY: In part (a) no horizontal force implies P_x is constant. In part (b) use the energy expression, Eq.(7.15), to find the potential energy initially in the spring.

SET UP: Initially both blocks are at rest.

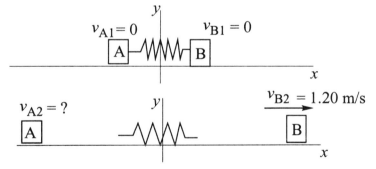

EXECUTE:

a)

$m_A v_{A1x} + m_B v_{B1x} = m_A v_{A2x} + m_B v_{B2x}$

$0 = m_A v_{A2x} + m_B v_{B2x}$

$v_{A2x} = -\left(\dfrac{m_B}{m_A}\right)v_{b2x} = -\left(\dfrac{3.00 \text{ kg}}{1.00 \text{ kg}}\right)(+1.20 \text{ m/s}) = -3.60 \text{ m/s}$

Block A has a final speed of 3.60 m/s, and moves off in the opposite direction to B.

b) Use energy conservation: $K_1 + U_1 + W_{\text{other}} = K_2 + U_2$

Only the spring force does work so $W_{\text{other}} = 0$ and $U = U_{\text{el}}$.

$K_1 = 0$ (the blocks initially are at rest)

$U_2 = 0$ (no potential energy is left in the spring)

$K_2 = \frac{1}{2} m_A v_{A2}^2 + \frac{1}{2} m_B v_{B2}^2 = \frac{1}{2}(1.00 \text{ kg})(3.60 \text{ m/s})^2 + \frac{1}{2}(3.00 \text{ kg})(1.20 \text{ m/s})^2 = 8.64 \text{ J}$

$U_1 = U_{1,\text{el}}$ the potential energy stored in the compresssed spring.

Thus $U_{1,\text{el}} = K_2 = 8.64 \text{ J}$

EVALUATE: The blocks have equal and opposite momenta as they move apart, since the total momentum is zero. The kinetic energy of each block is positive and doesn't depend on the direction of the block's velocity, just on its magnitude.

8.23   IDENTIFY and **SET UP:** Let the $+x$-direction be horizontal, along the direction the rock is thrown. There is no net horizontal force, so P_x is constant. Let object A be you and object B be the rock.

EXECUTE: $0 = -m_A v_A + m_B v_B \cos 35.0°$

$v_A = \dfrac{m_B v_B \cos 35.0°}{m_A} = 2.11 \text{ m/s}$

EVALUATE: P_y is not conserved because there is a net external force in the vertical direction; as you throw the rock the normal force exerted on you by the ice is larger than the total weight of the system.

8.25   IDENTIFY: There is no net external horizontal force so P_x is constant.

SET UP: Take $+x$ to be the direction of the initial velocity. Let Ken be object A and Kim be object B. Solve m_B.

EXECUTE:

P_x constant implies $m_A v_{A1x} + m_B v_{B1x} = m_A v_{A2x} + m_B v_{B2x}$

$m_A(v_{A1x} - v_{A2x}) = m_B(v_{B2x} - v_{B1x})$

$m_B = m_A \left(\dfrac{v_{A1x} - v_{A2x}}{v_{B2x} - v_{B1x}} \right)$; multiply by g and get $w_B = w_A \left(\dfrac{v_{A1x} - v_{A2x}}{v_{B2x} - v_{B1x}} \right)$

$w_B = (700 \text{ N}) \left(\dfrac{3.00 \text{ m/s} - 2.25 \text{ m/s}}{4.00 \text{ m/s} - 3.00 \text{ m/s}} \right) = (700 \text{ N})(0.75) = 525 \text{ N}$

EVALUATE: When Kim pushes away, by Newton's 3rd law they exert equal and opposite forces on each other. So, Kim speeds up and Ken slows down. The change in velocity each force produces depends on the mass of the object.

8.27 IDENTIFY: Since the pucks move on a smooth ice surface there is no net horizontal force and the horizontal momentum is conserved. In part (b) use $K = \frac{1}{2}mv^2$ to calculate the kinetic energy of A before and after the collision.

SET UP: There is final momentum in both the x and y directions. Momentum is a vector, so conservation of momentum is expressed in two separate equations: P_x is constant and P_y is constant. Take the x-axis to lie along the initial velocity of A.

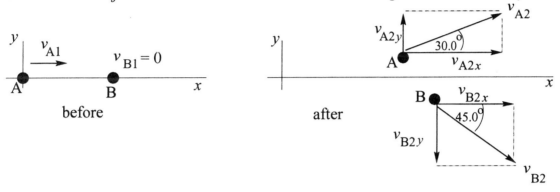

EXECUTE:

a) P_x is constant implies $m_A v_{A1x} + m_B v_{B1x} = m_A v_{A2x} + m_B v_{B2x}$

$mv_{A1} = mv_{A2}\cos 30.0° + mv_{B2}\cos 45.0°$

The mass m divides out and the equation becomes $40.0 \text{ m/s} = 0.8667v_{A2} + 0.7071v_{B2}$.

P_y is constant implies $m_A v_{A1y} + m_B v_{B1y} = m_A v_{A2y} + m_B v_{B2y}$

$0 = mv_{A2}\sin 30.0° - mv_{B2}\sin 45.0°$

The mass m divides out and the equation becomes $0 = 0.5000v_{A2} - 0.7071v_{B2}$.

Adding these two equations gives $1.366v_{A2} = 40.0 \text{ m/s}$ and $v_{A2} = 29.3 \text{ m/s}$.

Then $v_{B2} = \left(\dfrac{0.5000}{0.7071}\right) v_{A2} = \left(\dfrac{0.5000}{0.7071}\right)(29.3 \text{ m/s}) = 20.7 \text{ m/s}$.

b) $K = K_{A1} = \frac{1}{2}mv_{A1}^2$

$K_2 = K_{A2} + K_{B2} = \frac{1}{2}mv_{A2}^2 + \frac{1}{2}mv_{B2}^2$

$\Delta K = K_2 - K_1$

The fraction of the original kinetic energy of puck A dissipated during the collision is

$$-\frac{\Delta K}{K_1} = -\frac{K_2 - K_1}{K_1} = -\left(\frac{\frac{1}{2}mv_{A2}^2 + \frac{1}{2}mv_{B2}^2 - \frac{1}{2}mv_{A1}^2}{\frac{1}{2}mv_{A1}^2}\right) = 1 - \left(\frac{v_{A2}}{v_{A1}}\right)^2 - \left(\frac{v_{B2}}{v_{A1}}\right)^2$$

$-\dfrac{\Delta K}{K_1} = 1 - ((29.3 \text{ m/s})/(40.0 \text{ m/s}))^2 - ((20.7 \text{ m/s})/(40.0 \text{ m/s}))^2 =$

$1 - 0.5366 - 0.2679 = 0.196; \quad 19.6\%$.

EVALUATE: There is no net horizontal external force so the total horizontal momentum of the system doesn't change. But there are internal forces, the forces that one puck exerts on the other during the collision. These forces cause equal and opposite momentum changes for the pucks. They also do work and this changes the kinetic energy of each puck and also changes the total kientic energy of the system.

8.33 IDENTIFY: There is no net external horizontal force so the horizontal momentum of the system is conserved. The objects stick together so have a common final velocity.

SET UP: Let the automobile be object A and the truck be object B.

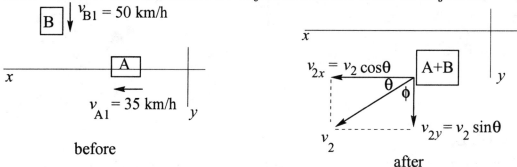

before   after

P_x is constant and P_y is constant, with x and y as shown in the sketch.

EXECUTE:

P_x is constant implies $m_A v_{A1x} + m_B v_{B1x} = (m_A + m_B)v_{2x}$

$m_A v_{A1x} = (m_A + m_B)v_{2x}$

$(1400 \text{ kg})(+35.0 \text{ km/h}) = (1400 \text{ kg} + 2800 \text{ kg})v_{2x}$

$v_{2x} = 11.67 \text{ km/h}$

P_y is constant implies $m_A v_{A1y} + m_B v_{B1y} = (m_A + m_B)v_{2y}$

$m_B v_{B1y} = (m_A + m_B)v_{2y}$

$(2800 \text{ kg})(+50.0 \text{ km/h}) = (1400 \text{ kg} + 2800 \text{ kg})v_{2y}$

$v_{2y} = 33.33 \text{ km/h}$

$v_2 = \sqrt{v_{2x}^2 + v_{2y}^2} = \sqrt{(11.67 \text{ km/h})^2 + (33.33 \text{ km/h})^2} = 35.3 \text{ km/h}$

$\tan\theta = \dfrac{v_{2y}}{v_{2x}} = \dfrac{33.33 \text{ km/h}}{11.67 \text{ km/h}}; \quad \theta = 70.7° \text{ and } \phi = 19.3°$

The wreckage is moving in the direction 19.3° west of south.

EVALUATE: The final velocity $\vec{v}_2$ has both x and y components and these are solved for separately and then combined to get $\vec{v}_2$.

8.35 IDENTIFY: Apply conservation of momentum to the collision.
Apply conservation of energy to the motion of the block after the collision.

SET UP: Conservation of momentum applied to the collision between the bullet and the block: Let object A be the bullet and object B be the block. Let v_A be the speed of the bullet before the collision and let V be the speed of the block with the bullet inside just after the collision.

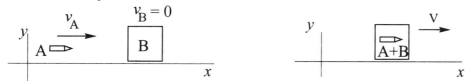

P_x is constant gives $m_A v_A = (m_A + m_B)V$

Conservation of energy applied to the motion of the block after the collision:

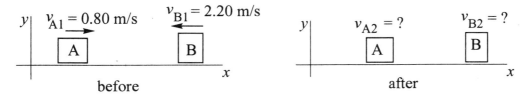

$K_1 + U_1 + W_{\text{other}} = K_2 + U_2$

EXECUTE:

Work is done by friction so $W_{\text{other}} = W_f = (f_k \cos \phi)s = -f_k s = -\mu_k mgs$

$U_1 = U_2 = 0$ (no work done by gravity)

$K_1 = \frac{1}{2}mV^2$; $K_2 = 0$ (block has come to rest)

Thus $\frac{1}{2}mV^2 - \mu_k mgs = 0$

$V = \sqrt{2\mu_k gs} = \sqrt{2(0.20)(9.80 \text{ m/s}^2)(0.230 \text{ m})} = 0.9495 \text{ m/s}$

Use this in the conservation of momentum equation

$$v_A = \left(\frac{m_A + m_B}{m_A}\right) V = \left(\frac{5.00 \times 10^{-3} \text{ kg} + 1.20 \text{ kg}}{5.00 \times 10^{-3} \text{ kg}}\right)(0.9495 \text{ m/s}) = 229 \text{ m/s}$$

EVALUATE: When we apply conservation of momentum to the collision we are ignoring the impulse of the friction force exerted by the surface during the collision. This is reasonable since this force is much smaller than the forces the bullet and block exert on each other during the collision. This force does work as the block moves after the collsion, and takes away all the kinetic energy.

8.39   IDENTIFY: No net external horizontal force so P_x is conserved. Elastic collision so $K_1 = K_2$ and can use Eq.(8.27).

SET UP:

EXECUTE:

From conservation of x-component of momentum:

$m_A v_{A1x} + m_B v_{B1x} = m_A v_{A2x} + m_B v_{B2x}$

$m_A v_{A1} - m_B v_{B1} = m_A v_{A2x} + m_B v_{B2x}$

$(0.150 \text{ kg})(0.80 \text{ m/s}) - (0.300 \text{ kg})(2.20 \text{ m/s}) = (0.150 \text{ kg})v_{A2x} + (0.300 \text{ kg})v_{B2x}$

$-3.60 \text{ m/s} = v_{A2x} + 2v_{B2x}$

From the relative velocity equation for an elastic collision (Eq.8-27):

$v_{B2x} - v_{A2x} = -(v_{B1x} - v_{A1x}) = -(-2.20 \text{ m/s} - 0.80 \text{ m/s}) = +3.00 \text{ m/s}$

$3.00 \text{ m/s} = -v_{A2x} + v_{B2x}$

Adding the two equations gives $-0.60 \text{ m/s} = 3v_{B2x}$ and $v_{B2x} = -0.20 \text{ m/s}$. Then $v_{A2x} = v_{B2x} - 3.00 \text{ m/s} = -3.20 \text{ m/s}$.

The 0.150 kg glider (A) is moving to the left at 3.20 m/s and the 0.300 kg glider (B) is moving to the left at 0.20 m/s.

EVALUATE: We can use our v_{A2x} and v_{B2x} to show that P_x is constant and $K_1 = K_2$

8.43   IDENTIFY: Elastic collision. Solve for mass and speed of target nucleus.

SET UP:

a) Let A be the proton and B be the target nucleus. The collision is elastic, all velocities lie along a line, and B is at rest before the collision. Hence the results of Eqs.(8.24) and (8.25) apply.

EXECUTE:

Eq.(8.24): $m_B(v_x + v_{Ax}) = m_A(v_x - v_{Ax})$, where v_x is the velocity component of A before the collision and v_{Ax} is the velocity component of A after the colliison.

Here, $v_x = 1.50 \times 10^7$ m/s (take direction of incident beam to be positive) and $v_{Ax} = -1.20 \times 10^7$ m/s (negative since traveling in direction opposite to incident beam).

$$m_B = m_A \left(\frac{v_x - v_{Ax}}{v_x + v_{Ax}} \right) = m \left(\frac{1.50 \times 10^7 \text{ m/s} + 1.20 \times 10^7 \text{ m/s}}{1.50 \times 10^7 \text{ m/s} - 1.20 \times 10^7 \text{ m/s}} \right) =$$

$$m \left(\frac{2.70}{0.30} \right) = 9.00m$$

b) Eq.(8.25): $v_{Bx} = \left(\frac{2m_A}{m_A + m_B} \right) v = \left(\frac{2m}{m + 9.00m} \right) (1.50 \times 10^7 \text{ m/s}) =$

3.00×10^6 m/s

EVALUATE: Can use our calculated v_{Bx} and m_B to show that P_x is constant and that $K_1 = K_2$.

8.45 IDENTIFY: Calculate x_{cm}.

SET UP: Apply Eq.(8.28) with the sun as mass 1 and Jupiter as mass 2.
Take the origin at the sun and let Jupiter lie on the positive x-axis.

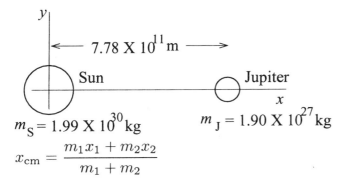

$$x_{cm} = \frac{m_1 x_1 + m_2 x_2}{m_1 + m_2}$$

EXECUTE: $x_1 = 0$ and $x_2 = 7.78 \times 10^{11}$ m

$$x_{cm} = \frac{(1.90 \times 10^{27} \text{ kg})(7.78 \times 10^{11} \text{ m})}{1.99 \times 10^{30} \text{ kg} + 1.90 \times 10^{27} \text{ kg}} = 7.42 \times 10^8 \text{ m}$$

The center of mass is 7.42×10^8 m from the center of the sun and is on the line
connecting the centers of the sun and Jupiter. The sun's radius is 6.96×10^8 m so
the center of mass lies just outside the sun.

EVALUATE: The mass of the sun is much greater than the mass of Jupiter so the
center of mass is much closer to the sun. For each object we have considered all the
mass as being at the center of mass (geometrical center) of the object.

8.47 a) IDENTIFY: Use Eq.(8.28).

SET UP: The target variable is m_1.

EXECUTE: $x_{cm} = 2.0$ m, $y_{cm} = 0$

$$x_{cm} = \frac{m_1 x_1 + m_2 x_2}{m_1 + m_2} = \frac{m_1(0) + (0.10 \text{ kg})(8.0 \text{ m})}{m_1 + (0.10 \text{ kg})} = \frac{0.80 \text{ kg} \cdot \text{m}}{m_1 + 0.10 \text{ kg}}$$

EVALUATE: The cm is closer to m_1 so its mass is larger than m_2.

$$x_{cm} = 2.0 \text{ m gives } 2.0 \text{ m} = \frac{0.80 \text{ kg} \cdot \text{m}}{m_1 + 0.10 \text{ kg}}$$

$$m_1 + 0.10 \text{ kg} = \frac{0.80 \text{ kg} \cdot \text{m}}{2.0 \text{ m}} = 0.40 \text{ kg}$$

$$m_1 = 0.30 \text{ kg}$$

b) IDENTIFY: Use Eq.(8.32) to calculate $\vec{P}$.

SET UP: $\vec{v}_{cm} = (5.0 \text{ m/s})\hat{j}$

$\vec{P} = M\vec{v}_{cm} = (0.10 \text{ kg} + 0.30 \text{ kg})(5.0 \text{ m/s})\hat{i} = (2.0 \text{ kg} \cdot \text{m/s})\hat{i}$

c) IDENTIFY: Use Eq.(8.31).

SET UP: $v_{cm} = \dfrac{m_1 \vec{v}_1 + m_2 \vec{v}_2}{m_1 + m_2}$. The target variable is $\vec{v}_1$. Particle 2 at rest says $v_2 = 0$.

EXECUTE:
$\vec{v}_1 = \left(\dfrac{m_1 + m_2}{m_1}\right)\vec{v}_{cm} = \left(\dfrac{0.30 \text{ kg} + 0.10 \text{ kg}}{0.30 \text{ kg}}\right)(5.00 \text{ m/s})\hat{i} = (6.7 \text{ m/s})\hat{i}$.

EVALUATE: Using the result of part (c) we can calculate $\vec{p}_1$ and $\vec{p}_2$ and show that $\vec{P}$ as calculated in part (b) does equal $\vec{p}_1 + \vec{p}_2$.

8.49 **a) IDENTIFY** and **SET UP:** Apply Eq.(8.28) and solve for m_1 and m_2.

EXECUTE:
$y_{cm} = \dfrac{m_1 y_1 + m_2 y_2}{m_1 + m_2}$

$m_1 + m_2 = \dfrac{m_1 y_1 + m_2 y_2}{y_{cm}} = \dfrac{m_1(0) + (0.50 \text{ kg})(6.0 \text{ m})}{2.4 \text{ m}} = 1.25 \text{ kg}$

EVALUATE: $m_1 = 0.75$ kg. y_{cm} is closer to m_1 since $m_1 > m_2$.

b) IDENTIFY and **SET UP:** Apply $\vec{a} = d\vec{v}/dt$ for the cm motion.

EXECUTE: $\vec{a}_{cm} = \dfrac{d\vec{v}_{cm}}{dt} = (1.5 \text{ m/s}^3)t\hat{i}$

c) IDENTIFY and **SET UP:** Apply Eq.(8.34).

EXECUTE: $\sum \vec{F}_{ext} = M\vec{a}_{cm} = (1.25 \text{ kg})(1.5 \text{ m/s}^3)t\hat{i}$

At $t = 3.0$ s, $\sum \vec{F}_{ext} = (1.25 \text{ kg})(1.5 \text{ m/s}^3)(3.0 \text{ s})\hat{i} = (5.6 \text{ N})\hat{i}$.

EVALUATE: v_{cm-x} is positive and increasing so a_{cm-x} is positive and $\vec{F}_{ext}$ is in the $+x$-direction. There is no motion and no force component in the y-direction.

8.53 **IDENTIFY** and **SET UP:** Apply Eq.(8.39): $a = -\dfrac{v_{ex}}{m}\dfrac{dm}{dt}$. Solve for dm/dt.

EXECUTE:
$\dfrac{dm}{dt} = -\dfrac{ma}{v_{ex}} = -\dfrac{(6000 \text{ kg})(25.0 \text{ m/s}^2)}{2000 \text{ m/s}} = -75.0 \text{ kg/s}$

So in 1 s the rocket must eject 75.0 kg of gas.

EVALUATE: We have approximated dm/dt by $\Delta m/\Delta t$. We have assumed that 25.0 m/s^2 is the average acceleration for the first second.

8.57 **IDENTIFY** and **SET UP:** Use Eq.(8.40): $v - v_0 = v_{ex} \ln(m_0/m)$

$v_0 = 0$ ("fired from rest"), so $v/v_{ex} = \ln(m_0/m)$

Thus $m_0/m = e^{v/v_{ex}}$, or $m/m_0 = e^{-v/v_{ex}}$

If v is the final speed then m is the mass left when all the fuel has been expended; m/m_0 is the fraction of the initial mass that is not fuel.

a) EXECUTE: $v = 1.00 \times 10^{-3}c = 3.00 \times 10^5$ m/s gives

$m/m_0 = e^{-(3.00 \times 10^5 \text{ m/s})/(2000 \text{ m/s})} = 7.2 \times 10^{-66}$

EVALUATE: This is clearly not feasible, for so little of the initial mass to not be fuel.

b) EXECUTE: $v = 3000$ m/s gives $m/m_0 = e^{-(3000 \text{ m/s})/(2000 \text{ m/s})} = 0.223$

EVALUATE: 22.3% of the total initial mass not fuel, so 77.7% is fuel; this is possible.

Problems

8.59 **IDENTIFY** and **SET UP:** Use Eq.(8.7) to calculate the x-and y-components of the impulse. Then apply Eq.(8.6) to find $\vec{p}_2$ and the components of the final velocity.

EXECUTE:

$\vec{F} = (\alpha t^2)\hat{i} + (\beta + \gamma t)\hat{j}$; $\alpha = 25.0$ N/s^2, $\beta = 30.0$ N, $\gamma = 5.0$ N/s

$J_x = \int_{t_1}^{t_2} F_x(t)\,dt = \int_0^{t_2}(\alpha t^2)\,dt = \frac{1}{3}\alpha t_2^3 = \frac{1}{3}(25.0 \text{ N/s}^2)(0.500 \text{ s})^3 = 1.042$ N $\cdot$ s

$J_y = \int_{t_1}^{t_2} F_y(t)\,dt = \int_0^{t_2}(\beta + \gamma t)\,dt = (\beta t_2 + \frac{1}{2}\gamma t_2^2)$

$J_y = (30.0 \text{ N})(0.500 \text{ s}) + \frac{1}{2}(5.00 \text{ N/s})(0.500 \text{ s})^2 = 15.62$ N $\cdot$ s

$J_x = p_{2x} - p_{1x} = m(v_{2x} - v_{1x}) = mv_{2x}$, since $v_{1x} = 0$

$v_{2x} = \dfrac{J_x}{m} = \dfrac{1.042 \text{ N} \cdot \text{s}}{2.00 \text{ kg}} = 0.521$ m/s

$J_y = p_{2y} - p_{1y} = m(v_{2y} - v_{1y}) = mv_{2y}$, since $v_{1y} = 0$

$v_{2y} = \dfrac{J_y}{m} = \dfrac{15.62 \text{ N} \cdot \text{s}}{2.00 \text{ kg}} = 7.81$ m/s

Thus $\vec{v}_2 = (0.521 \text{ m/s})\hat{i} + (7.81 \text{ m/s})\hat{j}$.

EVALUATE: The force has positive x and y components so the object accelerated in the $+x$ and $+y$ directions and v_x and v_y are both positive after 0.500 s.

8.65 **IDENTIFY:** Use a coordinate system attached to the ground. Take the

x-axis to be east (along the tracks) and the y-axis to be north (parallel to the ground and perpendicular to the tracks). Then P_x is conserved and P_y is <u>not</u> conserved, due to the sideways force exerted by the tracks, the force that keeps the handcar on the tracks.

a) SET UP: Let A be the 25.0 kg mass and B be the car (mass 175 kg). After the mass is thrown sideways relative to the car it still has the same eastward component of velocity, 5.00 m/s, as it had before it was thrown.

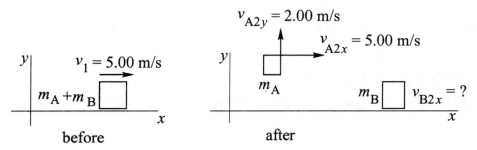

before   after

P_x is conserved so $(m_A + m_B)v_1 = m_A v_{A2x} + m_B v_{B2x}$

EXECUTE:

$(200 \text{ kg})(5.00 \text{ m/s}) = (25.0 \text{ kg})(5.00 \text{ m/s}) + (175 \text{ kg})v_{B2x}$

$$v_{B2x} = \frac{1000 \text{ kg} \cdot \text{m/s} - 125 \text{ kg} \cdot \text{m/s}}{175 \text{ kg}} = 5.00 \text{ m/s}.$$

The final velocity of the car is 5.00 m/s, east (unchanged).

EVALUATE: The thrower exerts a force on the mass in the y-direction and by Newton's 3rd law the mass exerts an equal and opposite force in the $-y$-direction on the thrower and car.

b) SET UP: We are applying $P_x = $ constant in coordinates attached to the ground, so we need the final velocity of A relative to the ground. Use the relative velocity addition equation. Then use $P_x = $ constant to find the final velocity of the car.

EXECUTE: $\vec{v}_{A/E} = \vec{v}_{A/B} + \vec{v}_{B/E}$

$v_{B/E} = +5.00 \text{ m/s}$

$v_{A/B} = -5.00 \text{ m/s}$ (minus since the mass is moving west relative to the car). This gives $v_{A/E} = 0$; the mass is at rest relative to the earth after it is thrown backwards from the car.

As in part (a), $(m_A + m_B)v_1 = m_A v_{A2x} + m_B v_{B2x}$.

Now $v_{A2x} = 0$, so $(m_A + m_B)v_1 = m_B v_{B2x}$.

$$v_{B2x} = \left(\frac{m_A + m_B}{m_B}\right)v_1 = \left(\frac{200 \text{ kg}}{175 \text{ kg}}\right)(5.00 \text{ m/s}) = 5.71 \text{ m/s}.$$

The final velocity of the car is 5.71 m/s, east.

EVALUATE: The thrower exerts a force in the $-x$-direction so the mass exerts a force on him in the $+x$-direction and he and the car speed up.

c) SET UP: Let A be the 25.0 kg mass and B be the car (mass $m_B = 200$ kg).

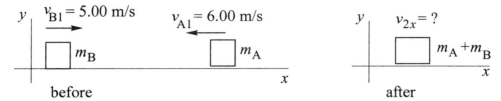

P_x is conserved so $m_A v_{A1x} + m_B v_{B1x} = (m_A + m_B)v_{2x}$

EXECUTE: $-m_A v_{A1} + m_B v_{B1} = (m_A + m_B)v_{2x}$

$$v_{2x} = \frac{m_B v_{B1} - m_A v_{A1}}{m_A + m_B} = \frac{(200 \text{ kg})(5.00 \text{ m/s}) - (25.0 \text{ kg})(6.00 \text{ m/s})}{200 \text{ kg} + 25.0 \text{ kg}} = 3.78 \text{ m/s}.$$

The final velocity of the car is 3.78 m/s, east.

EVALUATE: The mass has negative p_x so reduces the total P_x of the system and the car slows down.

8.69 IDENTIFY: Find k for the spring from the forces when the frame hangs at rest, use constant acceleration equations to find the speed of the putty just before it strikes the frame, apply conservation of momentum to the collision between the putty and the frame and then apply conservation of energy to the motion of the frame after the collision.

SET UP: Use the free-body diagram for the frame when it hangs at rest on the end of the spring to find the force constant k of the spring. Let s be the amount the spring is stretched.

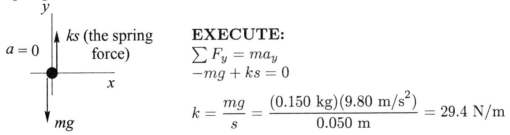

EXECUTE:
$\sum F_y = ma_y$
$-mg + ks = 0$

$$k = \frac{mg}{s} = \frac{(0.150 \text{ kg})(9.80 \text{ m/s}^2)}{0.050 \text{ m}} = 29.4 \text{ N/m}$$

SET UP: Next find the speed of the putty when it reaches the frame. The putty falls with acceleration $a = g$, downward.

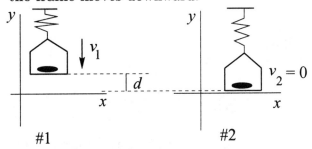

$$v_0 = 0$$
$$y - y_0 = 0.300 \text{ m}$$
$$a = +9.80 \text{ m/s}^2$$
$$v = ?$$

$$v^2 = v_0^2 + 2a(y - y_0)$$

EXECUTE: $v = \sqrt{2a(y - y_0)} = \sqrt{2(9.80 \text{ m/s}^2)(0.300 \text{ m})} = 2.425 \text{ m/s}$

SET UP: Apply conservation of momentum to the collision between the putty (A) and the frame (B):

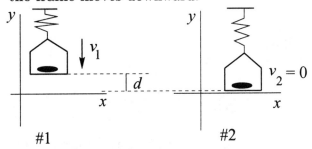

P_y is conserved, so $-m_A v_{A1} = -(m_A + m_B)v_2$

EXECUTE: $v_2 = \left(\dfrac{m_A}{m_A + m_B}\right) v_{A1} = \left(\dfrac{0.200 \text{ kg}}{0.350 \text{ kg}}\right)(2.425 \text{ m/s}) = 1.386 \text{ m/s}$

SET UP: Apply conservation of energy to the motion of the frame on the end of the spring after the collision. Let point 1 be just after the putty strikes and point 2 be when the frame has its maximum downward displcement. Let d be the amount the frame moves downward.

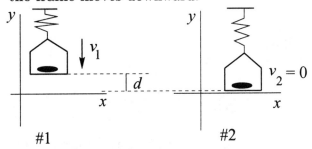

When the frame is at position 1 the spring is stretched a distance $x_1 = 0.050$ m. When the frame is at position 2 the spring is stretched a distance $x_2 = 0.050$ m $+ d$. Use coordinates with the y-direction upward and $y = 0$ at the lowest point reached by the frame, so that $y_1 = d$ and $y_2 = 0$. Work is done on the frame by gravity and by the spring force, so $W_{\text{other}} = 0$, and $U = U_{\text{el}} + U_{\text{gravity}}$.

EXECUTE: $K_1 + U_1 + W_{\text{other}} = K_2 + U_2$

$W_{\text{other}} = 0$

$K_1 = \frac{1}{2}mv_1^2 = \frac{1}{2}(0.350 \text{ kg})(1.386 \text{ m/s})^2 = 0.3362 \text{ J}$

$U_1 = U_{1,\text{el}} + U_{1,\text{grav}} = \frac{1}{2}kx_1^2 + mgy_1 = \frac{1}{2}(29.4 \text{ N/m})(0.050 \text{ m})^2 + (0.350 \text{ kg})(9.80 \text{ m/s}^2)d$

$U_1 = 0.03675 \text{ J} + (3.43 \text{ N})d$

$U_2 = U_{2,\text{el}} + U_{2,\text{grav}} = \frac{1}{2}kx_2^2 + mgy_2 = \frac{1}{2}(29.4 \text{ N/m})(0.050 \text{ m} + d)^2$

$U_2 = 0.03675 \text{ J} + (1.47 \text{ N})d + (14.7 \text{ N/m})d^2$

Thus $0.3362 \text{ J} + 0.03675 \text{ J} + (3.43 \text{ N})d = 0.03675 \text{ J} + (1.47 \text{ N})d + (14.7 \text{ N/m})d^2$

$(14.7 \text{ N/m})d^2 - (1.96 \text{ N})d - 0.3362 \text{ J} = 0$

$d = (1/29.4)[1.96 \pm \sqrt{(1.96)^2 - 4(14.7)(-0.3362)}] \text{ m} = 0.0667 \text{ m} \pm 0.1653 \text{ m}.$

The solution we want is a positive (downward) distance, so $d = 0.0667 \text{ m} + 0.1653 \text{ m} = 0.232 \text{ m}.$

EVALUATE: The collision is inelastic and mechanical energy is lost. Thus the decrease in gravitational potential energy is <u>not</u> equal to the increase in potential energy stored in the spring.

8.71 IDENTIFY: The horizontal components of momentum of the system of bullet plus stone are conserved. The collision is elastic if $K_1 = K_2$.

SET UP: Let A be the bullet and B be the stone.

a)

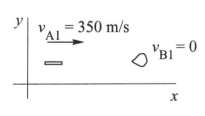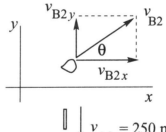

EXECUTE:

P_x is conserved so $m_A v_{A1x} + m_B v_{B1x} = m_A v_{A2x} + m_B v_{B2x}$

$m_A v_{A1} = m_B v_{B2x}$

$v_{B2x} = \left(\dfrac{m_A}{m_B}\right) v_{A1} = \left(\dfrac{6.00 \times 10^{-3} \text{ kg}}{0.100 \text{ kg}}\right)(350 \text{ m/s}) = 21.0 \text{ m/s}$

P_y is conserved so $m_A v_{A1y} + m_B v_{B1y} = m_A v_{A2y} + m_B v_{B2y}$

$0 = -m_A v_{A2} + m_B v_{B2y}$

$v_{B2y} = \left(\dfrac{m_A}{m_B}\right) v_{A2} = \left(\dfrac{6.00 \times 10^{-3} \text{ kg}}{0.100 \text{ kg}}\right)(250 \text{ m/s}) = 15.0 \text{ m/s}$

$v_{B2} = \sqrt{v_{B2x}^2 + v_{B2y}^2} = \sqrt{(21.0 \text{ m/s})^2 + (15.0 \text{ m/s})^2} = 25.8 \text{ m/s}$

$$\tan \theta = \frac{v_{B2y}}{v_{B2x}} = \frac{15.0 \text{ m/s}}{21.0 \text{ m/s}} = 0.7143; \quad \theta = 35.5° \text{ (defined in the sketch)}$$

b) To answer this question compare K_1 and K_2 for the system:

$K_1 = \frac{1}{2}m_A v_{A1}^2 + \frac{1}{2}m_B v_{B1}^2 = \frac{1}{2}(6.00 \times 10^{-3} \text{ kg})(350 \text{ m/s})^2 = 368 \text{ J}$

$K_2 = \frac{1}{2}m_A v_{A2}^2 + \frac{1}{2}m_B v_{B2}^2 = \frac{1}{2}(6.00 \times 10^{-3} \text{ kg})(250 \text{ m/s})^2 + \frac{1}{2}(0.100 \text{ kg})(25.8 \text{ m/s})^2 = 221 \text{ J}$

$\Delta K = K_2 - K_1 = 221 \text{ J} - 368 \text{ J} = -147 \text{ J}$

EVALUATE: The kinetic energy of the system decreases by 147 J as a result of the collision; the collision is <u>not</u> elastic. Momentum is conserved because $\sum F_{ext,x} = 0$ and $\sum F_{ext,y} = 0$. But there are internal forces between the bullet and the stone. These forces do negative work that reduces K.

8.73 IDENTIFY: Apply conservation of energy to the motion before and after the collision and apply conservation of momentum to the collision.

SET UP: Let v be the speed of the mass released at the rim just before it strikes the second mass. Let each object have mass m.

EXECUTE: Conservation of energy says $\frac{1}{2}mv^2 = mgR$; $\quad v = \sqrt{2gR}$

SET UP: This is speed v_1 for the collision. Let v_2 be the speed of the combined object just after the collision.

EXECUTE: Conservation of momentum applied to the collision gives

$mv_1 = 2mv_2$ so $v_2 = v_1/2 = \sqrt{gR/2}$

SET UP: Apply conservation of energy to the motion of the combined object after the collision. Let y_3 be the final height above the bottom of the bowl.

EXECUTE: $\frac{1}{2}(2m)v_2^2 = (2m)gy_3$

$$y_3 = \frac{v_2^2}{2g} = \frac{1}{2g}\left(\frac{gR}{2}\right) = R/4$$

EVALUATE: Mechanical energy is lost in the collision, so the final gravitational potential energy is less than the initial gravitational potential energy.

8.75 IDENTIFY: Apply conservation of energy to the motion before and after the collision. Apply conservation of momentum to the collision.

SET UP: First consider the motion after the collision. The combined object has mass $m_{tot} = 25.0$ kg. Apply $\sum \vec{F} = m\vec{a}$ to the object at the top of the circular loop, where the object has speed v_3. The acceleration is $a_{rad} = v_3^2/R$, downward.

EXECUTE: $T + mg = m\dfrac{v_3^2}{R}$

The minimum speed v_3 for the object not to fall out of the circle is given by setting $T = 0$. This gives $v_3 = \sqrt{Rg}$, where $R = 3.50$ m.

SET UP: Next, use conservation of energy with point 2 at the bottom of the loop and point 3 at the top of the loop. Take $y = 0$ at point 2. Only gravity does work, so $K_2 + U_2 = K_3 + U_3$

EXECUTE: $\frac{1}{2}m_{\text{tot}}v_2^2 = \frac{1}{2}m_{\text{tot}}v_3^2 + m_{\text{tot}}g(2R)$

Use $v_3 = \sqrt{Rg}$ and solve for v_2: $v_2 = \sqrt{5gR} = 13.1$ m/s

SET UP: Now apply conservation of momentum to the collision between the dart and the sphere. Let v_1 be the speed of the dart before the collision.

EXECUTE: $(5.00 \text{ kg})v_1 = (25.0 \text{ kg})(13.1 \text{ m/s})$

$v_1 = 65.5$ m/s

EVALUATE: The collision is inelastic and mechanical energy is removed from the system by the negative work done by the forces between the dart and the sphere.

8.77 IDENTIFY: Apply conservation of momentum to the collision between the bullet and the block and apply conservation of energy to the motion of the block after the collision.

a) SET UP: Collision between the bullet and the block: Let object A be the bullet and object B be the block. Apply momentum conservation to find the speed v_{B2} of the block just after the collision.

EXECUTE:

P_x is conserved so $m_A v_{A1x} + m_B v_{B1x} = m_A v_{A2x} + m_B v_{B2x}$

$m_A v_{A1} = m_A v_{A2} + m_B v_{B2x}$

$$v_{B2x} = \frac{m_A(v_{A1} - v_{A2})}{m_B} = \frac{4.00 \times 10^{-3} \text{ kg}(400 \text{ m/s} - 120 \text{ m/s})}{0.800 \text{ kg}} = 1.40 \text{ m/s}$$

SET UP: Motion of the block after the collision.

Let point 1 in the motion be just after the collision, where the block has the speed 1.40 m/s calculated above, and let point 2 be where the block has come to rest.

$K_1 + U_1 + W_{\text{other}} = K_2 + U_2$

EXECUTE: Work is done on the block by friction, so $W_{\text{other}} = W_f$.

$W_{\text{other}} = W_f = (f_k \cos \phi)s = -f_k s = -\mu_k mgs$, where $s = 0.450$ m

$U_1 = 0, \quad U_2 = 0$

$K_1 = \frac{1}{2}mv_1^2, \quad K_2 = 0$ (block has come to rest)

Thus $\frac{1}{2}mv_1^2 - \mu_k mgs = 0$.

$$\mu_k = \frac{v_1^2}{2gs} = \frac{(1.40 \text{ s})^2}{2(9.80 \text{ m/s}^2)(0.450 \text{ m})} = 0.222$$

b) For the bullet,

$K_1 = \frac{1}{2}mv_1^2 = \frac{1}{2}(4.00 \times 10^{-3} \text{ kg})(400 \text{ m/s})^2 = 320$ J

$K_2 = \frac{1}{2}mv_2^2 = \frac{1}{2}(4.00 \times 10^{-3} \text{ kg})(120 \text{ m/s})^2 = 28.8$ J

$\Delta K = K_2 - K_1 = 28.8 \text{ J} - 320 \text{ J} = -291$ J

The kinetic energy of the bullet decreases by 291 J.

c) Immediately after the collision the speed of the block is 1.40 m/s so its kinetic energy is $K = \frac{1}{2}mv^2 = \frac{1}{2}(0.800 \text{ kg})(1.40 \text{ m/s})^2 = 0.784$ J.

EVALUATE: The collision is highly inelastic. The bullet loses 291 J of kinetic energy but only 0.784 J is gained by the block. But momentum is conserved in the collision. All the momentum lost by the bullet is gained by the block.

8.81   IDENTIFY: Apply conservation of energy to the motion of the package before the collision and apply conservation of the horizontal component of momentum to the collision.

a) SET UP: Apply conservation of energy to the motion of the package from point 1 as it leaves the chute to point 2 just before it lands in the cart. Take $y = 0$ at point 2, so $y_1 = 4.00$ m. Only gravity does work, so

$K_1 + U_1 = K_2 + U_2$

EXECUTE: $\frac{1}{2}mv_1^2 + mgy_1 = \frac{1}{2}mv_2^2$

$v_2 = \sqrt{v_1^2 + 2gy_1} = 9.35$ m/s

b) SET UP: In the collision between the package and the cart momentum is conserved in the horizontal direction. (But not in the vertical direction, due to the vertical force the floor exerts on the cart.) Take $+x$ to be to the right. Let A be the package and B be the cart.

EXECUTE: P_x is constant gives $m_a v_{A1x} + m_B v_{B1x} = (m_A + m_B)v_{2x}$

$v_{B1x} = -5.00$ m/s

$v_{A1x} = (3.00 \text{ m/s}) \cos 37.0°$ (The horizontal velocity of the package is constant during its free-fall.)

Solving for v_{2x} gives $v_{2x} = -3.29$ m/s. The cart is moving to the left at 3.29 m/s after the package lands in it.

EVALUATE: The cart is slowed by its collision with the package, whose horizontal component of momentum is in the opposite direction to the motion of the cart.

8.83 a) IDENTIFY and **SET UP:** $K = \frac{1}{2}m_A v_A^2 + \frac{1}{2}m_B v_B^2$.

Use $\vec{v}_A = \vec{v}_A' + \vec{v}_{\text{cm}}$ and $\vec{v}_B = \vec{v}_B' + \vec{v}_{\text{cm}}$ to replace v_A and v_B in this equation. Note $\vec{v}_A'$ and $\vec{v}_B'$ as defined in the problem are the velocities of A and B in coordinates moving with the center of mass. Note also that $m_A \vec{v}_A' + m_B \vec{v}_B' = M \vec{v}_{\text{cm}}'$ where $\vec{v}_{\text{cm}}'$ is the velocity of the car in these coordinates. But that's zero, so $m_A \vec{v}_A' + m_B \vec{v}_B' = 0$; we can use this in the proof.

In part (b), use that $\vec{P}$ is conserved in a collision.

EXECUTE:

$\vec{v}_A = \vec{v}_A' + \vec{v}_{\text{cm}}$, so $v_A^2 = v_A'^2 + v_{\text{cm}}^2 + 2\vec{v}_A' \cdot \vec{v}_{\text{cm}}$

$\vec{v}_B = \vec{v}_B' + \vec{v}_{\text{cm}}$, so $v_B^2 = v_B'^2 + v_{\text{cm}}^2 + 2\vec{v}_B' \cdot \vec{v}_{\text{cm}}$

(We have used that for a vector $\vec{A}$, $A^2 = \vec{A} \cdot \vec{A}$.)

Thus $K = \frac{1}{2}m_A v_A'^2 + \frac{1}{2}m_A v_{\text{cm}}^2 + m_A \vec{v}_A' \cdot \vec{v}_{\text{cm}} + \frac{1}{2}m_B v_B'^2 + \frac{1}{2}m_B v_{\text{cm}}^2 + m_B \vec{v}_B' \cdot \vec{v}_{\text{cm}}$

$K = \frac{1}{2}(m_A + m_B)v_{\text{cm}}^2 + \frac{1}{2}(m_A v_A'^2 + m_B v_B'^2) + (m_A \vec{v}_A' + m_B \vec{v}_B') \cdot \vec{v}_{\text{cm}}$

But $m_A + m_B = M$ and as noted earlier $m_A \vec{v}_A' + m_B \vec{v}_B' = 0$, so

$K = \frac{1}{2}M v_{\text{cm}}^2 + \frac{1}{2}(m_A v_A'^2 + m_B v_B'^2)$. This is the result the problem asked us to derive.

b) EVALUATE: In the collision $\vec{P} = M\vec{v}_{\text{cm}}$ is constant, so $\frac{1}{2}M v_{\text{cm}}^2$ stays constant. The asteroids can lose all their relative kinetic energy but the $\frac{1}{2}M v_{\text{cm}}^2$ must remain.

8.91 IDENTIFY and **SET UP:**

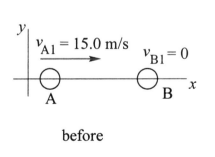
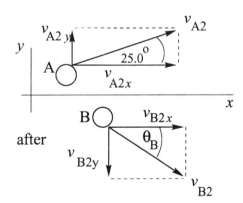

P_x and P_y are conserved in the collision since there is no external horizontal force.

The result of Problem 8.90 part (d) applies here since the collision is elastic This says that $25.0° + \theta_B = 90.0°$, so that $\theta_B = 65.0°$. (A and B move off in perpendicular directions.)

EXECUTE:

P_x is conserved so $m_A v_{A1x} + m_B v_{B1x} = m_A v_{A2x} + m_B v_{B2x}$

But $m_A = m_B$ so $v_{A1} = v_{A2} \cos 25.0° + v_{B2} \cos 65.0°$

P_y is conserved so $m_A v_{A1y} + m_B v_{B1y} = m_A v_{A2y} + m_B v_{B2y}$

$0 = v_{A2y} + v_{B2y}$

$0 = v_{A2} \sin 25.0° + v_{B2} \sin 65.0°$

$v_{B2} = (\sin 25.0° / \sin 65.0°)v_{A2}$

This result in the first equation gives $v_{A1} = v_{A2} \cos 25.0° + \left(\dfrac{\sin 25.0° \cos 65.0°}{\sin 65.0°} \right) v_{A2}$

$v_{A1} = 1.103 v_{A2}$

$v_{A2} = v_{A1}/1.103 = 15.0 \text{ m/s}/1.103 = 13.6 \text{ m/s}$

And then $v_{B2} = (\sin 25.0° / \sin 65.0°)(13.6 \text{ m/s}) = 6.34 \text{ m/s}.$

EVALUATE: We can use our numerical results to show that $K_1 = K_2$ and that $P_{x1} = P_{x2}$ and $P_{y1} = P_{y2}$.

8.95 **IDENTIFY:** Take as the system you and the slab. There is no horizontal force, so horizontal momentum is conserved. By Eq.(8.32), $\vec{P}$ is constant $\vec{v}_{cm}$ is constant (for a system of constant mass). Use coordinates fixed to the ice, with the direction you walk as the x-direction. $\vec{v}_{cm}$ is constant and initially $\vec{v}_{cm} = 0$.

$\vec{v}_{cm} = \dfrac{m_p \vec{v}_p + m_s \vec{v}_s}{m_p + m_s} = 0$

$m_p \vec{v}_p + m_s \vec{v}_s = 0$

$m_p v_{px} + m_s v_{sx} = 0$

$v_{sx} = -(m_p/m_s)v_{px} = -(m_p/5m_p)2.00 \text{ m/s} = -0.400 \text{ m/s}$

The slab moves at 0.400 m/s, in the direction opposite to the direction you are walking.

EVALUATE: The initial momentum of the system is zero. You gain momentum in the $+x$-direction so the slab gains momentum in the $-x$-direction. The slab exerts a force on you in the $+x$-direction so you exert a force on the slab in the $-x$-direction.

8.97 **IDENTIFY:** The rocket moves in projectile motion before the explosion and its fragments move in projectile motion after the explosion. Apply conservation of energy and conservation of momentum to the explosion.

SET UP: Apply conservation of energy to the explosion. Just before the explosion the shell is at its maximum height and has zero kinetic energy. Let A be the piece with mass 1.40 kg and B be the piece with mass 0.28 kg. Let v_A and v_B be the speeds of the two pieces immediately after the collision.

EXECUTE: $\frac{1}{2}m_A v_A^2 + \frac{1}{2}m_B v_B^2 = 860$ J

SET UP: Since the two fragments reach the ground at the same time, their velocities just after the explosion must be horizontal. The initial momentum of the shell before the explosion is zero, so after the explosion the pieces must be moving in opposite horizontal directions and have equal magnitude of momentum: $m_A v_A = m_B v_B$.

EXECUTE: Use this to eliminate v_A in the first equation and solve for v_B:
$\frac{1}{2}m_B v_B^2(1 + m_B/m_A) = 860$ J and $v_B = 71.6$ m/s.
Then $v_A = (m_B/m_A)v_B = 14.3$ m/s.

b) SET UP: Use the vertical motion from the maximum height to the ground to find the time it takes the pieces to fall to the ground after the explosion. Take $+y$ downward.

$v_{0y} = 0$, $a_y = +9.80$ m/s^2, $y - y_0 = 80.0$ m, $t = ?$

EXECUTE: $y - y_0 = v_{0y}t + \frac{1}{2}a_y t^2$ gives $t = 4.04$ s.

During this time the horizontal distance each piece moves is

$x_A = v_A t = 57.8$ m and $x_B = v_B t = 289.1$ m.

They move in opposite directions, so they are $x_A + x_B = 347$ m apart when they land.

EVALUATE: Fragment A has more mass so it is moving slower right after the collision, and it travels horizontally a smaller distance as it falls to the ground.

8.99 IDENTIFY: No external force, so $\vec{P}$ is conserved in the collision.

SET UP: Apply momentum conservation in the x and y directions:

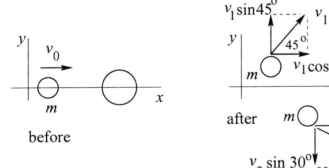 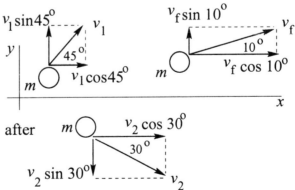

Solve for v_1 and v_2.

EXECUTE: P_x is conserved so $mv_0 = m(v_1 \cos 45° + v_f \cos 10^0 + v_2 \cos 30°)$

$v_0 - v_f \cos 10° = v_1 \cos 45° + v_2 \cos 30°$

1030.4 m/s $= v_1 \cos 45° + v_2 \cos 30°$

P_x is conserved so $0 = m(v_1 \sin 45° - v_2 \sin 30° + v_f \sin 10°)$

$v_1 \sin 45° = v_2 \sin 30° - 347.3$ m/s

$\sin 45° = \cos 45°$ so

$1030.4 \text{ m/s} = v_2 \sin 30° - 347.3 \text{ m/s} + v_2 \cos 30°$

$$v_2 = \frac{1030.4 \text{ m/s} + 347.3 \text{ m/s}}{\sin 30° + \cos 30.0°} = 1010 \text{ m/s}$$

And then $v_1 = \dfrac{v_2 \sin 30° - 347.3 \text{ m/s}}{\sin 45°} = 223 \text{ m/s}.$

The two emitted neutrons have speeds of 223 m/s and 1010 m/s.

The speeds of the Ba and Kr nuclei are related by P_z conservation. P_z is constant implies that $0 = m_{Ba}v_{Ba} - m_{Kr}v_{Kr}$

$$v_{Kr} = \left(\frac{m_{Ba}}{m_{Kr}}\right) v_{Ba} = \left(\frac{2.3 \times 10^{-25} \text{ kg}}{1.5 \times 10^{-25} \text{ kg}}\right) v_{Ba} = 1.5 v_{Ba}.$$

We can't say what these speeds are but they must satisfy this relation. The value of v_{Ba} depends on energy considerations.

EVALUATE: $K_1 = \frac{1}{2}m_n(3.0 \times 10^3 \text{ m/s})^2 = (4.5 \times 10^6 \text{ J/kg})m_n.$ $K_2 = \frac{1}{2}m_n(2.0 \times 10^3 \text{ m/s})^2 + \frac{1}{2}m_n(223 \text{ m/s})^2 + \frac{1}{2}m_n(1010 \text{ m/s})^2 + K_{Ba} + K_{Kr} = (2.5 \times 10^6 \text{ J/kg})m_n + K_{Ba} + K_{Kr}.$ We don't know what K_{Ba} and K_{Kr} are, but they are positive. We will study such nuclear reactions further in Chapter 41 and will find that energy is released in this process; $K_2 > K_1$. Some of the potential energy stored in the ^{235}U nucleus is released as kinetic energy and shared by the collision fragments.

8.101   IDENTIFY and **SET UP:** Apply conservation of energy to find the total energy before and after the collision with the floor from the initial and final maximum heights.

EXECUTE:

a) Objects stick together says that the relative speed after the collision is zero, so $\epsilon = 0$.

b) In an elastic collision the relative velociy of the two bodies has the same magnitude before and after the collision, so $\epsilon = 1$.

c) Speed of ball just before collsion: $mgh = \frac{1}{2}mv_1^2$

$v_1 = \sqrt{2gh}$

Speed of ball just after collision: $mgH_1 = \frac{1}{2}mv_2^2$

$v_2 = \sqrt{2gH_1}$

The second object (the surface) is stationary, so $\epsilon = v_2/v_1 = \sqrt{H_1/h}.$

d) $\epsilon = \sqrt{H_1/h}$ implies $H_1 = h\epsilon^2 = (1.2 \text{ m})(0.85)^2 = 0.87 \text{ m}$

e) $H_1 = h\epsilon^2$

$H_2 = h_1\epsilon^2 = h\epsilon^4$

$H_3 = H_2\epsilon^2 = (h\epsilon^4)\epsilon^2 = h\epsilon^6$

Generalize to $H_n = H_{n-1}\epsilon^2 = h\epsilon^{2(n-1)}\epsilon^2 = h\epsilon^{2n}$

f) 8th bounce implies $n = 8$

$H_8 = h\epsilon^{16} = 1.2 \text{ m}(0.85)^{16} = 0.089 \text{ m}$

EVALUATE: ϵ is a measure of the kinetic energy lost in the collision. The collision here is between a ball and the earth. Momentum lost by the ball is gained by the earth, but the velocity gained by the earth is very small and can be taken to be zero.

8.105 IDENTIFY and **SET UP:** Eq.(8.40) to the single-stage rocket and to each stage of the two-stage rocket.

a) EXECUTE: $v - v_0 = v_{\text{ex}}\ln(m_0/m)$ **a)** $v_0 = 0$ so $v = v_{\text{ex}}\ln(m_0/m)$

The total initial mass of the rocket is $m_0 = 12,000 \text{ kg} + 1000 \text{ kg} = 13,000 \text{ kg}$. Of this, $9000 \text{ kg} + 700 \text{ kg} = 9700 \text{ kg}$ is fuel, so the mass m left after all the fuel is burned is $13,000 \text{ kg} - 9700 \text{ kg} = 3300 \text{ kg}$.

$v = v_{\text{ex}}\ln(13,000 \text{ kg}/3300 \text{ kg}) = 1.37v_{\text{ex}}$

b) First stage: $v = v_{\text{ex}}\ln(m_0/m)$

$m_0 = 13,000 \text{ kg}$

The first stage has 9000 kg of fuel, so the mass left after the first stage fuel has burned is $13,000 \text{ kg} - 9000 \text{ kg} = 4000 \text{ kg}$.

$v = v_{\text{ex}}\ln(13,000 \text{ kg}/4000 \text{ kg}) = 1.18v_{\text{ex}}$

c) Second stage:

$m_0 = 1000 \text{ kg}, \quad m = 1000 \text{ kg} - 700 \text{ kg} = 300 \text{ kg}$

$v = v_0 + v_{\text{ex}}\ln(m_0/m) = 1.18v_{\text{ex}} + v_{\text{ex}}\ln(1000 \text{ kg}/300 \text{ kg}) = 2.38v_{\text{ex}}$

d) $v = 7.00 \text{ km/s}$

$v_{\text{ex}} = v/2.38 = (7.00 \text{ km/s})/2.38 = 2.94 \text{ km/s}$

EVALUATE: The two-stage rocket achieves a greater final speed because it jetisons the left-over mass of the first stage before the second-state fires and this reduces the final m and increases m_0/m.

CHAPTER 9
ROTATION OF RIGID BODIES

Exercises 5, 7, 11, 13, 19, 21, 25, 27, 31, 35, 37, 41, 47, 49, 51, 53, 55, 59
Problems 63, 65, 67, 69, 71, 73, 81, 83, 85, 87, 91, 97

Exercises

9.5 **IDENTIFY** and **SET UP:** Use Eq.(9.3) to calculate the angular velocity and Eq.(9.2) to calculate the average angular velocity for the specified time internval.

EXECUTE:

$\theta = \gamma t + \beta t^3$; $\gamma = 0.400$ rad/s, $\beta = 0.0120$ rad/s^3

a) $\omega_z = \dfrac{d\theta}{dt} = \gamma + 3\beta t^2$

b) At $t = 0$, $\omega_z = \gamma = 0.400$ rad/s

c) At $t = 5.00$ s, $\omega_z = 0.400$ rad/s $+ 3(0.0120$ rad/s$^2)(5.00$ s$)^2 = 1.30$ rad/s

$\omega_{\text{av}-z} = \dfrac{\Delta\theta}{\Delta t} = \dfrac{\theta_2 - \theta_1}{t_2 - t_1}$

For $t_1 = 0$, $\theta_1 = 0$.

For $\theta_2 = 5.00$ s, $\theta_2 = (0.400$ rad/s$)(5.00$ s$) + (0.012$ rad/s$^3)(5.00$ s$)^3 = 3.50$ rad

So $\omega_{\text{av}-z} = \dfrac{3.50 \text{ rad} - 0}{5.00 \text{ s} - 0} = 0.700$ rad/s.

EVALUATE: The average of the instantaneous angular velocities at the beginning and end of the time interval is $\frac{1}{2}(0.400$ rad/s $+ 1.30$ rad/s$) = 0.850$ rad/s. This is larger than $\omega_{\text{av}-z}$, because $\omega_z(t)$ is increasing faster than linearly.

9.7 **IDENTIFY** and **SET UP:** Use Eq.(9.3) to calculate $\omega_z(t)$ and Eq.(9.6) to calculate $\alpha_z(t)$.

EXECUTE:

a) $\theta = a + bt^2 - ct^3$

$\omega_z = \dfrac{d\theta}{dt} = 2bt - 3ct^2$

$\alpha_z = \dfrac{d\omega}{dt} = 2b - 6ct$

b) SET UP: The angular velocity is instantaneously not changing when $\alpha_z = 0$.

EXECUTE: $\alpha_z = 0$ implies $2b - 6ct = 0$ and $t = 2b/6c = b/3c$.

EVALUATE: α_z is the time rate of change of ω_z.

9.11 IDENTIFY: Apply the constant angular acceleration equations to the motion of the fan.

a) SET UP: $\omega_{0z} = (500 \text{ rev/min})(1 \text{ min}/60 \text{ s}) = 8.333 \text{ rev/s}$, $\omega_z = (200 \text{ rev/min})(1 \text{ min}/60 \text{ s}) = 3.333 \text{ rev/s}$, $t = 4.00 \text{ s}$, $\alpha_z = ?$

$\omega_z = \omega_{0z} + \alpha_z t$

EXECUTE: $\alpha_z = \dfrac{\omega_z - \omega_{0z}}{t} = \dfrac{3.333 \text{ rev/s} - 8.333 \text{ rev/s}}{4.00 \text{ s}} = -1.25 \text{ rev/s}^2$

$\theta - \theta_0 = ?$

$\theta - \theta_0 = \omega_{0z} t + \frac{1}{2}\alpha_z t^2 = (8.333 \text{ rev/s})(4.00 \text{ s}) + \frac{1}{2}(-1.25 \text{ rev/s}^2)(4.00 \text{ s})^2 = 23.3 \text{ rev}$

b) SET UP: $\omega_z = 0$ (comes to rest); $\omega_{0z} = 3.333 \text{ rev/s}$; $\alpha_z = -1.25 \text{ rev/s}^2$; $t = ?$

$\omega_z = \omega_{0z} + \alpha_z t$

EXECUTE: $t = \dfrac{\omega_z - \omega_{0z}}{\alpha_z} = \dfrac{0 - 3.333 \text{ rev/s}}{-1.25 \text{ rev/s}^2} = 2.67 \text{ s}$

EVALUATE: The angular acceleration is negative because the angular velocity is decreasing. The average angular velocity during the 4.00 s time interval is 350 rev/min and $\theta - \theta_0 = \omega_{av-z} t$ gives $\theta - \theta_0 = 23.3 \text{ rev}$, which checks.

9.13 IDENTIFY: Apply the constant angular acceleration equations to the motion. The target variables arc t and $\theta - \theta_0$.

SET UP:

a) $\alpha_z = 1.50 \text{ rad/s}^2$; $\omega_{0z} = 0$ (starts from rest); $\omega_z = 36.0 \text{ rad/s}$; $t = ?$

$\omega_z = \omega_{0z} + \alpha_z t$

EXECUTE: $t = \dfrac{\omega_z - \omega_{0z}}{\alpha_z} = \dfrac{36.0 \text{ rad/s} - 0}{1.50 \text{ rad/s}^2} = 24.0 \text{ s}$

b) $\theta - \theta_0 = ?$

$\theta - \theta_0 = \omega_{0z} t + \frac{1}{2}\alpha_z t^2 = 0 + \frac{1}{2}(1.50 \text{ rad/s}^2)(24.0 \text{ s})^2 = 432 \text{ rad}$

$\theta - \theta_0 = 432 \text{ rad}(1 \text{ rev}/2\pi \text{ rad}) = 68.8 \text{ rev}$

EVALUATE: We could use $\theta - \theta_0 = \frac{1}{2}(\omega_z + \omega_{0z})t$ to calculate $\theta - \theta_0 = \frac{1}{2}(0 + 36.0 \text{ rad/s})(24.0 \text{ s}) = 432 \text{ rad}$, which checks.

9.19 IDENTIFY: Apply the constant angular equations separately to the time intervals 0 to 2.00 s and 2.00 s until the wheel stops.

a) SET UP: Consider the motion from $t = 0$ to $t = 2.00$ s:

$\theta - \theta_0 = ?$; $\omega_{0z} = 24.0$ rad/s; $\alpha_z = 30.0$ rad/s^2; $t = 2.00$ s

EXECUTE: $\theta - \theta_0 = \omega_{0z}t + \frac{1}{2}\alpha_z t^2 = (24.0 \text{ rad/s})(2.00 \text{ s}) + \frac{1}{2}(30.0 \text{ rad/s}^2)(2.00 \text{ s})^2$

$\theta - \theta_0 = 48.0 \text{ rad} + 60.0 \text{ rad} = 108 \text{ rad}$

Total angular displacement from $t = 0$ until stops: $108 \text{ rad} + 432 \text{ rad} = 540 \text{ rad}$

Note: At $t = 2.00$ s, $\omega_z = \omega_{0z} + \alpha_z t = 24.0 \text{ rad/s} + (30.0 \text{ rad/s}^2)(2.00 \text{ s}) = 84.0$ rad/s; angular speed when breaker trips.

b) SET UP: Consider the motion from when the circuit trips until the wheel stops. For this calculation let $t = 0$ when the breaker trips.

$t = ?$; $\theta - \theta_0 = 432$ rad; $\omega_z = 0$; $\omega_{0z} = 84.0$ rad/s (from part (a))

$\theta - \theta_0 = \left(\dfrac{\omega_{0z} + \omega_z}{2}\right) t$

EXECUTE: $t = \dfrac{2(\theta - \theta_0)}{\omega_{0z} + \omega_z} = \dfrac{2(432 \text{ rad})}{84.0 \text{ rad/s} + 0} = 10.3$ s

The wheel stops 10.3 s after the breaker trips so $2.00 \text{ s} + 10.3 \text{ s} = 12.3 \text{ s}$ from the beginning.

c) SET UP: $\alpha_z = ?$; consider the same motion as in part (b):

$\omega_z = \omega_{0z} + \alpha_z t$

EXECUTE: $\alpha_z = \dfrac{\omega_z - \omega_{0z}}{t} = \dfrac{0 - 84.0 \text{ rad/s}}{10.3 \text{ s}} = -8.16$ rad/s^2

EVALUATE: The angular acceleration is positive while the wheel is speeding up and negative while it is slowing down. We could also use $\omega_z^2 = \omega_{0z}^2 + 2\alpha_z(\theta - \theta_0)$

to calculate $\alpha_z = \dfrac{\omega_z^2 - \omega_{0z}^2}{2(\theta - \theta_0)} = \dfrac{0 - (84.0 \text{ rad/s})^2}{2(432 \text{ rad})} = -8.16$ rad/s^2 for the accelration after the breaker trips.

9.21 IDENTIFY and **SET UP:** The blade tip moves in a circular path and its tangential speed is related to its angular speed by Eq.(9.13). In part (b) its resultant velocity is the vector sum of its tangential and vertical velocities.

EXECUTE:

The tangential speed of a blade tip is $v = r\omega$.

$\omega = (90.0 \text{ rev/min}) \left(\dfrac{2\pi \text{ rad}}{1 \text{ rev}}\right) \left(\dfrac{1 \text{ min}}{60 \text{ s}}\right) = 9.425$ rad/s

$v = r\omega = (5.0 \text{ m})(9.425 \text{ rad/s}) = 47.1$ m/s

The upward velocity of the entire blade has magnitude 4.00 m/s. The tangential velocity and the upward velocity of a blade tip are perpendicular, so their resultant has magnitude

$$v_{\rm res} = \sqrt{(47.1 \text{ m/s})^2 + (4.00 \text{ m/s})^2} = 47.3 \text{ m/s}$$

EVALUATE: $v = r\omega$ requires that ω be in rad/s to give v in m/s. The tangential velocity is much larger than the vertical velocity so the speed (magnitude of the resultant velocity) is not much larger than the magnitude of the tangential velocity.

9.25 **IDENTIFY** and **SET UP:** Use constant acceleration equations to find ω and α after each displacement. Then use Eqs.(9.14) and (9.15) to find the components of the linear acceleration.

EXECUTE:

a) at the start $t = 0$

flywheel starts from rest so $\omega_z = \omega_{0z} = 0$

$a_{\rm tan} = r\alpha = (0.300 \text{ m})(0.600 \text{ rad/s}^2) = 0.180 \text{ m/s}^2$

$a_{\rm rad} = r\omega^2 = 0$

$a = \sqrt{a_{\rm rad}^2 + a_{\rm tan}^2} = 0.180 \text{ m/s}^2$

b) $\theta - \theta_0 = 60°$

$a_{\rm tan} = r\alpha = 0.180 \text{ m/s}^2$

Calculate ω:

$\theta - \theta_0 = 60°(\pi \text{ rad}/180°) = 1.047 \text{ rad}; \quad \omega_{0z} = 0; \quad \alpha_z = 0.600 \text{ rad/s}^2; \quad \omega_z = ?$

$\omega_z^2 = \omega_{0z}^2 + 2\alpha_z(\theta - \theta_0)$

$\omega_z = \sqrt{2\alpha_z(\theta - \theta_0)} = \sqrt{2(0.600 \text{ rad/s}^2)(1.047 \text{ rad})} = 1.121 \text{ rad/s}$ and $\omega = \omega_z$.

Then $a_{\rm rad} = r\omega^2 = (0.300 \text{ m})(1.121 \text{ rad/s})^2 = 0.377 \text{ m/s}^2$.

$a = \sqrt{a_{\rm rad}^2 + a_{\rm tan}^2} = \sqrt{(0.377 \text{ m/s}^2)^2 + (0.180 \text{ m/s}^2)^2} = 0.418 \text{ m/s}^2$

c) $\theta - \theta_0 = 120°$

$a_{\rm tan} = r\alpha = 0.180 \text{ m/s}^2$

Calculate ω:

$\theta - \theta_0 = 120°(\pi \text{ rad}/180°) = 2.094 \text{ rad}; \quad \omega_{0z} = 0; \quad \alpha_z = 0.600 \text{ rad/s}^2; \quad \omega_z = ?$

$\omega_z^2 = \omega_{0z}^2 + 2\alpha_z(\theta - \theta_0)$

$\omega_z = \sqrt{2\alpha_z(\theta - \theta_0)} = \sqrt{2(0.600 \text{ rad/s}^2)(2.094 \text{ rad})} = 1.585 \text{ rad/s}$ and $\omega = \omega_z$.

Then $a_{rad} = r\omega^2 = (0.300 \text{ m})(1.585 \text{ rad/s})^2 = 0.754 \text{ m/s}^2$.

$a = \sqrt{a_{rad}^2 + a_{tan}^2} = \sqrt{(0.754 \text{ m/s}^2)^2 + (0.180 \text{ m/s}^2)^2} = 0.775 \text{ m/s}^2$

EVALUATE: α is constant so α_{tan} is constant. ω increases so a_{rad} increases.

9.27 **IDENTIFY:** Use Eq.(9.15) and solve for r.

SET UP: $a_{rad} = r\omega^2$ so $r = a_{rad}/\omega^2$, where ω must be in rad/s

EXECUTE: $a_{rad} = 3000g = 3000(9.80 \text{ m/s}^2) = 29,400 \text{ m/s}^2$

$$\omega = (5000 \text{ rev/min})\left(\frac{1 \text{ min}}{60 \text{ s}}\right)\left(\frac{2\pi \text{ rad}}{1 \text{ rev}}\right) = 523.6 \text{ rad/s}$$

Then $r = \dfrac{a_{rad}}{\omega^2} = \dfrac{29,400 \text{ m/s}^2}{(523.6 \text{ rad/s})^2} = 0.107 \text{ m}$.

EVALUATE: The diameter is then 0.214 m, which is larger than 0.127 m, so the claim is <u>not</u> realistic.

9.31 **IDENTIFY** and **SET UP:** Use Eq.(9.15) to relate ω to a_{rad} and $\sum \vec{F} = m\vec{a}$ to relate a_{rad} to F_{rad}. Use Eq.(9.13) to relate ω amd v, where v is the tangential speed.

EXECUTE:

a) $a_{rad} = r\omega^2$ and $F_{rad} = ma_{rad} = mr\omega^2$

$$\frac{F_{rad,2}}{F_{rad,1}} = \left(\frac{\omega_2}{\omega_1}\right)^2 = \left(\frac{640 \text{ rev/min}}{423 \text{ rev/min}}\right)^2 = 2.29$$

b) $v = r\omega$

$$\frac{v_2}{v_1} = \frac{\omega_2}{\omega_1} = \frac{640 \text{ rev/min}}{423 \text{ rev/min}} = 1.51$$

c) $v = r\omega$

$$\omega = (640 \text{ rev/min})\left(\frac{1 \text{ min}}{60 \text{ s}}\right)\left(\frac{2\pi \text{ rad}}{1 \text{ rev}}\right) = 67.0 \text{ rad/s}$$

Then $v = r\omega = (0.235 \text{ m})(67.0 \text{ rad/s}) = 15.7 \text{ m/s}$.

$a_{rad} = r\omega^2 = (0.235 \text{ m})(67.0 \text{ rad/s})^2 = 1060 \text{ m/s}^2$

$$\frac{a_{rad}}{g} = \frac{1060 \text{ m/s}^2}{9.80 \text{ m/s}^2} = 108; \quad a = 108g$$

EVALUATE: In parts (a) and (b), since a ratio is used the untis cancel and there is no need to convert ω to rad/s. In part (c), v and a_{rad} are calculated from ω, and ω must be in rad/s.

9.35 IDENTIFY and **SET UP:** According to Eq.(9.16), I for the entire object equals the sum of I for each piece, the rod plus the end caps.

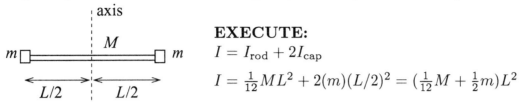

EXECUTE:

$I = I_{rod} + 2I_{cap}$

$I = \frac{1}{12}ML^2 + 2(m)(L/2)^2 = (\frac{1}{12}M + \frac{1}{2}m)L^2$

EVALUATE: Table 9.2 was used for I_{rod} and $I = mr^2$ for the end caps, since they are treated as point particles.

9.37 IDENTIFY and **SET UP:** Use Eq.(9.16). Treat the spheres as point **EXECUTE:**
masses and ignore I of the light rods.

a)

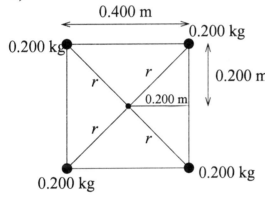

$r = \sqrt{(0.200 \text{ m})^2 + (0.200 \text{ m})^2} = 0.2828 \text{ m}$

$I = \sum m_i r_i^2 = 4(0.200 \text{ kg})(0.2828 \text{ m})^2$

$I = 0.0640 \text{ kg} \cdot \text{m}^2$

b)

0.200 kg   0.200 kg

0.200 m

axis

0.200 m

0.200 kg   0.200 kg

$r = 0.200 \text{ m}$

$I = \sum m_i r_i^2 = 4(0.200 \text{ kg})(0.200 \text{ m})^2$

$I = 0.0320 \text{ kg} \cdot \text{m}^2$

c)

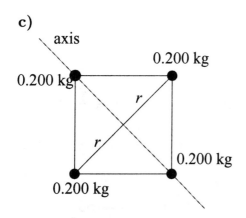

axis

0.200 kg

0.200 kg

0.200 kg

0.200 kg

$r = 0.2828$ m

$I = \sum m_i r_i^2 = 2(0.200 \text{ kg})(0.2828 \text{ m})^2$

$I = 0.0320 \text{ kg} \cdot \text{m}^2$

EVALUATE: In general I depends on the axis and our answer for part(a) is larger than for parts (b) and (c). It just happens that I is the same in parts (b) and (c).

9.41 **IDENTIFY** and **SET UP:** $I = \sum m_i r_i^2$ implies $I = I_{\text{rim}} + I_{\text{spokes}}$

EXECUTE:

$I_{\text{rim}} = MR^2 = (1.40 \text{ kg})(0.300 \text{ m})^2 = 0.126 \text{ kg} \cdot \text{m}^2$

Each spoke can be treated as a slender rod with the axis through one end, so

$I_{\text{spokes}} = 8(\frac{1}{3}ML^2) = \frac{8}{3}(0.280 \text{ kg})(0.300 \text{ m})^2 = 0.0672 \text{ kg} \cdot \text{m}^2$

$I = I_{\text{rim}} + I_{\text{spokes}} = 0.126 \text{ kg} \cdot \text{m}^2 + 0.0672 \text{ kg} \cdot \text{m}^2 = 0.193 \text{ kg} \cdot \text{m}^2$

EVALUATE: Our result is smaller than $m_{\text{tot}} R^2 = (3.64 \text{ kg})(0.300 \text{ m})^2 = 0.328$ kg·m², since the mass of each spoke is distributed between $r = 0$ and $r = R$.

9.47 **IDENTIFY** and **SET UP:** Combine eqs.(9.17) and (9.15) to solve for K. Use Table 9.2 to get I.

EXECUTE: $K = \frac{1}{2}I\omega^2$

$a_{\text{rad}} = R\omega^2$, so $\omega = \sqrt{a_{\text{rad}}/R} = \sqrt{(3500 \text{ m/s}^2)/1.20 \text{ m}} = 54.0 \text{ rad/s}$

For a disk, $I = \frac{1}{2}MR^2 = \frac{1}{2}(70.0 \text{ kg})(1.20 \text{ m})^2 = 50.4 \text{ kg} \cdot \text{m}^2$

Thus $K = \frac{1}{2}I\omega^2 = \frac{1}{2}(50.4 \text{ kg} \cdot \text{m}^2)(54.0 \text{ rad/s})^2 = 7.35 \times 10^4$ J

EVALUATE: The limit on a_{rad} limits ω which in turn limits K.

9.49 **IDENTIFY** and **SET UP:** $\omega = 2\pi/T$ since 2π rad is the angular displacement in time T. Combine this with Eq.(9.17) to express K in terms of T. Use this expression to calculate dK/dt in terms of dT/dt.

EXECUTE:

a) $K = \frac{1}{2}I\omega^2, \quad \omega = 2\pi/T$

$$K = \tfrac{1}{2}I\left(\frac{2\pi}{T}\right)^2 = \frac{2\pi^2 I}{T^2}$$

b) $\dfrac{dK}{dt} = 2I\pi^2\left(\dfrac{dT^{-2}}{dt}\right) = 2I\pi^2\left(-\dfrac{2}{T^3}\right)\dfrac{dT}{dt} = -\dfrac{4\pi^2 I}{T^3}\dfrac{dT}{dt}$

c) $K = \dfrac{2\pi^2 I}{T^2} = \dfrac{2\pi^2(8.0 \text{ kg} \cdot \text{m}^2)}{(1.5 \text{ s})^2} = 70$ J

d) $\dfrac{dK}{dt} = -\dfrac{4\pi^2 I}{T^3}\dfrac{dT}{dt} = -\dfrac{4\pi^2(8.0 \text{ kg}\cdot\text{m}^2)}{(1.5 \text{ s})^3}(0.0060) = -0.56$ J/s

EVALUATE: The greater T is the slower the rotation and the smaller K is. If dT/dt is positive then T is increasing and K is decreasing, so dK/dt is negative.

9.51 IDENTIFY and **SET UP:** Eq.(9.18) says $U = Mgy_{cm}$

The positive work done by the wrestler must equal in magnitude the negative work done by gravity.

EXECUTE: $W = -W_{grav} = U_2 - U_1 = Mg(\Delta y_{cm})$

$W = (120 \text{ kg})(9.80 \text{ m/s}^2)(0.700 \text{ m}) = 823$ J

EVALUATE: The force the wrestler exerts is upward and the displcement of the opponent is upward, so the force does positive work.

9.53 IDENTIFY: Use Eq.(9.19) to relate I for the wood sphere about the desired axis to I for an axis along a diameter.

SET UP: For a thin-walled hollow sphere, axis along a diameter, $I = \tfrac{2}{3}MR^2$.

For a solid sphere with mass M and radius R, $I_{cm} = \tfrac{2}{5}MR^2$, for an axis along a diameter.

EXECUTE: Find d such that $I_P = I_{cm} + Md^2$ with $I_P = \tfrac{2}{3}MR^2$:

$\tfrac{2}{3}MR^2 = \tfrac{2}{5}MR^2 + Md^2$

The factors of M divide out and the equation bercomes $\left(\tfrac{2}{3} - \tfrac{2}{5}\right)R^2 = d^2$

$d = \sqrt{(10-6)/15}\,R = 2R/\sqrt{15} = 0.516R$.

The axis is parallel to a diameter and is $0.516R$ from the center.

EVALUATE: $I_{cm}(\text{lead}) > I_{cm}(\text{wood})$ even though M and R are the same since for a hollow sphere all the mass is a distance R from the axis. Eq.(9.19) says $I_P > I_{cm}$, so there must be a d where $I_P(\text{wood}) = I_{cm}(\text{lead})$.

9.55 IDENTIFY and **SET UP:** Use Eq.(9.19). The cm of the sheet is at its geometrical center.

EXECUTE: $I_P = I_{cm} + Md^2$

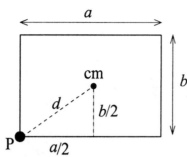

From part (c) of Fig.9-2,
$I_{cm} = \frac{1}{12}M(a^2 + b^2)$.

The distance d of P from the cm is
$d = \sqrt{(a/2)^2 + (b/2)^2}$.

Thus $I_P = I_{cm} + Md^2 = \frac{1}{12}M(a^2 + b^2) + M(\frac{1}{4}a^2 + \frac{1}{4}b^2) = (\frac{1}{12} + \frac{1}{4})M(a^2 + b^2) = \frac{1}{3}M(a^2 + b^2)$

EVALUATE: $I_P = 4I_{cm}$. For an axis through P mass is farther from the axis.

9.59 IDENTIFY: Eq.(9.20), $I = \int r^2 \, dm$

SET UP:

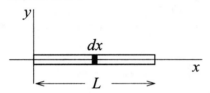

Take the x-axis to lie along the rod, with the origin at the left end. Consider a thin slice at coordinate x and width dx. The mass per unit length for this rod is M/L, so the mass of this slice is $dm = (M/L)dx$.

EXECUTE: $I = \int_0^L x^2(M/L) \, dx = (M/L)\int_0^L x^2 \, dx = (M/L)(L^3/3) = \frac{1}{3}ML^2$

EVALUATE: This result agrees with Table 9-2.

Problems

9.63 IDENTIFY and **SET UP:** Apply $v = r\omega$. v is the tangential speed of a point on the rim of the wheel and equals the linear speed of the car.

EXECUTE:

a) $v = 60$ mph $= 26.82$ m/s

$r = 12$ in. $= 0.3048$ m

$\omega = \dfrac{v}{r} = 88.0$ rad/s $= 14.0$ rev/s $= 840$ rpm

b) same ω as in part (a) since speedometer reads same

$r = 15$ in. $= 0.381$ m

$v = r\omega = (0.381 \text{ m})(88.0 \text{ rad/s}) = 33.5$ m/s $= 75$ mph

c) $v = 50$ mph $= 22.35$ m/s

$r = 10$ in. $= 0.254$ m

$\omega = \dfrac{v}{r} = 88.0$ rad/s; this is the same as for 60 mph with correct tires, so speedometer read 60 mph.

EVALUATE: For a given ω, v increases when r increases.

9.65 IDENTIFY and **SET UP:** Use Eqs.(9.3) amd (9.5). As long as $\alpha_Z > 0$, ω_z increases. At the t when $\alpha_z = 0$, ω_z is at its maximum positive value and then starts to decrease when α_z becomes negative.

$\theta(t) = \gamma t^2 - \beta t^3; \quad \gamma = 3.20$ rad/s^2, $\beta = 0.500$ rad/s^3

EXECUTE:

a) $\omega_z(t) = \dfrac{d\theta}{dt} = \dfrac{d(\gamma t^2 - \beta t^3)}{dt} = 2\gamma t - 3\beta t^2$

b) $\alpha_z(t) = \dfrac{d\omega_z}{dt} = \dfrac{d(2\gamma t - 3\beta t^2)}{dt} = 2\gamma - 6\beta t$

c) The maximum angular velocity occurs when $\alpha_z = 0$.

$2\gamma - 6\beta t = 0$ implies $t = \dfrac{2\gamma}{6\beta} = \dfrac{\gamma}{3\beta} = \dfrac{3.20 \text{ rad/s}^2}{3(0.500 \text{ rad/s}^3)} = 2.133$ s

At this t, $\omega_z = 2\gamma t - 3\beta t^2 = 2(3.20 \text{ rad/s}^2)(2.133 \text{ s}) - 3(0.500 \text{ rad/s}^3)(2.133 \text{ s})^2 = 6.83$ rad/s

The maximum positive angular velocity is 6.83 rad/s and it occurs at 2.13 s.

EVALUATE: For large t both ω_z and α_z are negative and ω_z increases in magnitude. In fact, $\omega_z \to -\infty$ at $t \to \infty$. So the answer in (c) is not the largest angular speed, just the largest positive angular velocity.

9.67 IDENTIFY and **SET UP:** The translational kinetic energy is $K = \frac{1}{2}mv^2$ and the kinetic energy of the rotating flywheel is $K = \frac{1}{2}I\omega^2$. Use the scale speed to calculate the actual speed v. From that calculate K for the car and then solve for ω that gives this K for the flywheel.

EXECUTE:

a) $\dfrac{v_{\text{toy}}}{v_{\text{scale}}} = \dfrac{L_{\text{toy}}}{L_{\text{real}}}$

$v_{\text{toy}} = v_{\text{scale}}\left(\dfrac{L_{\text{toy}}}{L_{\text{real}}}\right) = (700 \text{ km/h})\left(\dfrac{0.150 \text{ m}}{3.0 \text{ m}}\right) = 35.0$ km/h

$v_{\text{toy}} = (35.0 \text{ km/h})(1000 \text{ m}/1 \text{ km})(1 \text{ h}/3600 \text{ s}) = 9.72$ m/s

b) $K = \frac{1}{2}mv^2 = \frac{1}{2}(0.180 \text{ kg})(9.72 \text{ m/s})^2 = 8.50$ J

c) $K = \frac{1}{2}I\omega^2$ gives that $\omega = \sqrt{\dfrac{2K}{I}} = \sqrt{\dfrac{2(8.50 \text{ J})}{4.00 \times 10^{-5} \text{ kg} \cdot \text{m}^2}} = 652$ rad/s

EVALUATE: $K = \frac{1}{2}I\omega^2$ gives ω in rad/s. 652 rad/s = 6200 rev/min so the rotation rate of the flywheel is very large.

9.69 **IDENTIFY** and **SET UP:** Use the work-energy theorem (Eq.6.6) to relate the work to the kinetic energy.

EXECUTE:

a) $W_{\text{tot}} = K_2 - K_1$ so $K_2 = K_1 + W_{\text{tot}}$

$W_{\text{tot}} = -4000$ J (the amount of energy given up by the flywheel)

$K_1 = \frac{1}{2}I\omega_1^2$, but ω_1 must be in rad/s.

$\omega_1 = (300 \text{ rev/min})(2\pi \text{ rad/1 rev})(1 \text{ min/60 s}) = 31.42$ rad/s

$K_1 = \frac{1}{2}(16.0 \text{ kg} \cdot \text{m}^2)(31.42 \text{ rad/s})^2 = 7898$ J

Then $K_2 = K_1 + W_{\text{tot}} = 7898$ J $- 4000$ J $= 3898$ J

and $K_2 = \frac{1}{2}I\omega_2^2$ gives $\omega_2 = \sqrt{\dfrac{2K_2}{I}} = \sqrt{\dfrac{2(3898 \text{ J})}{16.0 \text{ kg} \cdot \text{m}^2}} = 22.1$ rad/s

$\omega_2 = 22.1$ rad/s $(1 \text{ rev}/2\pi \text{ rad})(60 \text{ s}/1 \text{ min}) = 211$ rev/min.

b) The 4000 J of energy must be restored to the flywheel,

$$P_{\text{av}} = \frac{\Delta W}{\Delta t} = \frac{4000 \text{ J}}{5.00 \text{ s}} = 800 \text{ W}$$

EVALUATE: The flywheel gives up energy when it does work.

9.71 **IDENTIFY** and **SET UP:** All points on the belt move with the same speed. Since the belt doesn't slip, the speed of the belt is the same as the speed of a point on the rim of the shaft and on the rim of the wheel, and these speeds are related to the angular speed of each circular object by $v = r\omega$.

EXECUTE:

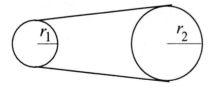

a) $v_1 = r_1\omega_1$

$\omega_1 = (60.0 \text{ rev/s})(2\pi \text{ rad/1 rev}) = 377$ rad/s

$v_1 = r_1\omega_1 = (0.45 \times 10^{-2} \text{ m})(377 \text{ rad/s}) = 1.70$ m/s

b) $v_1 = v_2$

$r_1\omega_1 = r_2\omega_2$

$\omega_2 = (r_1/r_2)\omega_1 = (0.45 \text{ cm}/2.00 \text{ cm})(377 \text{ rad/s}) = 84.8$ rad/s

EVALUATE: The wheel has a larger radius than the shaft so turns slower to have the same tangential speed for points on the rim.

9.73 IDENTIFY and **SET UP:** Use Eq.(9.15) to relate a_{rad} to ω and then use a constant acceleration equation to replace ω.

EXECUTE:

a) $a_{rad} = r\omega^2$, $\quad a_{rad,1} = r\omega_1^2$, $\quad a_{rad,2} = r\omega_2^2$

$\Delta a_{rad} = a_{rad,2} - a_{rad,1} = r(\omega_2^2 - \omega_1^2)$

One of the constant acceleration equations can be written

$\omega_{2z}^2 = \omega_{1z}^2 + 2\alpha(\theta_{2z} - \theta_1)$, or $\omega_{2z}^2 - \omega_{1z}^2 = 2\alpha_z(\theta_2 - \theta_1)$

Thus $\Delta a_{rad} = r2\alpha_z(\theta_2 - \theta_1) = 2r\alpha_z(\theta_2 - \theta_1)$, as was to be shown.

b) $\alpha_z = \dfrac{\Delta a_{rad}}{2r(\theta_2 - \theta_1)} = \dfrac{85.0 \text{ m/s}^2 - 25.0 \text{ m/s}^2}{2(0.250 \text{ m})(15.0 \text{ rad})} = 8.00 \text{ rad/s}^2$

Then $a_{tan} = r\alpha = (0.250 \text{ m})(8.00 \text{ rad/s}^2) = 2.00 \text{ m/s}^2$

EVALUATE: ω^2 is proportional to α_z and $(\theta - \theta_0)$ so a_{rad} is also proportional to these quantities. a_{rad} increases while r stays fixed, ω_z increases, and α_z is positive.

IDENTIFY and **SET UP:** Use Eq.(9.17) to relate K and ω and then use a constant acceleration equation to replace ω.

EXECUTE:

c) $K = \frac{1}{2}I\omega^2$; $K_2 = \frac{1}{2}I\omega_2^2$, $K_1 = \frac{1}{2}I\omega_1^2$

$\Delta K = K_2 - K_1 = \frac{1}{2}I(\omega_2^2 - \omega_1^2) = \frac{1}{2}I(2\alpha(\theta_2 - \theta_1)) = I\alpha(\theta_2 - \theta_1)$, as was to be shown.

d) $I = \dfrac{\Delta K}{\alpha(\theta_2 - \theta_1)} = \dfrac{45.0 \text{ J} - 20.0 \text{ J}}{(8.00 \text{ rad/s}^2)(15.0 \text{ rad})} = 0.208 \text{ kg} \cdot \text{m}^2$

EVALUATE: α_z is positive, ω increases, and K increases.

9.81 IDENTIFY: Use Eq.(9.20) to calculate I. Then use $K = \frac{1}{2}I\omega^2$ to calculate K.

a) SET UP:

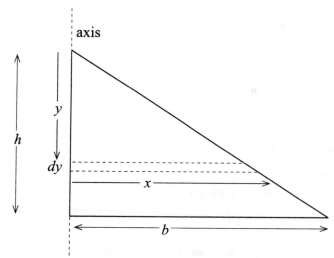

Consider a small strip of width dy and a distance y below the top of the triangle.

The length of the strip is $x = (y/h)b$.

EXECUTE:

The strip has area $x\,dy$ and the area of the sign is $\frac{1}{2}bh$, so the mass of the strip is

$$dm = M\left(\frac{x\,dy}{\frac{1}{2}bh}\right) = M\left(\frac{yb}{h}\right)\left(\frac{2\,dy}{bh}\right) = \left(\frac{2M}{h^2}\right)y\,dy$$

$$dI = \tfrac{1}{3}(dm)x^2 = \frac{2Mb^2}{3h^4}y^3\,dy$$

$$I = \int_0^h dI = \frac{2Mb^2}{3h^4}\int_0^h y^3\,dy = \frac{2Mb^2}{3h^4}\left(\frac{1}{4}y^4\Big|_0^h\right) = \frac{1}{6}Mb^2$$

b) $I = \tfrac{1}{6}Mb^2 = 2.304 \text{ kg} \cdot \text{m}^2$

$\omega = 2.00 \text{ rev/s} = 4.00\pi \text{ rad/s}$

$K = \tfrac{1}{2}I\omega^2 = 182 \text{ J}$

EVALUATE: From Table (9.2), if the sign were rectangular, with length b, then $I = \tfrac{1}{3}Mb^2$. Our result is one-half this, since mass is closer to the axis for the triangular than for the rectangular shape.

9.83 IDENTIFY: Use conservation of energy. The stick rotates about a fixed axis so $K = \tfrac{1}{2}I\omega^2$. Once we have ω use $v = r\omega$ to calculate v for the end of the stick.

SET UP:

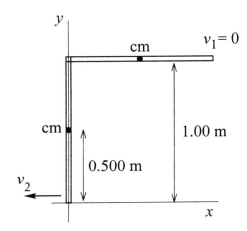

Take the origin of coordinates at the lowest point reached by the stick and take the positive y-direction to be upward.

EXECUTE:

a) Use Eq.(9.18): $U = Mgy_{\text{cm}}$

$\Delta U = U_2 - U_1 = Mg(y_{\text{cm2}} - y_{\text{cm1}})$

The center of mass of the meter stick is at its geometrical center, so

$y_{\text{cm1}} = 1.00$ m and $y_{\text{cm2}} = 0.50$ m

Then $\Delta U = (0.160 \text{ kg})(9.80 \text{ m/s}^2)(0.50 \text{ m} - 1.00 \text{ m}) = -0.784$ J

b) Use conservation of energy: $K_1 + U_1 + W_{\text{other}} = K_2 + U_2$

Gravity is the only force that does work on the meter stick, so $W_{\text{other}} = 0$.

$K_1 = 0$.

Thus $K_2 = U_1 - U_2 = -\Delta U$, where ΔU was calculated in part (a).

$K_2 = \frac{1}{2}I\omega_2^2$ so $\frac{1}{2}I\omega_2^2 = -\Delta U$ and $\omega_2 = \sqrt{2(-\Delta U)/I}$

For stick pivoted about one end, $I = \frac{1}{3}ML^2$ where $L = 1.00$ m, so

$$\omega_2 = \sqrt{\frac{6(-\Delta U)}{ML^2}} = \sqrt{\frac{6(0.784 \text{ J})}{(0.160 \text{ kg})(1.00 \text{ m})^2}} = 5.42 \text{ rad/s}$$

c) $v = r\omega = (1.00 \text{ m})(5.42 \text{ rad/s}) = 5.42$ m/s

d) For a particle in free-fall, with $+y$ upward,

$v_{0y} = 0; \quad y - y_0 = -1.00 \text{ m}; \quad a_y = -9.80 \text{ m/s}^2; \quad v = ?$

$v^2 = v_{0y}^2 + 2a_y(y - y_0)$

$v = -\sqrt{2a_y(y - y_0)} = -\sqrt{2(-9.80 \text{ m/s}^2)(-1.00 \text{ m})} = -4.43$ m/s

EVALUATE: The magnitude of the answer in part (c) is larger. $U_{1,\text{grav}}$ is the same for the stick as for a particle falling from a height of 1.00 m. For the stick $K = \frac{1}{2}I\omega_2^2 = \frac{1}{2}(\frac{1}{3}ML^2)(v/L)^2 = \frac{1}{6}Mv^2$. For the stick and for the particle, K_2 is the same but the same K gives a larger v for the end of the stick than for the particle.

The reason is that all the other points along the stick are moving slower than the end opposite the axis.

9.85 IDENTIFY: Apply conservation of energy to the system consisting of blocks A and B and the pulley.

SET UP:

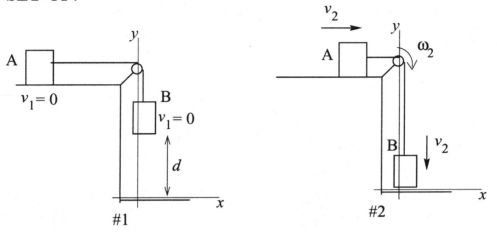

#1   #2

Use the work-energy relation $K_1 + U_1 + W_{\text{other}} = K_2 + U_2$. Use coordinates where $+y$ is upward and where the origin is at the position of block B after it has descended.

The tension in the rope does positive work on block A and negative work of the same magnitude on block B, so the net work done by the tension in the rope is zero.

Both blocks have the same speed.

EXECUTE: Gravity does work on block B and kinetic friction does work on block A. Therefore $W_{\text{other}} = W_f = -\mu_k m_A g d$.

$K_1 = 0$ (system is released from rest)

$U_1 = m_B g y_{B1} = m_B g d; \quad U_2 = m_B g y_{B2} = 0$

$K_2 = \frac{1}{2} m_A v_2^2 + \frac{1}{2} m_B v_2^2 + \frac{1}{2} I \omega_2^2.$

But $v(\text{blocks}) = R\omega(\text{pulley})$, so $\omega_2 = v_2/R$ and

$K_2 = \frac{1}{2}(m_A + m_B)v_2^2 + \frac{1}{2}I(v_2/R)^2 = \frac{1}{2}(m_A + m_B + I/R^2)v_2^2$

Putting all this into the work-energy relation gives

$m_B g d - \mu_k m_A g d = \frac{1}{2}(m_A + m_B + I/R^2)v_2^2$

$(m_A + m_B + I/R^2)v_2^2 = 2gd(m_B - \mu_k m_A)$

$$v_2 = \sqrt{\frac{2gd(m_B - \mu_k m_A)}{m_A + m_B + I/R^2}}$$

EVALUATE: If $m_B \gg m_A$ and I/R^2, then $v_2 = \sqrt{2gd}$; block B falls freely. If I is very large, v_2 is very small. Must have $m_B > \mu_k m_A$ for motion, so the weight of B will be larger than the friction force on A. I/R^2 has units of mass and is in a sense the "effective mass" of the pulley.

9.87 IDENTIFY and **SET UP:** Apply conservation of energy to the motion of the hoop. Use Eq.(9.18) to calculate U_{grav}. Use $K = \frac{1}{2}I\omega^2$ for the kinetic energy of the hoop. Solve for ω. The center of mass of the hoop is at its geometrical center.

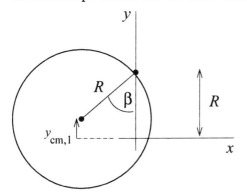

Take the origin to be at the original location of the center of the hoop, before it is rotated to one side.

$y_{\text{cm1}} = R - R\cos\beta = R(1 - \cos\beta)$

$y_{\text{cm2}} = 0$ (at equilibrium position hoop is at original position)

EXECUTE: $K_1 + U_1 + W_{\text{other}} = K_2 + U_2$

$W_{\text{other}} = 0$ (only gravity does work)

$K_1 = 0$ (released from rest), $K_2 = \frac{1}{2}I\omega_2^2$

For a hoop, $I_{\text{cm}} = MR^2$, so $I = Md^2 + MR^2$ with $d = R$ and $I = 2MR^2$, for an axis at the edge. Thus $K_2 = \frac{1}{2}(2MR^2)\omega_2^2 = MR^2\omega_2^2$.

$U_1 = Mgy_{\text{cm1}} = MgR(1 - \cos\beta)$, $U_2 = mgy_{\text{cm2}} = 0$

Thus $K_1 + U_1 + W_{\text{other}} = K_2 + U_2$ gives

$MgR(1 - \cos\beta) = MR^2\omega_2^2$ and $\omega_2 = \sqrt{g(1 - \cos\beta)/R}$

EVALUATE: If $\beta = 0$, then $\omega_2 = 0$. As β increases, ω_2 increases.

9.91 IDENTIFY: Apply conservation of energy to relate the height of the mass to the kinetic energy of the cylinder.

SET UP: First use $K(\text{cylinder}) = 250$ J to find ω for the cylinder and v for the mass.

EXECUTE: $I = \frac{1}{2}MR^2 = \frac{1}{2}(10.0 \text{ kg})(0.150 \text{ m})^2 = 0.1125 \text{ kg} \cdot \text{m}^2$

$K = \frac{1}{2}I\omega^2$ so $\omega = \sqrt{2K/I} = 66.67$ rad/s

$v = R\omega = 10.0$ m/s

SET UP: Use conservation of energy $K_1 + U_1 = K_2 + U_2$ to solve for the distance the mass descends. Take $y = 0$ at lowest point of the mass, so $y_2 = 0$ and $y_1 = h$, the distance the mass descends.

EXECUTE: $K_1 = U_2 = 0$ so $U_1 = K_2$.

$mgh = \frac{1}{2}mv^2 + \frac{1}{2}I\omega^2$, where $m = 12.0$ kg

For the cylinder, $I = \frac{1}{2}MR^2$ and $\omega = v/R$, so $\frac{1}{2}I\omega^2 = \frac{1}{4}Mv^2$.

$mgh = \frac{1}{2}mv^2 + \frac{1}{4}Mv^2$

$h = \dfrac{v^2}{2g}\left(1 + \dfrac{M}{2m}\right) = 7.23$ m

EVALUATE: For the cylinder $K_{cyl} = \frac{1}{2}I\omega^2 = \frac{1}{2}(\frac{1}{2}MR^2)(v/R)^2 = \frac{1}{4}Mv^2$.

$K_{mass} = \frac{1}{2}mv^2$, so $K_{mass} = (2m/M)K_{cyl} = [2(12.0$ kg$)/10.0$ kg$](250$ J$) = 600$ J. The mass has 600 J of kinetic energy when the cylinder has 250 J of kinetic energy and at this point the system has total energy 850 J since $U_2 = 0$. Initially the total energy of the system is $U_1 = mgy_1 = mgh = 850$ J, so the total energy is shown to be conserved.

9.97   Use Eq.(9.20) to calculate I.

a) SET UP: Let L be the length of the cylinder. Divide the cylinder into thin cylindrical shells of inner radius r and outer radius $r + dr$. An end view is

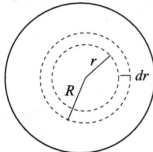

$\rho = \alpha r$

The mass of the thin cylindrical shell is
$dm = \rho\, dV = \rho(2\pi r\, dr)L = 2\pi\alpha Lr^2\, dr$

EXECUTE: $I = \int r^2\, dm = 2\pi\alpha L \int_0^R r^4\, dr = 2\pi\alpha L(\frac{1}{5}R^5) = \frac{2}{5}\pi\alpha LR^5$

Relate M to α:
$M = \int dm = 2\pi\alpha L \int_0^R r^2\, dr = 2\pi\alpha L(\frac{1}{3}R^3) = \frac{2}{3}\pi\alpha LR^3$, so $\pi\alpha LR^3 = 3M/2$.

Using this in the above result for I gives
$I = \frac{2}{5}(3M/2)R^2 = \frac{3}{5}MR^2$.

b) EVALUATE: For a cylinder of uniform density $I = \frac{1}{2}MR^2$. The answer in (a) is larger than this. Since the density increases with distance from the axis the cylinder in (a) has more mass farther from the axis than for a cylinder of uniform density.

CHAPTER 10
DYNAMICS OF ROTATIONAL MOTION

Exercises 1, 3, 5, 11, 13, 19, 23, 25, 27, 33, 35, 37, 39, 43, 45, 49
Problems 55, 57, 61, 63, 65, 67, 69, 75, 79, 85, 87, 89, 91, 93, 97

Exercises

10.1 **IDENTIFY:** Use Eq.(10.2) to calculate the magnitude of the torque and use the right-hand rule illustrated in Fig.(10.4) to calculate the torque direction.

a) SET UP:

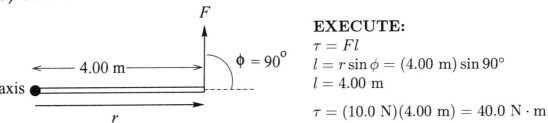

EXECUTE:
$\tau = Fl$
$l = r \sin \phi = (4.00 \text{ m}) \sin 90°$
$l = 4.00 \text{ m}$
$\tau = (10.0 \text{ N})(4.00 \text{ m}) = 40.0 \text{ N} \cdot \text{m}$

This force tends to produce a counterclockwise rotation about the axis; by the right-hand rule the vector $\vec{\tau}$ is directed out of the plane of the figure.

b) SET UP:

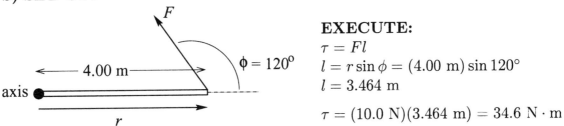

EXECUTE:
$\tau = Fl$
$l = r \sin \phi = (4.00 \text{ m}) \sin 120°$
$l = 3.464 \text{ m}$
$\tau = (10.0 \text{ N})(3.464 \text{ m}) = 34.6 \text{ N} \cdot \text{m}$

This force tends to produce a counterclockwise rotation about the axis; by the right-hand rule the vector $\vec{\tau}$ is directed out of the plane of the figure.

c) SET UP:

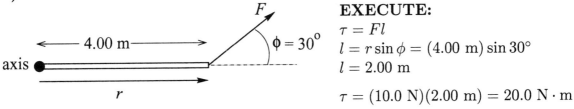

EXECUTE:
$\tau = Fl$
$l = r \sin \phi = (4.00 \text{ m}) \sin 30°$
$l = 2.00 \text{ m}$
$\tau = (10.0 \text{ N})(2.00 \text{ m}) = 20.0 \text{ N} \cdot \text{m}$

This force tends to produce a counterclockwise rotation about the axis; by the right-hand rule the vector $\vec{\tau}$ is directed out of the plane of the figure.

d) SET UP:

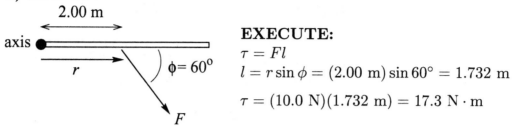

EXECUTE:

$\tau = Fl$

$l = r \sin \phi = (2.00 \text{ m}) \sin 60° = 1.732 \text{ m}$

$\tau = (10.0 \text{ N})(1.732 \text{ m}) = 17.3 \text{ N} \cdot \text{m}$

This force tends to produce a clockwise rotation about the axis; by the right-hand rule the vector $\vec{\tau}$ is directed into the plane of the figure.

e) SET UP:

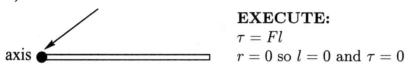

EXECUTE:

$\tau = Fl$

$r = 0$ so $l = 0$ and $\tau = 0$

f) SET UP:

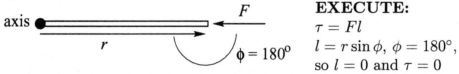

EXECUTE:

$\tau = Fl$

$l = r \sin \phi, \ \phi = 180°,$

so $l = 0$ and $\tau = 0$

EVALUATE: The torque is zero in parts (e) and (f) because the moment arm is zero; the line of action of the force passes through the axis.

10.3 **IDENTIFY** and **SET UP:** Use Eq.(10.2) to calculate the magnitude of each torque and use the right-hand rule (Fig.10.4) to determine the direction.

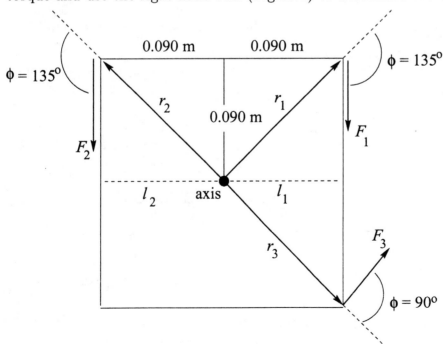

Let counterclockwise be the positive sense of rotation.

EXECUTE:

$r_1 = r_2 = r_3 = \sqrt{(0.090\text{ m})^2 + (0.090\text{ m})^2} = 0.1273\text{ m}$

$\tau_1 = -F_1 l_1$

$l_1 = r_1 \sin_1 = (0.1273\text{ m})\sin 135° = 0.0900\text{ m}$

$\tau_1 = -(18.0\text{ N})(0.0900\text{ m}) = -1.62\text{ N}\cdot\text{m}$

$\vec{\tau}_1$ is directed into paper

$\tau_2 = +F_2 l_2$

$l_2 = r_2 \sin_2 = (0.1273\text{ m})\sin 135° = 0.0900\text{ m}$

$\tau_2 = +(26.0\text{ N})(0.0900\text{ m}) = +2.34\text{ N}\cdot\text{m}$

$\vec{\tau}_2$ is directed out of paper

$\tau_3 = +F_3 l_3$

$l_3 = r_3 \sin_3 = (0.1273\text{ m})\sin 90° = 0.1273\text{ m}$

$\tau_3 = +(14.0\text{ N})(0.1273\text{ m}) = +1.78\text{ N}\cdot\text{m}$

$\vec{\tau}_3$ is directed out of paper

$\sum\tau = \tau_1 + \tau_2 + \tau_3 = -1.62\text{ N}\cdot\text{m} + 2.34\text{ N}\cdot\text{m} + 1.78\text{ N}\cdot\text{m} = 2.50\text{ N}\cdot\text{m}$

EVALUATE: The net torque is positive, which means it tends to produce a counterclockwise rotation; the vector torque is directed out of the plane of the paper. In summing the torques it is important to include + or − signs to show direction.

10.5 **IDENTIFY** and **SET UP:** Calculate the torque using Eq.(10.3) and also determine the direction of the torque using the right-hand rule.

a) $\vec{r} = (-0.450\text{ m})\hat{i} + (0.150\text{ m})\hat{j}$; $\vec{F} = (-5.00\text{ N})\hat{i} + (4.00\text{ N})\hat{j}$

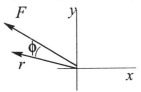

EXECUTE:

b) When the fingers of your right hand curl from the direction of $\vec{r}$ into the direction of $\vec{F}$ (through the smaller of the two angles, angle ϕ) your thumb points into the page (the direction of $\vec{\tau}$, the $-z$-direction).

c) $\vec{\tau} = \vec{r}\times\vec{F} = [(-0.450\text{ m})\hat{i} + (0.150\text{ m})\hat{j}]\times[(-5.00\text{ N})\hat{i} + (4.00\text{ N})\hat{j}]$

$\vec{\tau} = + (2.25\text{ N}\cdot\text{m})\hat{i}\times\hat{i} - (1.80\text{ N}\cdot\text{m})\hat{i}\times\hat{j} - (0.750\text{ N}\cdot\text{m})\hat{j}\times\hat{i} + (0.600\text{ N}\cdot\text{m})\hat{j}\times\hat{j}$

$\hat{i}\times\hat{i} = \hat{j}\times\hat{j} = 0$

$\hat{i}\times\hat{j} = \hat{k},\quad \hat{j}\times\hat{i} = -\hat{k}$

Thus $\vec{\tau} = -(1.80 \text{ N} \cdot \text{m})\hat{k} - (0.750 \text{ N} \cdot \text{m})(-\hat{k}) = (-1.05 \text{ N} \cdot \text{m})\hat{k}$.

EVALUATE: The calculation gives that $\vec{\tau}$ is in the $-z$-direction. This agrees with what we got from the right-hand rule.

10.11 IDENTIFY: Apply $\sum \vec{F} = m\vec{a}$ to the cylinder and calculate the normal force exerted on it by the axle.

a) SET UP:

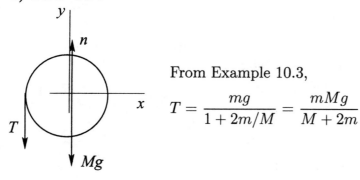

From Example 10.3,

$$T = \frac{mg}{1 + 2m/M} = \frac{mMg}{M + 2m}$$

EXECUTE: $\sum F_y = ma_y$

$n - T - Mg = 0$

$$n = T + Mg = \frac{mMg}{M + 2m} + Mg = \frac{mMg + M(M + 2m)g}{M + 2m} = \frac{M(M + 3m)g}{M + 2m}$$

$$n = \left(\frac{M + 3m}{M + 2m}\right)Mg = \left(\frac{M + 3m}{1 + 2m/M}\right)g$$

b) Re-write the expression for n:

$$n = \frac{mMg}{M + 2m} + Mg = Mg + mg\left(\frac{M}{M + 2m}\right) = Mg + mg + mg\left(\frac{M}{M + 2m} - 1\right)$$

$$n = (M + m)g - mg\left(\frac{M + 2m - M}{M + 2m}\right) = (M + m)g - \left(\frac{2m}{2m + M}\right)mg; \quad N \text{ is less}$$
than $(M + m)g$.

EVALUATE: The block is accelerating downward, so the tension is less than its weight. Or, you could say that part of the system is accelerating downward and the rest has no vertical acceleration, so the total upward force n on the system must be less than the total downward force $(M + m)g$ on the system.

c) The force diagrams in Example 10.3 are unchanged, so this has no effect on T and n.

10.13 IDENTIFY: Use the kinematic information to solve for the angular acceleration of the grindstone. Assume that the grindstone is rotating counterclockwise and let that be the positive sense of rotation. Then apply Eq.(10.6) to calculate the friction force and use $f_k = \mu_k n$ to calculate μ_k.

SET UP: $\omega_{0z} = 850$ rev/min$(2\pi$ rad/1 rev)(1 min/60 s) $= 89.0$ rad/s

$t = 7.50$ s; $\omega_z = 0$ (comes to rest); $\alpha_z = ?$

EXECUTE: $\omega_z = \omega_{0z} + \alpha_z t$

$$\alpha_z = \frac{0 - 89.0 \text{ rad/s}}{7.50 \text{ s}} = -11.9 \text{ rad/s}^2$$

SET UP: Apply $\sum \tau_z = I\alpha_z$ to the grindstone.

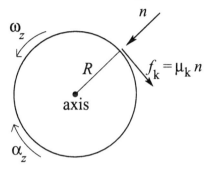

The normal force has zero moment arm for rotation about an axis at the center of the grindstone, and therefore zero torque. The only torque on the grindstone is that due to the friction force f_k exerted by the ax; for this force the moment arm is $l = R$ and the torque is negative.

EXECUTE: $\sum \tau_z = -f_k R = -\mu_k n R$

$I = \frac{1}{2} M R^2$ (solid disk, axis through center)

Thus $\sum \tau_z = I\alpha_z$ gives $-\mu_k n R = (\frac{1}{2} M R^2)\alpha_z$

$$\mu_k = -\frac{M R \alpha_z}{2n} = -\frac{(50.0 \text{ kg})(0.260 \text{ m})(-11.9 \text{ rad/s}^2)}{2(160 \text{ N})} = 0.483$$

EVALUATE: The friction torque is clockwise and slows down the counterclockwise rotation of the grindstone.

10.19 IDENTIFY: Apply Eq.(10.6) to the rotation about the cm to calculate α, and then use $a_{cm} = R\alpha$. Use constant acceleration equations to calculate time and final ω.

a) SET UP:

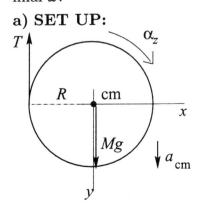

For translational motion of the center of mass of the hoop take the origin of coordinates at the cm and take $+y$ downward.

EXECUTE:

$\sum F_y = ma_y$ gives $Mg - T = Ma_{cm}$.

SET UP: Apply $\sum \tau_z = I_{cm}\alpha_z$ for rotation about the center of mass with the clockwise sense of rotation positive. The weight Mg has zero moment arm and therefore zero torque.

EXECUTE: $TR = I_{cm}\alpha_z$

For a hoop $I_{cm} = MR^2$ so $TR = MR^2\alpha_z$

$T = MR\alpha_z$

$a_{cm} = R\alpha_z$ so the equation becomes $T = Ma_{cm}$

Combine these two equations to eliminate T: $Mg - Ma_{cm} = Ma_{cm}$

The mass M divides out and we get $2a_{cm} = g$, so $a_{cm} = g/2$.

Then $T = M(g/2) = \frac{1}{2}Mg = \frac{1}{2}(0.180 \text{ kg})(9.80 \text{ m/s}^2) = 0.882 \text{ N}$

b) SET UP: Apply the constant acceleration kinematic equations to the motion of the center of mass:

$v_{0y} = 0$; $y - y_0 = 0.750 \text{ m}$; $a_y = g/2 = 4.90 \text{ m/s}^2$; $t = ?$

$y - y_0 = v_{0y}t + \frac{1}{2}a_y t^2$

EXECUTE: $t = \sqrt{\dfrac{2(y - y_0)}{a_y}} = \sqrt{\dfrac{2(0.750 \text{ m})}{4.90 \text{ m/s}^2}} = 0.553 \text{ s}$

c) SET UP: We can use the constant angular acceleration equations for the rotational motion:

$t = 0.553 \text{ s}$; $\omega_{0z} = 0$; $\alpha_z = a_{cm}/R = \frac{1}{2}(9.80 \text{ m/s}^2)/0.0800 \text{ m} = 61.25 \text{ rad/s}^2$;

$\omega_z = ?$

EXECUTE: $\omega_z = \omega_{0z} + \alpha_z t = 0 + (61.25 \text{ m/s}^2)(0.553 \text{ s}) = 33.9 \text{ rad/s}$

EVALUATE: We can alternatively find v_{cm} at this point in the motion and then use $\omega = v_{cm}/R$:

$v_{0y} = 0$; $v_y = ?$; $a_y = 4.90 \text{ m/s}^2$; $y - y_0 = 0.750 \text{ m}$

$v_y^2 = v_{0y}^2 + 2a_y(y - y_0)$

$v_y = \sqrt{2a_y(y - y_0)} = \sqrt{2(4.90 \text{ m/s}^2)(0.750 \text{ m})} = 2.71 \text{ m/s}$

Then $\omega = v_{cm}/R = (2.71 \text{ m/s})/0.0800 \text{ m} = 33.9 \text{ rad/s}$, the same as before.

T is less than the weight Mg of the hoop; the net force on the hoop is downward since the hoop accelerates downward. Part (c) could also be done using conservation of energy.

10.23 IDENTIFY: Apply $\sum \vec{F}_{ext} = m\vec{a}_{cm}$ and $\sum \tau_z = I_{cm}\alpha_z$ to the motion of the ball.

a) SET UP:

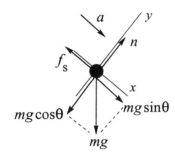

EXECUTE:

$\sum F_y = ma_y$

$n = mg\cos\theta$ and $f_s = \mu_s mg\cos\theta$

$\sum F_x = ma_x$

$mg\sin\theta - \mu_s mg\cos\theta = ma$

$g(\sin\theta - \mu_s\cos\theta) = a$ (eq. 1)

SET UP:

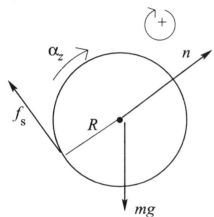

n and mg act at the center of the ball and provide no torque

EXECUTE:

$\sum\tau = \tau_f = \mu_s mg\cos\theta R;\ I = \frac{2}{5}mR^2$

$\sum\tau_z = I_{cm}\alpha_z$ gives $\mu_s mg\cos\theta = \frac{2}{5}mR^2\alpha$

No slipping means $\alpha = a/R$, so $\mu_s g\cos\theta = \frac{2}{5}a$ (eq.2)

We have two equations in the two unknowns a and μ_s. Solving gives

$a = \frac{5}{7}g\sin\theta$ and $\mu_s = \frac{2}{7}\tan\theta = \frac{2}{7}\tan 65.0° = 0.613$

b) Repeat the calculation of part (a), but now $I = \frac{2}{3}mR^2$.

$a = \frac{3}{5}g\sin\theta$ and $\mu_s = \frac{2}{5}\tan\theta = \frac{2}{5}\tan 65.0° = 0.858$

The value of μ_s calculated in part (a) is not large enough to prevent slipping for the hollow ball.

c) EVALUATE: There is no slipping at the point of contact.

More friction is required for a hollow ball since for a given m and R it has a larger I and more torque is needed to provide the same α. Note that the required μ_s is independent of the mass or radius of the ball and only depends on how that mass is distributed.

10.25 IDENTIFY: Apply conservation of energy to the motion of the wheel.

SET UP:

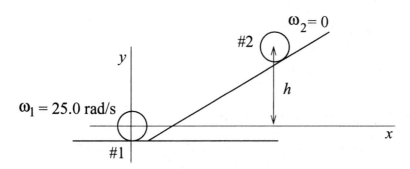

Take $y = 0$ at the center of the wheel when it is at the bottom of the hill.

The wheel has both translational and rotational motion so its kinetic energy is $K = \frac{1}{2}I_{cm}\omega^2 + \frac{1}{2}Mv_{cm}^2$.

EXECUTE: $K_1 + U_1 + W_{other} = K_2 + U_2$

$W_{other} = W_{fric} = -3500$ J (the friction work is negative)

$K_1 = \frac{1}{2}I\omega_1^2 + \frac{1}{2}Mv_1^2$; $v = R\omega$ and $I = 0.800MR^2$ so

$K_1 = \frac{1}{2}(0.800)MR^2\omega_1^2 + \frac{1}{2}MR^2\omega_1^2 = 0.900MR^2\omega_1^2$

$K_2 = 0$, $U_1 = 0$, $U_2 = Mgh$

Thus $0.900MR^2\omega_1^2 + W_{fric} = Mgh$

$M = w/g = 392$ N$/(9.80$ m/s$^2) = 40.0$ kg

$$h = \frac{0.900MR^2\omega_1^2 + W_{fric}}{Mg}$$

$$h = \frac{(0.900)(40.0 \text{ kg})(0.600 \text{ m})^2(25.0 \text{ rad/s})^2 - 3500 \text{ J}}{(40.0 \text{ kg})(9.80 \text{ m/s}^2)} = 11.7 \text{ m}$$

EVALUATE: Friction does negative work and reduces h.

10.27 a) IDENTIFY: Use Eq.(10.6) to find α_z and then use a constant angular acceleration equation to find ω_z.

SET UP:

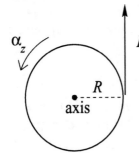

EXECUTE:

Apply $\sum \tau_z = I\alpha_z$ to find the angular acceleration:

$FR = I\alpha_z$

$$\alpha_z = \frac{FR}{I} = \frac{(18.0 \text{ N})(2.40 \text{ m})}{2100 \text{ kg} \cdot \text{m}^2} = 0.02057 \text{ rad/s}^2$$

SET UP: Use the constant α_z kinematic equations to find ω_z.

$\omega_z = ?$; ω_{0z} (initially at rest); $\alpha_z = 0.02057$ rad/s^2; $t = 15.0$ s

EXECUTE: $\omega_z = \omega_{0z} + \alpha_z t = 0 + (0.02057 \text{ rad/s}^2)(15.0 \text{ s}) = 0.309$ rad/s

b) IDENTIFY and **SET UP:** Calculate the work from Eq.(10.24),using a constant angular acceleration equation to calculate $\theta - \theta_0$, or use the work-energy theorem (Eq.10.25). We will do it both ways.

EXECUTE:

(1) $W = \tau_z \Delta\theta$ (Eq.(10.24))

$\Delta\theta = \theta - \theta_0 = \omega_{0z}t + \frac{1}{2}\alpha_z t^2 = 0 + \frac{1}{2}(0.02057 \text{ rad/s}^2)(15.0 \text{ s})^2 = 2.314 \text{ rad}$

$\tau_z = FR = (18.0 \text{ N})(2.40 \text{ m}) = 43.2 \text{ N} \cdot \text{m}$

Then $W = \tau_z \Delta\theta = (43.2 \text{ N} \cdot \text{m})(2.314 \text{ rad}) = 100 \text{ J}$.

or

(2) $W_{\text{tot}} = K_2 - K_1$ (the work-energy relation from chapter 6)

$W_{\text{tot}} = W$, the work done by the child

$K_1 = 0; \quad K_2 = \frac{1}{2}I\omega^2 = \frac{1}{2}(2100 \text{ kg} \cdot \text{m}^2)(0.309 \text{ rad/s})^2 = 100 \text{ J}$

Thus $W = 100$ J, the same as before.

EVALUATE: Either method yields the same result for W.

c) IDENTIFY and **SET UP:** Use Eq.(6.15) to calculate P_{av}

EXECUTE: $P_{\text{av}} = \dfrac{\Delta W}{\Delta t} = \dfrac{100 \text{ J}}{15.0 \text{ s}} = 6.67 \text{ W}$

EVALUATE: Work is in joules, power is in watts.

10.33 a) IDENTIFY and **SET UP:** Use Eq.(10-26) and solve for τ_z.
$P = \tau_z \omega_z$, where ω_z must be in rad/s.

EXECUTE:

$\omega_z = (4000 \text{ rev/min})(2\pi \text{ rad/1 rev})(1 \text{ min/ } 60 \text{ s}) = 418.9 \text{ rad/s}$

$\tau_z = \dfrac{P}{\omega_z} = \dfrac{1.50 \times 10^5 \text{ W}}{418.9 \text{ rad/s}} = 358 \text{ N} \cdot \text{m}$

b) IDENTIFY and **SET UP:** Apply $\sum \vec{F} = m\vec{a}$ to the drum. Find the tension T in the rope using τ_z from part (a).

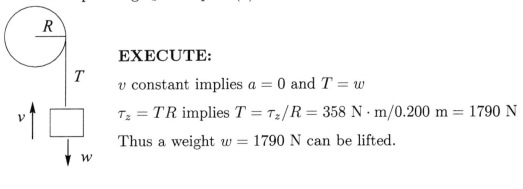

EXECUTE:

v constant implies $a = 0$ and $T = w$

$\tau_z = TR$ implies $T = \tau_z/R = 358 \text{ N} \cdot \text{m}/0.200 \text{ m} = 1790 \text{ N}$

Thus a weight $w = 1790$ N can be lifted.

c) IDENTIFY and **SET UP:** Use $v = R\omega$.

EXECUTE: The drum has $\omega = 418.9$ rad/s, so $v = (0.200$ m$)(418.9$ rad/s$) = 83.8$ m/s

EVALUATE: The rate at which T is doing work on the drum is $P = Tv = (1790$ N$)(83.8$ m/s$) = 150$ kW. This agrees with the work output of the motor.

10.35 a) IDENTIFY: Use $L = mvr \sin \phi$ (Eq.(10.28):

SET UP:

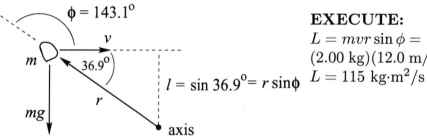

EXECUTE:
$L = mvr \sin \phi =$
$(2.00$ kg$)(12.0$ m/s$)(8.00$ m$) \sin 143.1°$
$L = 115$ kg·m²/s

To find the direction of $\vec{L}$ apply the right-hand rule by turning $\vec{r}$ into the direction of $\vec{v}$ by pushing on it with the fingers of your right hand. Your thumb points into the page, in the direction of $\vec{L}$.

b) IDENTIFY and **SET UP:** By Eq.(10.29) the rate of change of the angular momentum of the rock equals the torque of the net force acting on it.

EXECUTE: $\tau = mg(8.00$ m$) \cos 36.9° = 125$ kg·m²/s²

To find the direction of $\vec{\tau}$ and hence of $d\vec{L}/dt$, apply the right-hand rule by turning $\vec{r}$ into the direction of the gravity force by pushing on it with the fingers of your right hand. Your thumb points out of the page, in the direction of $d\vec{L}/dt$.

EVALUATE: $\vec{L}$ and $d\vec{L}/dt$ are in opposite directions, so L is decreasing. The gravity force is accelerating the rock downward, toward the axis. Its horizontal velocity is constant but the distance l is decreasing and hence L is decreasing.

10.37 IDENTIFY and **SET UP:** Use $L = I\omega$

EXECUTE: The second hand makes 1 revolution in 1 minute, so
$\omega = (1.00$ rev/min$)(2\pi$ rad/1 rev$)(1$ min/60 s$) = 0.1047$ rad/s
For a slender rod, with the axis about one end,
$I = \frac{1}{3}ML^2 = \frac{1}{3}(6.00 \times 10^{-3}$ kg$)(0.150$ m$)^2 = 4.50 \times 10^{-5}$ kg·m²

Then $L = I\omega = (4.50 \times 10^{-5}$ kg · m²$)(0.1047$ rad/s$) = 4.71 \times 10^{-6}$ kg·m²/s.
EVALUATE: $\vec{L}$ is clockwise.

10.39 IDENTIFY and **SET UP:** $\vec{L}$ is conserved if there is no net external torque.

Use conservation of angular momentum to find ω at the new radius and use $K = \frac{1}{2}I\omega^2$ to find the change in kinetic energy, which is equal to the work done on the block.

EXECUTE:

a) Yes, angular momentum is conserved. The moment arm for the tension in the cord is zero so this force exerts no torque and there is no net torque on the block.

b) $L_1 = L_2$ so $I_1\omega_1 = I_2\omega_2$. Block treated as a point mass, so $I = mr^2$, where r is the distance of the block from the hole.

$mr_1^2\omega_1 = mr_2^2\omega_2$

$\omega_2 = \left(\dfrac{r_1}{r_2}\right)^2 \omega_1 = \left(\dfrac{0.300\ \text{m}}{0.150\ \text{m}}\right)^2 (1.75\ \text{rad/s}) = 7.00\ \text{rad/s}$

c) $K_1 = \frac{1}{2}I_1\omega_1^2 = \frac{1}{2}mr_1^2\omega_1^2 = \frac{1}{2}mv_1^2$

$v_1 = r_1\omega_1 = (0.300\ \text{m})(1.75\ \text{rad/s}) = 0.525\ \text{m/s}$

$K_1 = \frac{1}{2}mv_1^2 = \frac{1}{2}(0.0250\ \text{kg})(0.525\ \text{m/s})^2 = 0.00345\ \text{J}$

$K_2 = \frac{1}{2}mv_2^2$

$v_2 = r_2\omega_2 = (0.150\ \text{m})(7.00\ \text{rad/s}) = 1.05\ \text{m/s}$

$K_2 = \frac{1}{2}mv_2^2 = \frac{1}{2}(0.0250\ \text{kg})(1.05\ \text{m/s})^2 = 0.01378\ \text{J}$

$\Delta K = K_2 - K_1 = 0.01378\ \text{J} - 0.00345\ \text{J} = 0.0103\ \text{J}$

d) $W_{\text{tot}} = \Delta K$

But $W_{\text{tot}} = W$, the work done by the tension in the cord, so $W = 0.0103\ \text{J}$

EVALUATE: Smaller r means smaller I. $L = I\omega$ is constant so ω increases and K increases. The work done by the tension is positive since it is directed inward and the block moves inward, toward the hole.

10.43 IDENTIFY and **SET UP:** There is no net external torque about the rotation axis so the angular momentum $L = I\omega$ is conserved.

EXECUTE:

a) $L_1 = L_2$ $I_1\omega_1 = I_2\omega_2$, so $\omega_2 = (I_1/I_2)\omega_1$

$I_1 = I_{\text{tt}} = \frac{1}{2}MR^2 = \frac{1}{2}(120\ \text{kg})(2.00\ \text{m})^2 = 240\ \text{kg} \cdot \text{m}^2$

$I_2 = I_{\text{tt}} + I_{\text{p}} = 240\ \text{kg} \cdot \text{m}^2 + mR^2 = 240\ \text{kg} \cdot \text{m}^2 + (70\ \text{kg})(2.00\ \text{m})^2 = 520\ \text{kg} \cdot \text{m}^2$

$\omega_2 = (I_1/I_2)\omega_1 = (240\ \text{kg} \cdot \text{m}^2/520\ \text{kg} \cdot \text{m}^2)(3.00\ \text{rad/s}) = 1.38\ \text{rad/s}$

b) $K_1 = \frac{1}{2}I_1\omega_1^2 = \frac{1}{2}(240\ \text{kg} \cdot \text{m}^2)(3.00\ \text{rad/s})^2 = 1080\ \text{J}$

$K_2 = \frac{1}{2}I_2\omega_2^2 = \frac{1}{2}(520\ \text{kg} \cdot \text{m}^2)(1.38\ \text{rad/s})^2 = 495\ \text{J}$

EVALUATE: The kinetic energy decreases because of the negative work done on the turntable and the parachutist by the friction force between these two objects.

The angular speed decreases because I increases when the parachutist is added to the system.

10.45 a) IDENTIFY and SET UP: Apply conservation of angular momentum $\vec{L}$, with the axis at the nail. Let object A be the bug and object B be the bar.

Initially, all objects are at rest and $L_1 = 0$.

Just after the bug jumps, it has angular momentum in one direction of rotation and the bar is rotating with angular velocity ω_B in the opposite direction.

EXECUTE:

$L_2 = m_A v_A r - I_B \omega_B$ where $r = 1.00$ m and $I_B = \frac{1}{3} m_B r^2$

$L_1 = L_2$ gives $m_A v_A r = \frac{1}{3} m_B r^2 \omega_B$

$\omega_B = \dfrac{3 m_A v_A}{m_B r} = 0.120$ rad/s

b) $K_1 = 0$; $K_2 = \frac{1}{2} m_A v_A^2 + \frac{1}{2} I_B \omega_B^2 =$
$\frac{1}{2}(0.0100 \text{ kg})(0.200 \text{ m/s})^2 + \frac{1}{2}(\frac{1}{3}[0.0500 \text{ kg}][1.00 \text{ m}]^2)(0.120 \text{ rad/s})^2 = 3.2 \times 10^{-4}$ J.
The increase in kinetic energy comes from work done by the bug when it pushes against the bar in order to jump.

EVALUATE: There is no external torque applied to the system and the total angular momentum of the system is constant. There are internal forces, forces the bug and bar exert on each other. The forces exert torques and change the angular momentum of the bug and the bar, but these changes are equal in magnitude and opposite in direction. These internal forces do positve work on the two objects and the kinetic energy of each object and of the system increases.

10.49 a) IDENTIFY and SET UP: By the work-energy relation $W = \Delta K$, the work done on the gyroscope equals its increase in kinetic energy.

Thus $P_{\text{av}} = W/t = \Delta K/t$ and $t = \Delta K/P$

EXECUTE: Since $K_1 = 0$, $\Delta K = K_2$, the amount of energy stored in the gyroscope when it is up to speed.

Thus $\Delta K = K_2 = \frac{1}{2} I \omega^2$.

For a solid disk $I = \frac{1}{2} M R^2 = \frac{1}{2}(60,000 \text{ kg})(2.00 \text{ m})^2 = 1.20 \times 10^5$ kg $\cdot$ m^2

$\omega = 500$ rev/min$(2\pi \text{ rad}/1 \text{ rev})(1 \text{ min}/60 \text{ s}) = 52.36$ rad/s

Thus $\Delta K = \frac{1}{2} I \omega^2 = \frac{1}{2}(1.20 \times 10^5 \text{ kg} \cdot \text{m}^2)(52.36 \text{ rad/s})^2 = 1.645 \times 10^8$ J

and $t = \dfrac{\Delta K}{P} = \dfrac{1.645 \times 10^8 \text{ J}}{7.46 \times 10^4 \text{ W}} = 2205$ s $= 36.8$ min.

b) IDENTIFY and SET UP: Use Eq.(10.36).

EXECUTE: $\Omega = \tau_z/L = \tau_z/I\omega_z$, so $\tau_z = I\omega_z\Omega$.

$\Omega = 1.00°/s(\pi \text{ rad}/180°) = 0.01745 \text{ rad/s}$

$\tau_z = I\omega_z\Omega = (1.20 \times 10^5 \text{ kg} \cdot \text{m}^2)(52.36 \text{ rad/s})(0.01745 \text{ rad/s}) = 1.10 \times 10^5 \text{ N} \cdot \text{m}$

EVALUATE: The larger P is the less time it takes to add the required energy to the system. The larger τ_z is the greater is the precession rate Ω. $\tau_z = wr = mgr$; this gives $r = 19$ cm. This is the distance from the center of mass of the gyroscope to the pivot; this value seems reasonable.

Problems

10.55 IDENTIFY and **SET UP:** Use Eq.(10.26) for P. Use a constant acceleration equation to replace ω_z and Eq.(10.6) to replace α_z.

EXECUTE:

a) $P = \tau_z\omega_z$ Since α_z is constant, $\omega_z = \omega_{0z} + \alpha_z t$. $\omega_{0z} = 0$.

And $\tau_z = I\alpha_z$ says $\alpha_z = \tau_z/I$ and then $\omega_z = \tau_z t/I$.

Thus $P = \tau_z(\tau_z t/I) = \tau_z^2 t/I$

b) $\dfrac{P}{\tau_z^2} = \dfrac{t}{I} = \text{constant}$, so $\dfrac{P_1}{\tau_{z1}^2} = \dfrac{P_2}{\tau_{z2}^2}$

$P_2 = P_1\left(\dfrac{\tau_{z2}}{\tau_{z1}}\right)^2 = 500 \text{ W}\left(\dfrac{60.0 \text{ N} \cdot \text{m}}{20.0 \text{ N} \cdot \text{m}}\right)^2 = 4500 \text{ W}$

EVALUATE: At a given t the instantaneous power is proportional to the square of the torque.

IDENTIFY and **SET UP:** Repeat the above but now write ω_z in terms of $\theta - \theta_0$.

c) EXECUTE: $P = \tau_z\omega_z$

$\omega_z^2 = \omega_{0z}^2 + 2\alpha_z(\theta - \theta_0) = 2(\tau_z/I)(\theta - \theta_0)$

$\omega_z = \sqrt{(2\tau_z/I)(\theta - \theta_0)}$

Thus $P = \tau_z\omega_z = \tau_z\sqrt{(2\tau_z/I)(\theta - \theta_0)} = (\sqrt{2(\theta - \theta_0)/I})\tau_z^{3/2}$.

d) $\dfrac{P}{\tau_z^{3/2}} = \sqrt{2(\theta - \theta_0)/I} = \text{constant}$, so $\dfrac{P_1}{\tau_{z1}^{3/2}} = \dfrac{P_2}{\tau_{z2}^{3/2}}$

$P_2 = P_1\left(\dfrac{\tau_{z2}}{\tau_{z1}}\right)^{3/2} = 500 \text{ W}\left(\dfrac{60.0 \text{ N} \cdot \text{m}}{20.0 \text{ N} \cdot \text{m}}\right)^{3/2} = 2600 \text{ W}$

e) EVALUATE: There is no contradiction. When the net torque is varied, keeping t constant is different from keeping $\theta - \theta_0$ constant.

10.57 IDENTIFY: Use $\sum\tau_z = I\alpha_z$ to find α_z, and then use the constant α_z kinematic equations to solve for t.

SET UP:

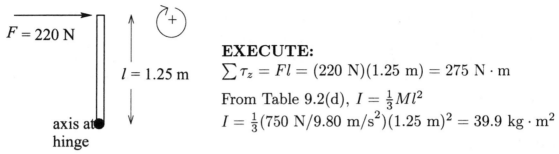

EXECUTE:

$$\sum \tau_z = Fl = (220\ \text{N})(1.25\ \text{m}) = 275\ \text{N} \cdot \text{m}$$

From Table 9.2(d), $I = \frac{1}{3} Ml^2$

$$I = \frac{1}{3}(750\ \text{N}/9.80\ \text{m/s}^2)(1.25\ \text{m})^2 = 39.9\ \text{kg} \cdot \text{m}^2$$

$$\sum \tau_z = I\alpha_z \ \text{so} \ \alpha_z = \frac{\sum \tau_z}{I} = \frac{275\ \text{N} \cdot \text{m}}{39.9\ \text{kg} \cdot \text{m}^2} = 6.89\ \text{rad/s}^2$$

SET UP: $\alpha_z = 6.89\ \text{rad/s}^2$; $\theta - \theta_0 = 90°(\pi\ \text{rad}/180°) = \pi/2\ \text{rad}$; $\omega_{0z} = 0$ (door initially at rest); $t = ?$

$$\theta - \theta_0 = \omega_{0z}t + \frac{1}{2}\alpha_z t^2$$

EXECUTE: $t = \sqrt{\dfrac{2(\theta - \theta_0)}{\alpha_z}} = \sqrt{\dfrac{2(\pi/2\ \text{rad})}{6.89\ \text{rad/s}^2}} = 0.675\ \text{s}$

EVALUATE: The forces and the motion are connected through the angular acceleration.

10.61 IDENTIFY: Apply $\sum \vec{F} = m\vec{a}$ to the crate and $\sum \tau_z = I\alpha_z$ to the cylinder. The motions are connected by $a(\text{crate}) = R\alpha(\text{cylinder})$.

SET UP: Force diagram for the crate

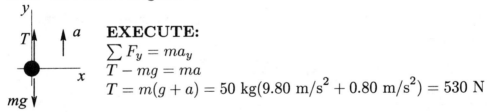

EXECUTE:

$$\sum F_y = ma_y$$
$$T - mg = ma$$
$$T = m(g + a) = 50\ \text{kg}(9.80\ \text{m/s}^2 + 0.80\ \text{m/s}^2) = 530\ \text{N}$$

SET UP: Force diagram for the cylinder:

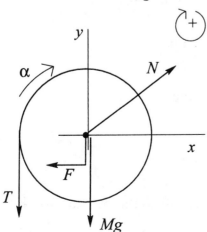

EXECUTE:

$$\sum \tau_z = I\alpha_z$$
$$Fl - TR = I\alpha_z, \ \text{where} \ l = 0.12\ \text{m and} \ R = 0.25\ \text{m}$$
$$a = R\alpha \ \text{so} \ \alpha_z = a/R$$
$$Fl = TR + Ia/R$$

$$F = T\left(\frac{R}{l}\right) + \frac{Ia}{Rl} = 530 \text{ N}\left(\frac{0.25 \text{ m}}{0.12 \text{ m}}\right) + \frac{(2.9 \text{ kg} \cdot \text{m}^2)(0.80 \text{ m/s}^2)}{(0.25 \text{ m})(0.12 \text{ m})} = 1200 \text{ N}$$

EVALUATE: The tension in the rope is greater than the weight of the crate since the crate accelerates upward. If F were applied to the rim of the cylinder ($l = 0.25$ m), it would have the value $F = 576$ N. This is greater than T because it must accelerate the cylinder as well as the crate. And F is larger than this because it is applied closer to the axis than R so has a smaller moment arm and must be larger to give the same torque.

10.63 IDENTIFY: Apply $\sum \tau_z = I\alpha_z$ to the flywheel and $\sum \vec{F} = m\vec{a}$ to the block. The target variables are the tension in the string and the acceleration of the block.

a) SET UP: Apply $\sum \tau_z = I\alpha_z$ to the rotation of the flywheel about the axis.

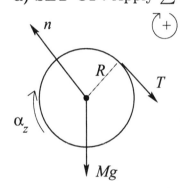

EXECUTE:

The forces n and Mg act at the axis so have zero torque.

$$\sum \tau_z = TR$$
$$TR = I\alpha_z$$

SET UP: Apply $\sum \vec{F} = m\vec{a}$ to the translational motion of the block:

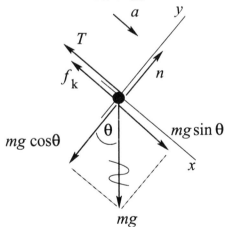

EXECUTE:

$$\sum F_y = ma_y$$
$$n - mg\cos 36.9° = 0$$
$$n = mg\cos 36.9°$$

$$f_k = \mu_k n = \mu_k mg\cos 36.9°$$

$$\sum F_x = ma_x$$
$$mg\sin 36.9° - T - \mu_k mg\cos 36.9° = ma$$
$$mg(\sin 36.9° - \mu_k \cos 36.9°) - T = ma$$

But we also know that $a_{\text{block}} = R\alpha_{\text{wheel}}$, so $\alpha = a/R$.

Using this in the $\sum \tau_z = I\alpha_z$ equation gives $TR = Ia/R$ and $T = (I/R^2)a$.

Use this to replace T in the $\sum F_y = ma_y$ equation:

$$mg(\sin 36.9° - \mu_k \cos 36.9°) - (I/R^2)a = ma$$

$$a = \frac{mg(\sin 36.9° - \mu_k \cos 36.9°)}{m + I/R^2}$$

$$a = \frac{(5.00 \text{ kg})(9.80 \text{ m/s}^2)(\sin 36.9° - (0.25)\cos 36.9°)}{5.00 \text{ kg} + 0.500 \text{ kg} \cdot \text{m}^2/(0.200 \text{ m})^2} = 1.12 \text{ m/s}^2$$

b) $T = \dfrac{0.500 \text{ kg} \cdot \text{m}^2}{(0.200 \text{ m})^2}(1.12 \text{ m/s}^2) = 14.0 \text{ N}$

EVALUATE: If the string is cut the block will slide down the incline with $a = g\sin 36.9° - \mu_k g \cos 36.9° = 3.92 \text{ m/s}^2$. The actual acceleration is less than this because $mg\sin 36.9°$ must also accelerate the flywheel. $mg\sin 36.9° - f_k = 19.6 \text{ N}$. T is less than this; there must be more force on the block directed down the incline than up then incline since the block accelerates down the incline.

10.65 IDENTIFY: Apply both $\sum \vec{F} = m\vec{a}$ and $\sum \tau_z = I\alpha_z$ to the motion of the roller. Rolling without slipping means $a_{cm} = R\alpha$. Target variables are a_{cm} and f.
SET UP:

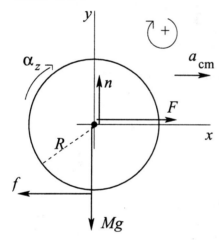

EXECUTE:

Apply $\sum \vec{F} = m\vec{a}$ to the translational motion of the center of mass:
$\sum F_x = ma_x$
$F - f = Ma_{cm}$

Apply $\sum \tau_z = I\alpha_z$ to the rotation about the center of mass:
$\sum \tau_z = fR$
thin-walled hollow cylinder: $I = MR^2$
Then $\sum \tau_z = I\alpha_z$ implies $fR = MR^2\alpha$.
But $a_{cm} = R\alpha$, so $f = Ma_{cm}$.
Using this in the $\sum F_x = ma_x$ equation gives $F - Ma_{cm} = Ma_{cm}$
$a_{cm} = F/2M$, and then $f = Ma_{cm} = M(F/2M) = F/2$.
EVALUATE: If the surface were frictionless the object would slide without rolling and the acceleration would be $a_{cm} = F/M$. The acceleration is less when the object rolls.

10.67 IDENTIFY: This problem can be done either with conservation of energy or with $\sum \vec{F}_{\text{ext}} = m\vec{a}$. We will do it both ways.

a) SET UP: (1) <u>Conservation of energy:</u> $K_1 + U_1 + W_{\text{other}} = K_2 + U_2$

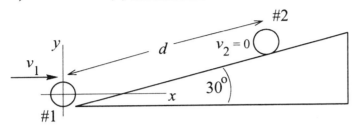

Take position 1 to be the location of the disk at the base of the ramp and 2 to be where the disk momentarily stops before rolling back down.

Take the origin of coordinates at the center of the disk at position 1 and take $+y$ to be upward. Then $y_1 = 0$ and $y_2 = d\sin 30°$, where d is the distance that the disk rolls up the ramp.

"rolls without slipping" and neglect rolling friction says $W_f = 0$; only gravity does work on the disk, so $W_{\text{other}} = 0$

EXECUTE: $U_1 = Mgy_1 = 0$

$K_1 = \frac{1}{2}Mv_1^2 + \frac{1}{2}I_{\text{cm}}\omega_1^2$ (Eq.10-11). But $\omega_1 = v_1/R$ and $I_{\text{cm}} = \frac{1}{2}MR^2$, so

$\frac{1}{2}I_{\text{cm}}\omega_1^2 = \frac{1}{2}(\frac{1}{2}MR^2)(v_1/R)^2 = \frac{1}{4}Mv_1^2$.

Thus $K_1 = \frac{1}{2}Mv_1^2 + \frac{1}{4}Mv_1^2 = \frac{3}{4}Mv_1^2$.

$U_2 = Mgy_2 = Mgd\sin 30°$

$K_2 = 0$ (disk is at rest at point 2).

Thus $\frac{3}{4}Mv_1^2 = Mgd\sin 30°$

$$d = \frac{3v_1^2}{4g\sin 30°} = \frac{3(2.50 \text{ m/s})^2}{4(9.80 \text{ m/s}^2)\sin 30°} = 0.957 \text{ m}$$

SET UP: (2) <u>force and acceleration</u>

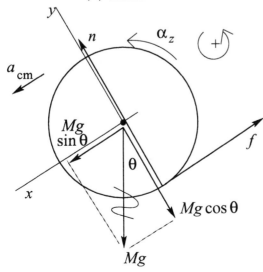

EXECUTE:

Apply $\sum F_x = ma_x$ to the translational motion of the center of mass:

$Mg\sin\theta - f = Ma_{\text{cm}}$

Apply $\sum \tau_z = I\alpha_z$ to the rotation about the center of mass:

$fR = (\frac{1}{2}MR^2)\alpha_z$

$f = \frac{1}{2}MR\alpha_z$

But $a_{cm} = R\alpha$ in this equation gives $f = \frac{1}{2}Ma_{cm}$.

Use this in the $\sum F_x = ma_x$ equation to eliminate f.

$Mg\sin\theta - \frac{1}{2}Ma_{cm} = Ma_{cm}$

M divides out and $\frac{3}{2}a_{cm} = g\sin\theta$. $a_{cm} = \frac{2}{3}g\sin\theta = \frac{2}{3}(9.80 \text{ m/s}^2)\sin 30° = 3.267 \text{ m/s}^2$

SET UP: Apply the constant acceleration equations to the motion of the center of mass. Note that in our coordinates the positive x-direction is down the incline.

$v_{0x} = -2.50$ m/s (directed up the incline); $\quad a_x = +3.267$ m/s^2;

$v_x = 0$ (momentarily comes to rest); $\quad x - x_0 = ?$

$v_x^2 = v_{0x}^2 + 2a_x(x - x_0)$

EXECUTE: $x - x_0 = -\dfrac{v_{0x}^2}{2a_x} = -\dfrac{(-2.50 \text{ m/s})^2}{2(3.267 \text{ m/s}^2)} = -0.957$ m

b) EVALUATE: The results from the two methods agree; the disk rolls 0.957 m up the ramp before it stops.

The mass M enters both in the linear inertia and in the gravity force so divides out. The mass M and radius R enter in both the rotational inertia and the gravitational torque so divide out.

10.69 IDENTIFY: Use conservation of energy to relate h to the speed of the marble at the top of the loop. The marble has both translational and rotational kinetic energy. Then apply $\sum \vec{F} = m\vec{a}$ to the marble when it is at the top of the loop to find the minimum speed there required if it is to stay in contact with the track. Combine these two results to find the minimum h.

a) SET UP:

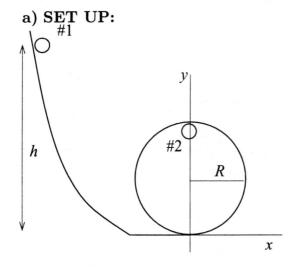

Take the origin at the lowest point in the track.

$y_{cm,1} = h$
$y_{cm,2} = 2R - r$
$I_{cm} = \frac{2}{5}mr^2$

EXECUTE:

$K_1 + U_1 + W_{other} = K_2 + U_2$

$W_{\text{other}} = 0; \quad K_1 = 0$

$K_2 = \frac{1}{2}mv^2 + \frac{1}{2}I_{\text{cm}}\omega^2 = \frac{1}{2}mv^2 + \frac{1}{2}(\frac{2}{5}mr^2)\omega^2$

Rolls without slipping implies $v = r\omega$ or $\omega = v/r$.

Thus $K_2 = \frac{1}{2}mv^2 + \frac{1}{2}(\frac{2}{5}mr^2)(v^2/r^2) = mv^2(\frac{1}{2} + \frac{1}{5}) = \frac{7}{10}mv^2$.

$U_1 = mgy_{\text{cm},1} = mgh$

$U_2 = mgy_{\text{cm},2} = mg(2R - r)$

Thus $K_1 + U_1 + W_{\text{other}} = K_2 + U_2$ gives

$mgh = \frac{7}{10}mv^2 + mg(2R - r)$

$\frac{7}{10}v^2 = g(h - 2R + r)$

SET UP: Force diagram for the marble at point 2:

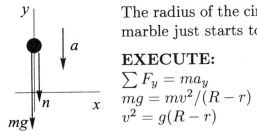

The radius of the circle in which the cm travels is $R - r$. When the marble just starts to leave the track (minimum h) $n \to 0$

EXECUTE:

$\sum F_y = ma_y$

$mg = mv^2/(R - r)$

$v^2 = g(R - r)$

Using this in the conservation of energy equation gives $\frac{7}{10}g(R - r) = g(h - 2R + r)$

and $h = \frac{7}{10}R + 2R - \frac{7}{10}r - r = (27R - 17r)/10$.

b) No friction implies that marble slides without rolling, so $K_1 = \frac{1}{2}mv^2$, no $\frac{1}{2}I_{\text{cm}}\omega^2$ term.

In the conservation of energy equation this gives $\frac{1}{2}v^2 = g(h - 2R + r)$.

Combining this with $v^2 = g(R - r)$ gives $\frac{1}{2}g(R - r) = g(h - 2R + r)$.

$h = \frac{5}{2}R - \frac{3}{2}r = (5R - 3r)/2$.

EVALUATE: Note that when $r = 0$ this result is the same as for Problem 7.46 part (a). The answer in part (a) is larger than this by $(R-r)/5$. A larger h is needed when the marble rolls because some of the decrease in gravitational potential energy goes into rotational kinetic energy and there is less that goes into translational kinetic energy, so v is less for the same h.

10.75 IDENTIFY: Apply conservation of energy to the motion of the ball as it rolls up the hill. After the ball leaves the edge of the cliff it moves in projectile motion and constant acceleration equations can be used.

a) SET UP: Use conservation of energy to find the speed v_2 of the ball just before it leaves the top of the cliff. Let point 1 be at the bottom of the hill and point 2 be at the top of the hill. Take $y = 0$ at the bottom of the hill, so $y_1 = 0$ and $y_2 = 28.0$ m.

EXECUTE: $K_1 + U_1 = K_2 + U_2$

$\frac{1}{2}mv_1^2 + \frac{1}{2}I\omega_1^2 = mgy_2 + \frac{1}{2}mv_2^2 + \frac{1}{2}I\omega_2^2$

Rolling without slipping means $\omega = v/r$ and $\frac{1}{2}I\omega^2 = \frac{1}{2}(\frac{2}{5}mr^2)(v/r)^2 = \frac{1}{5}mv^2$

$\frac{7}{10}mv_1^2 = mgy_2 + \frac{7}{10}mv_2^2$

$v_2 = \sqrt{v_1^2 - \frac{10}{7}gy_2} = 15.26$ m/s

SET UP: Consider the projectile motion of the ball, from just after it leaves the top of the cliff until just before it lands. Take $+y$ to be downward.

Use the vertical motion to find the time in the air:

$v_{0y} = 0,\quad a_y = 9.80$ m/s^2,$\quad y - y_0 = 28.0$ m,$\quad t = ?$

EXECUTE: $y - y_0 = v_{0y}t + \frac{1}{2}a_y t^2$ gives $t = 2.39$ s

During this time the ball travels horizontally

$x - x_0 = v_{0x}t = (15.26$ m/s$)(2.39$ s$) = 36.5$ m.

Just before it lands, $v_y = v_{0y} + a_y t = 23.4$ m/s and $v_x = v_{0x} = 15.3$ m/s

$v = \sqrt{v_x^2 + v_y^2} = 28.0$ m/s

b) EVALUATE: At the bottom of the hill, $\omega = v/r = (25.0$ m/s$)/r$. The rotation rate doesn't change while the ball is in the air, after it leaves the top of the cliff, so just before it lands $\omega = (15.3$ m/s$)/r$. The total kinetic energy is the same at the bottom of the hill and just before it lands, but just before it lands less of this energy is rotational kinetic energy, so the translational kinetic energy is greater.

10.79 IDENTIFY and **SET UP:** Apply conservation of energy to the motion of the ball. The ball ends up with both translational and rotational kinetic energy. Use Fig.(10.14) to relate the speed of different points on the ball to v_{cm}.

EXECUTE:

a) $U_{el} = \frac{1}{2}kx^2 = \frac{1}{2}(400$ N $\cdot$ m$)(0.15$ m$)^2 = 4.50$ J and $K_1 = 0.800U_{el} = 3.60$ J

$K_1 = \frac{1}{2}mv_{cm}^2 + \frac{1}{2}I_{cm}\omega^2$ rolling without slipping says $\omega = v_{cm}/R$

$I_{cm} = \frac{2}{5}mR^2$

Thus $K_1 = \frac{1}{2}mv_{cm}^2 + (\frac{2}{5}mR^2)(v_{cm}/R)^2 = mv_{cm}^2(\frac{1}{2} + \frac{1}{5}) = \frac{7}{10}mv_{cm}^2$

and $v_{cm} = \sqrt{\dfrac{10K_1}{7m}} = \sqrt{\dfrac{10(3.60 \text{ J})}{7(0.0590 \text{ kg})}} = 9.34$ m/s.

b)

From Fig.(10.14), at the top of the ball
$v = 2v_{cm} = 18.7$ m/s

c)

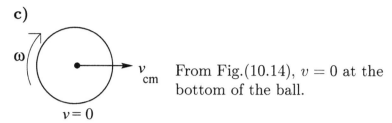

From Fig.(10.14), $v = 0$ at the bottom of the ball.

d) The problem says that $U_2 = 0.900K_1 = 3.24$ J. Thus $U_2 = mgh = 3.24$ J and

$$h = \frac{3.24 \text{ J}}{mg} = \frac{3.24 \text{ J}}{(0.0590 \text{ kg})(9.80 \text{ m/s}^2)} = 5.60 \text{ m}$$

EVALUATE: Not all the potential energy stored in the spring goes into kinetic energy at the base of the ramp or into gravitational potential energy at the top of the ramp because of loss of mechanical energy due to negative work done by friction. If the ball slides without rolling, then $K_1 = \frac{1}{2}mv_{cm}^2$ and $v_{cm} = 11.0$ m/s. v_{cm} is less than this when the ball rolls and some of its total kinetic energy is rotational.

10.85 IDENTIFY: Apply conservation of energy to the motion of the first ball before the collision and to the motion of the second ball after the collision. Apply conservation of angular momentum to the collision between the first ball and the bar.

SET UP: The speed of the ball just before it hits the bar is $v = \sqrt{2gy} = 15.34$ m/s. Use conservation of angular momentum to find the angular velocity ω of the bar just after the collision. Take the axis at the center of the bar.

EXECUTE:

$L_1 = mvr = (5.00 \text{ kg})(15.34 \text{ m/s})(2.00 \text{ m}) = 153.4 \text{ kg} \cdot \text{m}^2$

Immediately after the collsion the bar and both balls are rotating together.

$L_2 = I_{tot}\omega$

$I_{tot} = \frac{1}{12}Ml^2 + 2mr^2 = \frac{1}{12}(8.00 \text{ kg})(4.00 \text{ m})^2 + 2(5.00 \text{ kg})(2.00 \text{ m})^2 = 50.67 \text{ kg} \cdot \text{m}^2$

$L_2 = L_1 = 153.4 \text{ kg} \cdot \text{m}^2$

$\omega = L_2/I_{tot} = 3.027$ rad/s

Just after the collision the second ball has linear speed $v = r\omega = (2.00 \text{ m})(3.027 \text{ rad/s}) = 6.055$ m/s and is moving upward.

$\frac{1}{2}mv^2 = mgy$ gives $y = 1.87$ m for the height the second ball goes.

EVALUATE: Mechanical energy is lost in the inelastic collsion and some of the final energy is in the rotation of the bar with the first ball stuck to it. As a result, the second ball does not reach the height from which the first ball was dropped.

10.87 IDENTIFY: Apply conservation of angular momentum to the collision. Linear momentum is not conserved because of the force applied to the rod at the axis. But since this external force acts at the axis, it produces no torque and angular momentum is conserved.

SET UP:

a) $m_b = \frac{1}{4}m_{rod}$

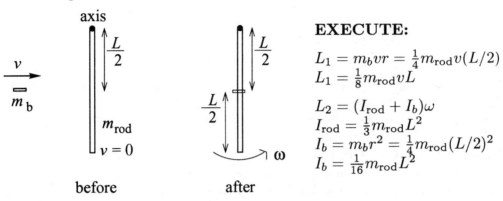

before   after

EXECUTE:

$L_1 = m_b v r = \frac{1}{4}m_{rod}v(L/2)$
$L_1 = \frac{1}{8}m_{rod}vL$

$L_2 = (I_{rod} + I_b)\omega$
$I_{rod} = \frac{1}{3}m_{rod}L^2$
$I_b = m_b r^2 = \frac{1}{4}m_{rod}(L/2)^2$
$I_b = \frac{1}{16}m_{rod}L^2$

Thus $L_1 = L_2$ gives $\frac{1}{8}m_{rod}vL = (\frac{1}{3}m_{rod}L^2 + \frac{1}{16}m_{rod}L^2)\omega$

$\frac{1}{8}v = \frac{19}{48}L\omega$

$\omega = \frac{6}{19}v/L$

b) $K_1 = \frac{1}{2}mv^2 = \frac{1}{8}m_{rod}v^2$

$K_2 = \frac{1}{2}I\omega^2 = \frac{1}{2}(I_{rod} + I_b)\omega^2 = \frac{1}{2}(\frac{1}{3}m_{rod}L^2 + \frac{1}{16}m_{rod}L^2)(6v/19L)^2$

$K_2 = \frac{1}{2}(\frac{19}{48})(\frac{6}{19})^2 m_{rod}v^2 = \frac{3}{152}m_{rod}v^2$

Then $\dfrac{K_2}{K_1} = \dfrac{\frac{3}{152}m_{rod}v^2}{\frac{1}{8}m_{rod}v^2} = 3/19$.

EVALUATE: The collision is inelastic and $K_2 < K_1$.

10.89 a) IDENTIFY: Apply conservation of angular momentum to the collision between the bullet and the board:

SET UP:

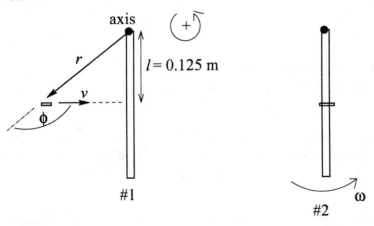

#1   #2

EXECUTE: $L_1 = L_2$

$L_1 = mvr\sin\phi = mvl = (1.90 \times 10^{-3}\text{ kg})(360\text{ m/s})(0.125\text{ m}) = 0.0855\text{ kg·m}^2/\text{s}$

$L_2 = I_2\omega_2$

$I_2 = I_{\text{board}} + I_{\text{bullet}} = \frac{1}{3}ML^2 + mr^2$

$I_2 = \frac{1}{3}(0.750 \text{ kg})(0.250 \text{ m})^2 + (1.90 \times 10^{-3} \text{ kg})(0.125 \text{ m})^2 = 0.01565 \text{ kg} \cdot \text{m}^2$

Then $L_1 = L_2$ gives that $\omega_2 = \dfrac{L_1}{I_2} = \dfrac{0.0855 \text{ kg} \cdot \text{m}^2/\text{s}}{0.01565 \text{ kg} \cdot \text{m}^2} = 5.46 \text{ rad/s}$

b) IDENTIFY: Apply conservation of energy to the motion of the board after the collision.

SET UP: Take the origin of coordinates at the center of the board and $+y$ to be upward, so $y_{\text{cm},1} = 0$ and $y_{\text{cm},2} - h$, the height being asked for.

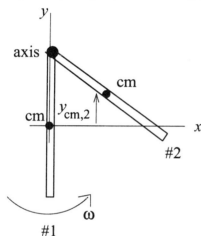

$K_1 + U_1 + W_{\text{other}} = K_2 + U_2$

EXECUTE:
Only gravity does work, so $W_{\text{other}} = 0$.
$K_1 = \frac{1}{2}I\omega^2$
$U_1 = mgy_{\text{cm},1} = 0$
$K_2 = 0$
$U_2 = mgy_{\text{cm},2} = mgh$

Thus $\frac{1}{2}I\omega^2 = mgh$.

$h = \dfrac{I\omega^2}{2mg} = \dfrac{(0.01565 \text{ kg} \cdot \text{m}^2)(5.46 \text{ rad/s})^2}{2(0.750 \text{ kg} + 1.90 \times 10^{-3} \text{ kg})(9.80 \text{ m/s}^2)} = 0.0317 \text{ m} = 3.17 \text{ cm}$

c) IDENTIFY and **SET UP:**

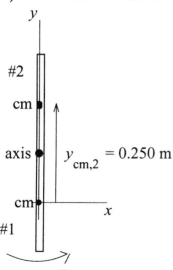

Apply conservation of energy as in part (b), except now we want $y_{\text{cm},2} = h = 0.250 \text{ m}$. Solve for the ω after the collision that is required for this to happen.

EXECUTE: $\frac{1}{2}I\omega^2 = mgh$

$$\omega = \sqrt{\frac{2mgh}{I}} = \sqrt{\frac{2(0.750 \text{ kg} + 1.90 \times 10^{-3} \text{ kg})(9.80 \text{ m/s}^2)(0.250 \text{ m})}{0.01565 \text{ kg} \cdot \text{m}^2}}$$

$\omega = 15.34 \text{ rad/s}$

Now go back to the equation that results from applying conservation of angular momentum to the collision and solve for the initial speed of the bullet.

$L_1 = L_2$ implies $m_{\text{bullet}} vl = I_2 \omega_2$

$$v = \frac{I_2 \omega_2}{m_{\text{bullet}} l} = \frac{(0.01565 \text{ kg} \cdot \text{m}^2)(15.34 \text{ rad/s})}{(1.90 \times 10^{-3} \text{ kg})(0.125 \text{ m})} = 1010 \text{ m/s}$$

EVALUATE: We have divided the motion into two separate events: the collision and the motion after the collision. Angular momentum is conserved in the collision because the collision happens quickly. The board doesn't move much until after the collision is over, so there is no gravity torque about the axis. The collision is inelastic and mechanical energy is lost in the collision. Angular momentum of the system is not conserved during this motion, due to the external gravity torque. Our answer to parts (b) and (c) say that a bullet speed of 360 m/s causes the board to swing up only a little and a speed of 1010 m/s causes it to swing all the way over.

10.91 IDENTIFY: There is no external torque on the system consisting of the two disks, so we can apply conservation of angular momentum, $L_1 = L_2$. Solve for the final angular velocity and then use conservation of energy, with $K = \frac{1}{2} I \omega^2$.

SET UP: $I_A = \frac{1}{3} I_B$, so $I_B = 3 I_A$

L_1 is the initial angular momentum of disk A, $L_1 = I_A \omega_0$.

L_2 is the final angular momentum of the two disks after they are connected and reach a common angular velocity ω, $L_2 = (I_A + I_B)\omega = 4 I_A \omega$.

EXECUTE: Then $L_1 = L_2$ implies $I_A \omega_0 = 4 I_A \omega$.

$\omega = \omega_0 / 4$

Energy considerations: $W_{\text{tot}} = K_2 - K_1$

$W_{\text{tot}} = W_f = -2400 \text{ J}$

$K_1 = \frac{1}{2} I_A \omega_0^2$

$K_2 = \frac{1}{2}(I_A + I_B)\omega^2 = \frac{1}{2}(4 I_A)(\omega_0/4)^2 = \frac{1}{4}(\frac{1}{2} I_A \omega_0^2) = \frac{1}{4} K_1$

Thus $-2400 \text{ J} = \frac{1}{4} K_1 - K_1$

$\frac{3}{4} K_1 = 2400 \text{ J}$ and $K_1 = \frac{4}{3}(2400 \text{ J}) = 3200 \text{ J}$

EVALUATE: W_f is negative and the kinetic energy decreases. Also, ω decreases when I increases.

10.93 IDENTIFY and **SET UP:** Apply conservation of angular momentum to the system consisting of the disk and train.

SET UP: $L_1 = L_2$, counterclockwise positive

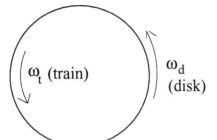

ω_d
(disk)

$L_1 = 0$ (before you switch on the train's engine; both the train and the platform are at rest)

$L_2 = L_{\text{train}} + L_{\text{disk}}$

EXECUTE: The train is $\frac{1}{2}(0.95\text{ m}) = 0.475$ m from the axis of rotation, so for it

$I_t = m_t R_t^2 = (1.20\text{ kg})(0.475\text{ m})^2 = 0.2708\text{ kg}\cdot\text{m}^2$

$\omega_{\text{rel}} = v_{\text{rel}}/R_t = (0.600\text{ m/s})/0.475\text{ s} = 1.263\text{ rad/s}$

This is the angular velocity of the train relative to the disk. Relative to the earth $\omega_t = \omega_{\text{rel}} + \omega_d$.

Thus $L_{\text{train}} = I_t\omega_t = I_t(\omega_{\text{rel}} + \omega_d)$.

$L_2 = L_1$ says $L_{\text{disk}} = -L_{\text{train}}$

$L_{\text{disk}} = I_d\omega_d$, where $I_d = \frac{1}{2}m_d R_d^2$

$\frac{1}{2}m_d R_d^2 \omega_d = -I_t(\omega_{\text{rel}} + \omega_d)$

$$\omega_d = -\frac{I_t\omega_{\text{rel}}}{\frac{1}{2}m_d R_d^2 + I_t} = -\frac{(0.2708\text{ kg}\cdot\text{m}^2)(1.263\text{ rad/s})}{\frac{1}{2}(7.00\text{ kg})(0.500\text{ m})^2 + 0.2708\text{ kg}\cdot\text{m}^2} = -0.30\text{ rad/s}.$$

EVALUATE: The minus sign tells us that the disk is rotating clockwise relative to the earth. The disk and train rotate in opposite directions, since the total angular momentum of the system must remain zero. Note that we applied $L_1 = L_2$ in an inertial frame attached to the earth.

10.97 IDENTIFY and **SET UP:** Eq.(10.36): $\Omega = \tau_z/I\omega_z$. The problem is asking for Ω.

EXECUTE:

The capsizing torque is $\tau_z = wr = (50.0\text{ kg})(9.80\text{ m/s}^2)(0.040\text{ m}) = 19.6\text{ N}\cdot\text{m}$.

$I = 0.085\text{ kg}\cdot\text{m}^2$

$v_{\text{cm}} = 6.00$ m/s, so the angular velocity of the wheel about its axle is

$\omega = v_{\text{cm}}/r = (6.00\text{ m/s})/0.33\text{ m} = 18.18\text{ rad/s}$

Thus $\Omega = \dfrac{\tau}{I\omega} = \dfrac{19.6\text{ N}\cdot\text{m}}{(0.085\text{ kg}\cdot\text{m}^2)(18.18\text{ rad/s})} = 12.7\text{ rad/s}.$

EVALUATE: The bicycle is more stable and the required Ω is smaller the greater ω is. ω is the angular velocity of the front wheel about the horizontal axle. Ω is the rate at which the direction of this axle is turning about a vertical axis.

CHAPTER 11
EQUILIBRIUM AND ELASTICITY

Exercises 1, 3, 7, 11, 13, 15, 17, 19, 21, 23, 27, 33, 35, 37
Problems 41, 45, 49, 53, 59, 67, 71, 73, 75, 77, 79, 81, 83, 85, 87, 89, 91

Exercises

11.1 **IDENTIFY:** Use Eq.(11.3) for x_{cm} and the fact that the center of gravity and the center of mass are the same point.

SET UP:

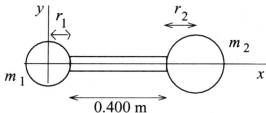

Use coordinates with the origin at the center of the 1.00 kg ball and the x-axis along the rod.

The x-coordinate of the center of gravity is given by $x_{cm} = \dfrac{m_1 x_1 + m_2 x_2}{m_1 + m_2}$.

EXECUTE: The center of gravity of ball 1 is at $x_1 = 0$ and the center of gravity of ball 2 is at $x_2 = 0.400 + r_1 + r_2 = 0.400\ \text{m} + 0.080\ \text{m} + 0.100\ \text{m} = 0.580\ \text{m}$.

Then $x_{cm} = \dfrac{0 + (2.00\ \text{kg})(0.580\ \text{m})}{1.00\ \text{kg} + 2.00\ \text{kg}} = 0.387\ \text{m}$.

The center of gravity is 0.387 m to the right of the center of the 1.00 kg ball.

EVALUATE: All the mass of each ball is taken to be at its cg. The cg is closer to the more massive ball.

11.3 **IDENTIFY:** As the child moves to the right the cg of the system moves to the right. When the plank is on the verge of tipping the cg of the system is at point S in Fig.11.4.

SET UP: The center of gravity of the plank, at the geometrical center of the plank, 0.75 m to the left of the support S. Use coordinates where the origin is at point S and $+x$ is to the right. Let the child be a distance d to the right of point S. In these coordinates, $x_{cm} = 0$. Solve for d.

EXECUTE: Eq.(11.3) gives $0 = \dfrac{(-0.75\ \text{m})(90\ \text{kg}) + d(60\ \text{kg})}{90\ \text{kg} + 60\ \text{kg}}$

$d = (0.75\ \text{m})(90\ \text{kg})/(60\ \text{kg}) = 1.125\ \text{m}$.

EVALUATE: The child is halfway from S to the end of the plank. In example 11.1 a 30-kg child could stand at the end of the plank without it tipping. This more massive child must stand closer to the support.

11.7 IDENTIFY: The board plus motor system is in vertical and rotational equilibrium, so $\sum F_y = 0$ and $\sum \tau_z = 0$.

SET UP: Free-body diagram for the board and motor

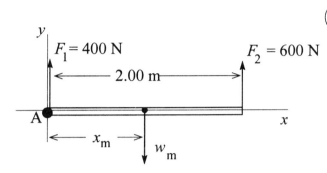

Let w_m be the weight of the. motor. Take the origin of coordinates at the end where the 400 N force is applied (point A).

Take all the weight of the motor to act at its cg, a distance x_{cm} to the right of A.

EXECUTE:

$\sum F_y = ma_y$

$F_1 + F_2 - w_m = 0$

$w_m = F_1 + F_2 = 400 \text{ N} + 600 \text{ N} = 1000 \text{ N}$

$\sum \tau_A = 0$

$+F_2(2.00 \text{ m}) - w_m x_m = 0$

$x_m = 2.00 \text{ m}(F_2/w_m) = 2.00 \text{ m}(600 \text{ N}/1000 \text{ N}) = 1.20 \text{ m}$

The center of gravity of the motor is 1.20 m from the end where the 400 N force is applied.

EVALUATE: The cg of the motor is closer to the person who lifts with the greater force.

11.11 IDENTIFY: The system of the person and diving board is at rest so the two conditions of equilibrium apply.

a) SET UP: Free-body diagram for the diving board:

Take the origin of coordinates at the left-hand end of the board (point A).

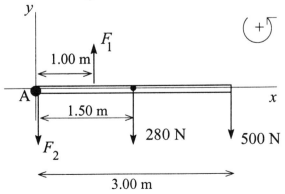

$\vec{F}_1$ is the force applied at the support point and $\vec{F}_2$ is the force at the end that is held down.

EXECUTE:

$\sum \tau_A = 0$ gives $+F_1(1.0 \text{ m}) - (500 \text{ N})(3.00 \text{ m}) - (280 \text{ N})(1.50 \text{ m}) = 0$

$$F_1 = \frac{(500 \text{ N})(3.00 \text{ m}) + (280 \text{ N})(1.50 \text{ m})}{1.00 \text{ m}} = 1920 \text{ N}$$

b) $\sum F_y = ma_y$

$F_1 - F_2 - 280 \text{ N} - 500 \text{ N} = 0$

$F_2 = F_1 - 280 \text{ N} - 500 \text{ N} = 1920 \text{ N} - 280 \text{ N} - 500 \text{ N} = 1140 \text{ N}$

EVALUATE: We can check our answers by calculating the net torque about some point and checking that $\tau_z = 0$ for that point also. Net torque about A:

$(1920 \text{ N})(1.00 \text{ m}) - (280 \text{ N})(1.50 \text{ m}) - (500 \text{ N})(3.00 \text{ m}) =$
$1920 \text{ N} \cdot \text{m} - 420 \text{ N} \cdot \text{m} - 1500 \text{ N} \cdot \text{m} = 0$, which checks.

11.13 IDENTIFY: Apply the first and second conditions of equilibrium to the strut.

a) SET UP: Free-body diagram for the strut. Take the origin of coordinates at the hinge (point A) and $+y$ upward. Let F_h and F_v be the horizontal and vertical components of the force $\vec{F}$ exerted on the strut by the pivot. The tension in the vertical cable is the weight w of the suspended object. The weight w of the strut can be taken to act at the center of the strut. Let L be the length of the strut.

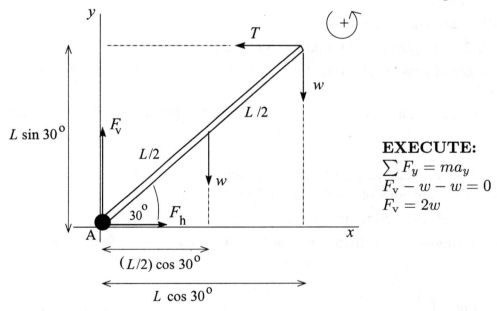

EXECUTE:
$\sum F_y = ma_y$
$F_v - w - w = 0$
$F_v = 2w$

Sum torques about point A. The pivot force has zero moment for this axis and so doesn't enter into the torque equation.

$\tau_A = 0$

$TL \sin 30.0° - w((L/2)\cos 30.0°) - w(L\cos 30.0°) = 0$

$T \sin 30.0° - (3w/2)\cos 30.0° = 0$

$T = \dfrac{3w \cos 30.0°}{2 \sin 30.0°} = 2.60w$

Then $\sum F_x = ma_x$ implies $T - F_h = 0$ and $F_h = 2.60w$.

We now have the components of $\vec{F}$ so can find its magnitude and direction:

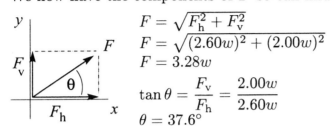

$$F = \sqrt{F_h^2 + F_v^2}$$
$$F = \sqrt{(2.60w)^2 + (2.00w)^2}$$
$$F = 3.28w$$
$$\tan\theta = \frac{F_v}{F_h} = \frac{2.00w}{2.60w}$$
$$\theta = 37.6°$$

b) SET UP: Free-body diagram for the strut

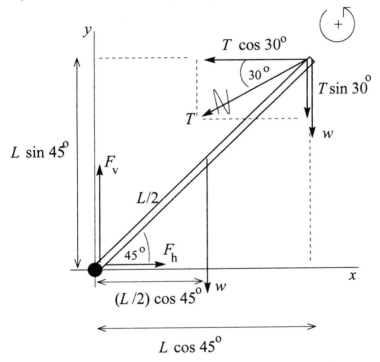

The tension T has been replaced by its x and y components. The torque due to T equals the sum of the torques of its components, and the latter are easier to calculate.

EXECUTE:

$$\sum\tau_A = 0 + (T\cos 30.0°)(L\sin 45.0°) - (T\sin 30.0°)(L\cos 45.0°) - w((L/2)\cos 45.0°) - w(L\cos 45.0°) = 0$$

The length L divides out of the equation. The equation can also be simplified by noting that $\sin 45.0° = \cos 45.0°$.

Then $T(\cos 30.0° - \sin 30.0°) = 3w/2$.

$$T = \frac{3w}{2(\cos 30.0° - \sin 30.0°)} = 4.10w$$

$$\sum F_x = ma_x$$

$F_h - T\cos 30.0° = 0$

$F_h = T\cos 30.0° = (4.10w)(\cos 30.0°) = 3.55w$

$\sum F_y = ma_y$

$F_v - w - w - T\sin 30.0° = 0$

$F_v = 2w + (4.10w)\sin 30.0° = 4.05w$

$F = \sqrt{F_h^2 + F_v^2}$

$F = \sqrt{(3.55w)^2 + (4.05w)^2} = 5.39w$

$\tan\theta = \dfrac{F_v}{F_h} = \dfrac{4.05w}{3.55w}$

$\theta = 48.8°$

EVALUATE: In each case the force exerted by the pivot does not act along the strut. Consider the net torque about the upper end of the strut. If the pivot force acted along the strut, it would have zero torque about this point. The two forces acting at this point also have zero torque and there would be one nonzero torque, due to the weight of the strut. The net torque about this point would then not be zero, violating the second condition of equilibrium.

11.15 IDENTIFY: Apply the first and second conditions of equilibrium to the door.

SET UP: Free-body diagram for the door. Let $\vec{H}_1$ and $\vec{H}_2$ be the forces exerted by the upper and lower hinges. Take the origin of coordinates at the bottom hinge (point A) and $+y$ upward.

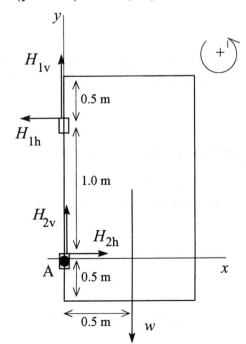

EXECUTE:

We are given that
$H_{1v} = H_{2v} = w/2 = 140$ N.

$\sum F_x = ma_x$

$H_{2h} - H_{1h} = 0$

$H_{1h} = H_{2h}$

The horizontal components of the hinge forces are equal in magnitude and opposite in direction.

Sum torques about point A. H_{1v}, H_{2v}, and H_{2h} all have zero moment arm and hence zero torque about an axis at this point.

Thus $\sum \tau_A = 0$ gives $H_{1h}(1.00 \text{ m}) - w(0.50 \text{ m}) = 0$

$$H_{1h} = w\left(\frac{0.50 \text{ m}}{1.00 \text{ m}}\right) = \tfrac{1}{2}(280 \text{ N}) = 140 \text{ N}.$$

The horizontal component of each hinge force is 140 N.

EVALUATE: The horizontal components of the force exerted by each hinge are the only horizontal forces so must be equal in magnitude and opposite in direction. With an axis at A, the torque due to the horizontal force exerted by the upper hinge must be counterclockwise to oppose the clockwise torque exerted by the weight of the door. So, the horizontal force exerted by the upper hinge must be to the left. You can also verify that the net torque is also zero if the axis is at the upper hinge.

11.17 **IDENTIFY:** Apply the first and second conditions of equilibrium to Clea.

SET UP: Consider the forces on Clea.

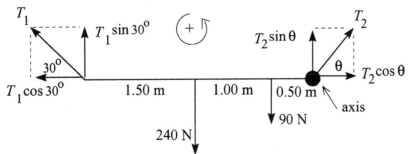

EXECUTE:

$n_r = 89 \text{ N}$, $n_f = 157 \text{ N}$

$n_r + n_f = w$ so $w = 246 \text{ N}$

$\sum \tau_z = 0$, axis at rear feet

Let x be the distance from the rear feet to the center of gravity.

$n_f(0.95 \text{ m}) - xw = 0$

$x = 0.606 \text{ m}$ from rear feet so 0.34 m from front feet.

EVALUATE: The normal force at her front feet is greater than at her rear feet, so her center of gravity is closer to her front feet.

11.19 **IDENTIFY:** Apply the first and second conditions of equilibrium to the rod.

SET UP: Force diagram for the rod.

EXECUTE:

$\sum \tau_z = 0$, axis at right end of rod, counterclockwise torques positive

$(240 \text{ N})(1.50 \text{ m}) + (90 \text{ N})(0.50 \text{ m}) - (T_1 \sin 30.0°)(3.00 \text{ m}) = 0$

$$T_1 = \frac{360 \text{ N} \cdot \text{m} + 45 \text{ N} \cdot \text{m}}{1.50 \text{ m}} = 270 \text{ N}$$

$\sum F_x = ma_x$

$T_2 \cos \theta - T_1 \cos 30° = 0$ and $T_2 \cos \theta = 234 \text{ N}$

$\sum F_y = ma_y$

$T_1 \sin 30° + T_2 \sin \theta - 240 \text{ N} - 90 \text{ N} = 0$

$T_2 \sin \theta = 330 \text{ N} - (270 \text{ N}) \sin 30° = 195 \text{ N}$

Then $\dfrac{T_2 \sin \theta}{T_2 \cos \theta} = \dfrac{195 \text{ N}}{234 \text{ N}}$ gives

$\tan \theta = 0.8333$ and $\theta = 40°$

And $T_2 = \dfrac{195 \text{ N}}{\sin 40°} = 303 \text{ N}.$

EVALUATE: The monkey is closer to the right rope than to the left one, so the tension is larger in the right rope. The horizontal components of the tensions must be equal in magnitude and opposite in direction. Since $T_2 > T_1$, the rope on the right must be at a greater angle above the horizontal to have the same horizontal component as the tension in the other rope.

11.21 a) IDENTIFY and **SET UP:** Use Eq.(10.3) to calculate the torque (magnitude and direction) for each force and add the torques as vectors.

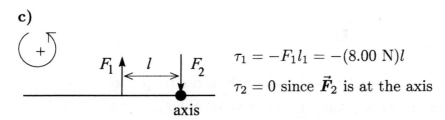

EXECUTE:

$\tau_1 = F_1 l_1 = +(8.00 \text{ N})(3.00 \text{ m})$

$\tau_1 = +24.0 \text{ N} \cdot \text{m}$

$\tau_2 = -F_2 l_2 = -(8.00 \text{ N})(l + 3.00 \text{ m})$

$\tau_2 = -24.0 \text{ N} \cdot \text{m} - (8.00 \text{ N})l$

$\sum \tau_z = \tau_1 + \tau_2 = +24.0 \text{ N} \cdot \text{m} - 24.0 \text{ N} \cdot \text{m} - (8.00 \text{ N})l = -(8.00 \text{ N})l$

Want l that makes $\sum \tau_z = -6.40 \text{ N} \cdot \text{m}$ (net torque must be clockwise)

$-(8.00 \text{ N})l = -6.40 \text{ N} \cdot \text{m}$

$l = (6.40 \text{ N} \cdot \text{m})/8.00 \text{ N} = 0.800 \text{ m}$

b) $|\tau_2| > |\tau_1|$ since F_2 has a larger moment arm; the net torque is clockwise.

c)

$\tau_1 = -F_1 l_1 = -(8.00 \text{ N})l$

$\tau_2 = 0$ since $\vec{F}_2$ is at the axis

$\sum \tau_z = -6.40$ N $\cdot$ m gives $-(8.00$ N$)l = -6.40$ N $\cdot$ m

$l = 0.800$ m, same as in part (a).

EVALUATE: The force couple gives the same magnitude of torque for the pivot at any point.

11.23 IDENTIFY and **SET UP:** Apply Eq.(11.10) and solve for A and then use $A = \pi r^2$ to get the radius and $d = 2r$ to calculate the diameter.

EXECUTE:

$Y = \dfrac{l_0 F_\perp}{A \, \Delta l}$ so $A = \dfrac{l_0 F_\perp}{Y \, \Delta l}$ (A is the cross-section area of the wire)

For steel, $Y = 2.0 \times 10^{11}$ Pa (Table 11-1)

Thus $A = \dfrac{(2.00 \text{ m})(400 \text{ N})}{(2.0 \times 10^{11} \text{ Pa})(0.25 \times 10^{-2} \text{ m})} = 1.6 \times 10^{-6}$ m^2.

$A = \pi r^2$, so $r = \sqrt{A/\pi} = \sqrt{1.6 \times 10^{-6} \text{ m}^2/\pi} = 7.1 \times 10^{-4}$ m

$d = 2r = 1.4 \times 10^{-3}$ m $= 1.4$ mm

EVALUATE: Steel wire of this diameter doesn't stretch much; $\Delta l/l_0 = 0.12\%$.

11.27 IDENTIFY Use the first condition of equilibrium to calculate the tensions T_1 and T_2 in the wires. Then use Eq.(11.10) to calculate the strain and elongation of each wire.

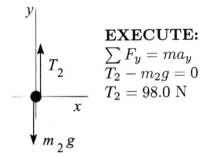

SET UP: Free-body-diagram for m_2

EXECUTE:

$\sum F_y = ma_y$

$T_2 - m_2 g = 0$

$T_2 = 98.0$ N

SET UP: Free-body-diagram for m_1

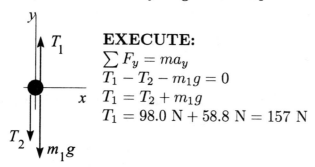

EXECUTE:
$$\sum F_y = ma_y$$
$$T_1 - T_2 - m_1 g = 0$$
$$T_1 = T_2 + m_1 g$$
$$T_1 = 98.0 \text{ N} + 58.8 \text{ N} = 157 \text{ N}$$

a) $Y = \dfrac{\text{stress}}{\text{stain}}$ so strain $= \dfrac{\text{stress}}{Y} = \dfrac{F_\perp}{AY}$

upper wire: strain $= \dfrac{T_1}{AY} = \dfrac{157 \text{ N}}{(2.5 \times 10^{-7} \text{ m}^2)(2.0 \times 10^{11} \text{ Pa})} = 3.1 \times 10^{-3}$

lower wire: strain $= \dfrac{T_2}{AY} = \dfrac{98 \text{ N}}{(2.5 \times 10^{-7} \text{ m}^2)(2.0 \times 10^{11} \text{ Pa})} = 2.0 \times 10^{-3}$

b) strain $= \Delta l / l_0$ so $\Delta l = l_0 (\text{strain})$

upper wire: $\Delta l = (0.50 \text{ m})(3.1 \times 10^{-3}) = 1.6 \times 10^{-4} \text{ m} = 1.6 \text{ mm}$

lower wire: $\Delta l = (0.50 \text{ m})(2.0 \times 10^{-3}) = 1.0 \times 10^{-3} \text{ m} = 1.0 \text{ mm}$

EVALUATE: The tension is greater in the upper wire because it must support both objects. The wires have the same length and diameter, so the one with the greater tension has the greater strain and elongation.

11.33 IDENTIFY and **SET UP:** Use Eqs.(11.13) and (11.14) to calculate B and k.
EXECUTE:
$$B = -\frac{\Delta p}{\Delta V/V_0} = -\frac{(3.6 \times 10^6 \text{ Pa})(600 \text{ cm}^3)}{(-0.45 \text{ cm}^3)} = +4.8 \times 10^9 \text{ Pa}$$
$$k = 1/B = 1/4.8 \times 10^9 \text{ Pa} = 2.1 \times 10^{-10} \text{ Pa}^{-1}$$
EVALUATE: k is the same as for glycerine (Table 11.2).

11.35 IDENTIFY and **SET UP:** Use Eq.(11.17). Same material implies same S
EXECUTE:
$S = \dfrac{\text{stress}}{\text{strain}}$ so strain $= \dfrac{\text{stress}}{S} = \dfrac{F_\parallel / A}{S}$ and same forces implies same $F_\parallel$.

For the smaller object, $(\text{strain})_1 = F_\parallel / A_1 S$

For the larger object, $(\text{strain})_2 = F_\parallel / A_2 S$

$$\frac{(\text{strain})_2}{(\text{strain})_1} = \left(\frac{F_\parallel}{A_2 S} \right) \left(\frac{A_1 S}{F_\parallel} \right) = \frac{A_1}{A_2}$$

Larger solid has triple each edge length, so $A_2 = 9A_1$, and $\dfrac{(\text{strain})_2}{(\text{strain})_1} = \dfrac{1}{9}$

EVALUATE: The larger object has a smaller deformation.

11.37 IDENTIFY and **SET UP:** Use Eq.(11.8).

EXECUTE:

Tensile stress $= \dfrac{F_\perp}{A} = \dfrac{F_\perp}{\pi r^2} = \dfrac{90.8 \text{ N}}{\pi(0.92 \times 10^{-3} \text{ m})^2} = 3.41 \times 10^7 \text{ Pa}$

EVALUATE: A modest force produces a very large stress because the cross-sectional area is small.

Problems

11.41 IDENTIFY: Apply the second condition of equilibrium to the car.

a) SET UP: Free-body diagram for the car

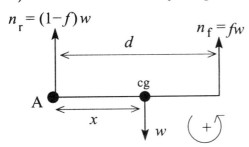

Let the cg be a distance x from the rear axle. Sum torques about the rear axle (point A).

EXECUTE: $\sum \tau_A = 0$ so $n_\mathrm{f}d - xw = 0$

$fwd - wx = 0$ and $x = fd$, as was to be shown.

b) EVALUATE: In Example 11.2, $n_\mathrm{f} = 0.53w$ so $f = 0.53$ and $d = 2.46$ m. The general result derived in part (a) gives $x = fd = (0.53)(2.46 \text{ m}) = 1.30$ m, the same as calculated in the example.

11.45 IDENTIFY: Apply $\sum \tau_z = 0$ to the system.

a) SET UP: Seesaw set so Karen exerts maximum torque about the pivot implies pivot moved as far as possible from Karen, so pivot is 0.20 m to left of center of seesaw.

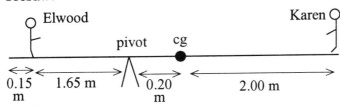

w_E is the weight of Elwood, w_K is the weight of Karen, and w_ss is the weight of the seesaw. n is the normal force exerted by the pivot

Free-body diagram for the seesaw:

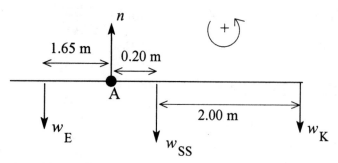

Apply $\sum \tau_z = 0$ with the axis at the pivot (point A) and counterclockwise torques positive.

EXECUTE:

$$w_E(1.65\ \text{m}) - w_{ss}(0.20\ \text{m}) - w_K(2.20\ \text{m}) = 0$$

$$w_E = \frac{w_K(2.20\ \text{m}) + w_{ss}(0.20\ \text{m})}{1.65\ \text{m}} = \frac{(420\ \text{N})(2.20\ \text{m}) + (240\ \text{N})(0.20\ \text{m})}{1.65\ \text{m}} = 589\ \text{N}$$

EVALUATE: The moment arm for Elwood is less than for Karen so his weight must be greater to give the same torque.

b) The center of gravity of each piece of the system (Elwood, Karen, and the seesaw) is located some distance above the pivot. When the seesaw is horizontal the force diagram that takes this account gives forces with lines of action and moment arms the same as in part (a):

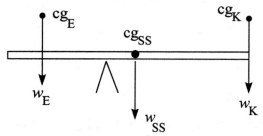

The center of gravity cg_{tot} of the total system (seesaw, Elwood, Karen) is located a distance y_{cm} directly above the pivot P. When the seesaw is horizontal the line of action of the total weight Mg and the normal n coincide and there is zero net torque.

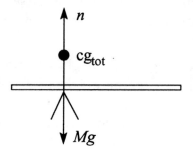

Now consider the force diagram when the system is rotated through an angle θ:

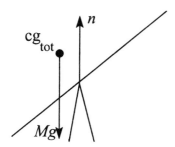

The line of action of the total weight force no longer passes through the pivot so now produces a counterclockwise torque about the pivot. The resultant torque in this direction rotates the system farther from equilibrium. (The same conclusion is reached if the seesaw is rotated clockwise from the horizontal.)

11.49 **IDENTIFY:** Apply the first and second conditions of equilibrium to the bar.
SET UP: Free-body diagram for the bar:

n is the normal force exerted on the bar by the surface. There is no friction force at this surface. H_h and H_v are the components of the force exerted on the bar by the hinge. The components of the force of the bar on the hinge will be equal in magnitude and opposite in direction.

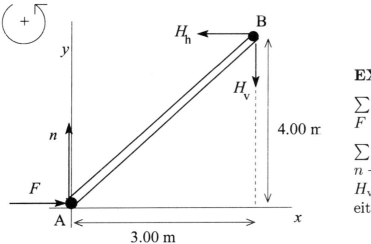

EXECUTE:

$\sum F_x = ma_x$
$F = H_h = 120 \text{ N}$

$\sum F_y = ma_y$
$n - H_v = 0$
$H_v = n$, but we don't know either of these forces.

$\sum \tau_B = 0$ gives $F(4.00 \text{ m}) - N(3.00 \text{ m}) = 0$

$n = (4.00 \text{ m}/3.00 \text{ m})F = \frac{4}{3}(120 \text{ N}) = 160 \text{ N}$ and then $H_v = 160 \text{ N}$

Force of bar on hinge:

horizontal component 120 N, to right

vertical component 160 N, upward

EVALUATE: $H_h/H_v = 120/160 = 3.00/4.00$, so the force the hinge exerts on the bar is directed along the bar. $\vec{n}$ and $\vec{F}$ have zero torque about point A, so the line of action of the hinge force $\vec{H}$ must pass through this point also if the net torque is to be zero.

11.53 **IDENTIFY** and **SET UP:** Use Eq.(10.3) to calculate the torque due to each force and add the forces as vectors.

a)

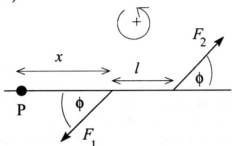

EXECUTE:
$F_1 = F_2 = F$

For an axis at P.
$$\tau_1 = -F_1 x \sin\phi = -Fx\sin\phi$$
$$\tau_2 = +F_2(x + l)\sin\phi = +F(x + l)\sin\phi$$

$\tau_P = \tau_1 + \tau_2 = F\sin\phi(x + l - x) = Fl\sin\phi$, which is independent of x. Therefore the resultant torque is the same for an axis at any point along the rod.

b)

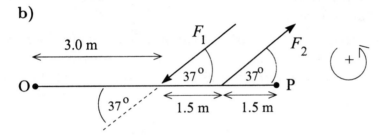

For an axis at O,
$$\tau_1 = -F_1(3.0 \text{ m})\sin 37° = -(14.0 \text{ N})(3.0 \text{ m})\sin 37° = -25.28 \text{ N} \cdot \text{m}$$
$$\tau_2 = +F_2(4.5 \text{ m})\sin 37° = +(14.0 \text{ N})(4.5 \text{ m})\sin 37° = +37.91 \text{ N} \cdot \text{m}$$
$$\sum \tau_O = \tau_1 + \tau_2 = -25.28 \text{ N} \cdot \text{m} + 37.91 \text{ N} \cdot \text{m} = +12.6 \text{ N} \cdot \text{m}.$$

For an axis at P,
$$\tau_1 = +F_1(3.0 \text{ m})\sin 37° = +(14.0 \text{ N})(3.0 \text{ m})\sin 37° = +25.28 \text{ N} \cdot \text{m}$$
$$\tau_2 = -F_2(1.5 \text{ m})\sin 37° = -(14.0 \text{ N})(1.5 \text{ m})\sin 37° = -12.64 \text{ N} \cdot \text{m}$$
$$\sum \tau_P = \tau_1 + \tau_2 = +25.28 \text{ N} \cdot \text{m} - 12.64 \text{ N} \cdot \text{m} = +12.6 \text{ N} \cdot \text{m}.$$

The general result derived in part (a) gives
$$\sum \tau = Fl\sin\phi = (14.0 \text{ N})(1.5 \text{ m})\sin 37° = +12.6 \text{ N} \cdot \text{m}.$$

EVALUATE: The torque about P is the same as about O, and agrees with the general result derived in part (a).

11.59 **IDENTIFY:** Apply the first and second conditions of equilibrium to the beam.
SET UP:

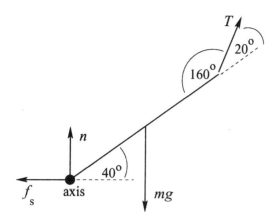

EXECUTE:

a) $\sum \tau_z = 0$, axis at lower end of beam

Let the length of the beam be L.

$$T(\sin 20°)L = -mg \left(\frac{L}{2}\right) \cos 40° = 0$$

$$T = \frac{\frac{1}{2}mg \cos 40°}{\sin 20°} = 2700 \text{ N}$$

b) Take $+y$ upward.

$\sum F_y = 0$ gives $n - w + T \sin 60° = 0$ so $n = 73.6$ N

$\sum F_x = 0$ gives $f_s = T \cos 60° = 1372$ N

$$f_s = \mu_s n, \quad \mu_s = \frac{f_s}{n} = \frac{1372 \text{ N}}{73.6 \text{ N}} = 19$$

EVALUATE: The floor must be very rough for the beam not to slip. The friction force exerted by the floor is to the left because T has a component that pulls the beam to the right.

11.67 IDENTIFY: Apply the first and second conditions of equilibrium to the crate.

SET UP: Free-body diagram for the crate

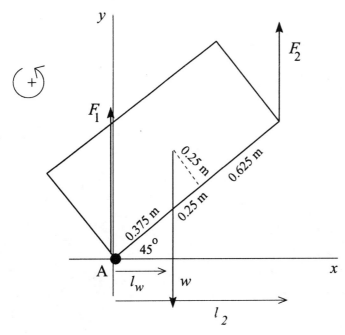

$l_w = (0.375 \text{ m}) \cos 45°$
$l_2 = (1.25 \text{ m}) \cos 45°$

Let $\vec{F}_1$ and $\vec{F}_2$ be the vertical forces exerted by you and your friend. Take the origin at the lower left-hand corner of the crate (point A).

EXECUTE:

$\sum F_y = ma_y$ gives $F_1 + F_2 - w = 0$

$F_1 + F_2 = w = (200 \text{ kg})(9.80 \text{ m/s}^2) = 1960 \text{ N}$

$\sum \tau_A = 0$ gives $F_2 l_2 - w l_w = 0$

$F_2 = w \left(\dfrac{l_w}{l_2} \right) = 1960 \text{ N} \left(\dfrac{0.375 \text{ m} \cos 45°}{1.25 \text{ m} \cos 45°} \right) = 590 \text{ N}$

Then $F_1 = w - F_2 = 1960 \text{ N} - 590 \text{ N} = 1370 \text{ N}$.

EVALUATE: The person below (you) applies a force of 1370 N. The person above (your friend) applies a force of 590 N. It is better to be the person above. As the sketch shows, the moment arm for $\vec{F}_1$ is less than for $\vec{F}_2$, so must have $F_1 > F_2$ to compensate.

11.71 IDENTIFY: Apply $\sum \tau_z = 0$ first to the roof and then to one wall.

a) SET UP: Consider the forces on the roof

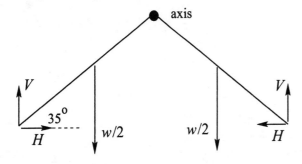

V and H are the vertical and horizontal forces each wall exerts on the roof.

$w = 20,000 \text{ N}$ is the total weight of the roof.

$2V = w$ so $V = w/2$

Apply $\sum \tau_z = 0$ to one half of the roof, with the axis along the line where the two halves join. Let each half have length L.

EXECUTE: $(w/2)(L/2)(\cos 35.0°) + HL \sin 35.0° - VL \cos 35° = 0$

L divides out, and use $V = w/2$

$H \sin 35.0° = \frac{1}{4} w \cos 35.0°$

$H = \dfrac{w}{4 \tan 35.0°} = 7140 \text{ N}$

EVALUATE: By Newton's 3rd law, the roof exerts a horizontal, outward force on the wall. For torque about an axis at the lower end of the wall, at the ground, this force has a larger moment arm and hence larger torque the taller the walls.

b) **SET UP:** Force Diagram for one wall

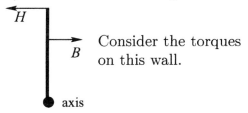

Consider the torques on this wall.

H is the horizontal force exerted by the roof, as considered in part (a). B is the horizontal force exerted by the buttress. Now the angle is $40°$, so

$H = \dfrac{w}{4 \tan 40°} = 5959 \text{ N}$

EXECUTE: $\sum \tau_z = 0$, axis at the ground

$H(40 \text{ m}) - B(30 \text{ m}) = 0$ and $B = 7900 \text{ N}$.

EVALUATE: The horizontal force exerted by the roof is larger as the roof becomes more horizontal, since for torques applied to the roof the moment arm for H decreases. The force B required from the buttress is less the higher up on the wall this force is applied.

11.73 IDENTIFY: Apply the first and second conditions of equilibrium to the gate.

SET UP: Free-body diagram for the gate

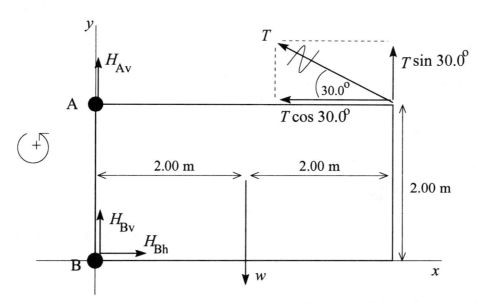

Use coordinates with the origin at B. Let $\vec{H}_A$ and $\vec{H}_B$ be the forces exerted by the hinges at A and B. The problem states that $\vec{H}_A$ has no horizontal component. Replace the tension $\vec{T}$ by its horizontal and vertical components.

EXECUTE:

a) $\sum \tau_B = 0$ gives $+(T \sin 30.0°)(4.00 \text{ m}) + (T \cos 30.0°)(2.00 \text{ m}) - w(2.00 \text{ m}) = 0$

$T(2 \sin 30.0° + \cos 30.0°) = w$

$$T = \frac{w}{2 \sin 30.0° + \cos 30.0°} = \frac{500 \text{ N}}{2 \sin 30.0° + \cos 30.0°} = 268 \text{ N}$$

b) $\sum F_x = ma_x$ says $H_{Bh} - T \cos 30.0° = 0$

$H_{Bh} = T \cos 30.0° = (268 \text{ N}) \cos 30.0° = 232 \text{ N}$

c) $\sum F_y = ma_y$ says $H_{Av} + H_{Bv} + T \sin 30.0° - w = 0$

$H_{Av} + H_{Bv} = w - T \sin 30.0° = 500 \text{ N} - (268 \text{ N}) \sin 30.0° = 366 \text{ N}$

EVALUATE: T has a horizontal component to the left so H_{Bh} must be to the right, as these are the only two horizontal forces. Note that we cannot determine H_{Av} and H_{Bv} separately, only their sum.

11.75 IDENTIFY: Apply the first and second conditions of equilibrium, first to both marbles considered as a composite object and then to the bottom marble.

a) SET UP:

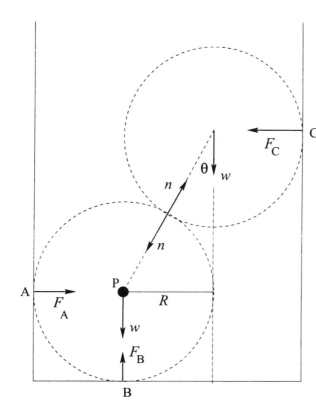

EXECUTE:

$F_B = 2w = 1.47$ N

$\sin\theta = R/2R$ so $\theta = 30°$

$\sum \tau_z = 0$, axis at P

$F_C(2R\cos\theta) - wR = 0$

$F_C = \dfrac{mg}{2\cos 30°} = 0.424$ N

$F_A = F_C = 0.424$ N

b) Consider the forces on the bottom marble.

The horizontal forces must sum to zero, so

$F_A = n\sin\theta$

$n = \dfrac{F_A}{\sin 30°} = 0.848$ N

Could use instead that the vertical forces sum to zero

$F_B - mg - n\cos\theta = 0$

$n = \dfrac{F_B - mg}{\cos 30°} = 0.848$ N, which checks.

EVALUATE: If we consider each marble separately, the line of action of every force passes through the center of the marble so there is clearly no torque about that point for each marble. We can use the results we obtained to show that $\sum F_x = 0$ and $\sum F_y = 0$ for the top marble.

11.77 IDENTIFY: Apply the first and second conditions of equilibrium to the bale.

a) SET UP: Find the angle where the bale starts to tip.

When starts to tip only the lower left-hand corner of the bale makes contact with the conveyor belt. Therefore the line of action of the normal force n passes through the left-hand edge of the bale. Consider $\sum \tau_A = 0$ with point A at the lower left-hand corner. Then $\tau_n = 0$ and $\tau_f = 0$, so it must be that $\tau_{mg} = 0$ also. This means that the line of action of the gravity must pass through point A. Thus the free-body diagram must be as follows:

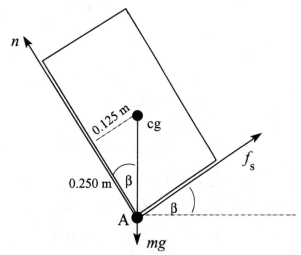

EXECUTE:

$$\tan \beta = \frac{0.125 \text{ m}}{0.250 \text{ m}}$$

$\beta = 27°$, angle where tips

SET UP: At the angle where the bale is ready to slip down the incline f_s has its maximum possible value, $f_s = \mu_s n$. Free-body diagram for the bale, with the origin of coordinates at the cg:

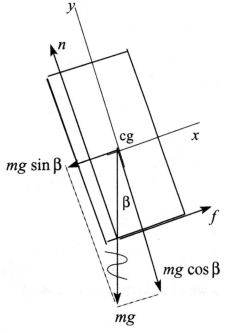

EXECUTE:

$\sum F_y = ma_y$
$n - mg \cos \beta = 0$
$n = mg \cos \beta$
$f_s = \mu_s mg \cos \beta$
(f_s has maximum value when bale ready to slip)

$\sum F_x = ma_x$
$f_s - mg \sin \beta = 0$
$\mu_s mg \cos \beta - mg \sin \beta = 0$
$\tan \beta = \mu_s$
$\mu_s = 0.60$ gives that $\beta = 31°$

$\beta = 27°$ to tip; $\beta = 31°$ to slip, so tips first

b) The magnitude of the friction force didn't enter into the calculation of the tipping angle; still tips at $\beta = 27°$.

For $\mu_s = 0.40$ tips at $\beta = \arctan(0.40) = 22°$

Now the bale will start to slide down the incline before it tips.

EVALUATE: With a smaller μ_s the slope angle β where the bale slips is smaller.

11.79 IDENTIFY: Apply the first and second conditions of equilibrium to the door.

a) SET UP: Free-body diagram for the door

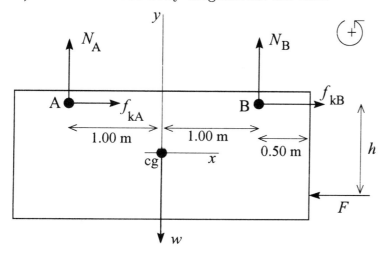

Take the origin of coordinates at the center of the door (at the cg). Let n_A f_{kA}, n_B, and f_{kB} be the normal and friction forces exerted on the door at each wheel.

EXECUTE:

$$\sum F_y = ma_y$$

$$n_A + n_B - w = 0$$

$$n_A + n_B = w = 950 \text{ N}$$

$$\sum F_x = ma_x$$

$$f_{kA} + f_{kB} - F = 0$$

$$F = f_{kA} + f_{kB}$$

$$f_{kA} = \mu_k n_A, \quad f_{kB} = \mu_k n_B,$$

so $F = \mu_k(n_A + n_B) = \mu_k w = (0.52)(950 \text{ N}) = 494 \text{ N}$

$$\sum \tau_B = 0$$

n_B, f_{kA}, and f_{kB} all have zero moment arms and hence zero torque about this point.

Thus $+w(1.00 \text{ m}) - n_A(2.00 \text{ m}) - F(h) = 0$

$$n_A = \frac{w(1.00 \text{ m}) - F(h)}{2.00 \text{ m}} = \frac{(950 \text{ N})(1.00 \text{ m}) - (494 \text{ N})(1.60 \text{ m})}{2.00 \text{ m}} = 80 \text{ N}$$

And then $n_B = 950 \text{ N} - n_A = 950 \text{ N} - 80 \text{ N} = 870 \text{ N}$.

b) SET UP: If h is too large the torque of F will cause wheel A to leave the track. When wheel A just starts to lift off the track n_A and f_{kA} both go to zero.

EXECUTE: The equations in part (a) still apply.

$n_A + n_B - w = 0$ gives $n_B = w = 950$ N

Then $f_{kB} = \mu_k n_B = 0.52(950$ N$) = 494$ N

$F = f_{kA} + f_{kB} = 494$ N

$+w(1.00$ m$) - n_A(2.00$ m$) - F(h) = 0$

$h = \dfrac{w(1.00 \text{ m})}{F} = \dfrac{(950 \text{ N})(1.00 \text{ m})}{494 \text{ N}} = 1.92$ m

EVALUATE: The result in part (b) is larger than the value of h in part (a). Increasing h increases the clockwise torque about B due to F and therefore decreases the clockwise torque that n_A must apply.

11.81 IDENTIFY: Apply the first and second conditions of equilibrium to the pole.

a) SET UP: Free-body diagram for the pole

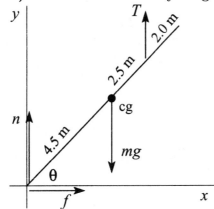

n and f are the vertical and horizontal components of the force the ground exerts on the pole.

$\sum F_x = ma_x$
$f = 0$
The force exerted by the ground has no horizontal component.

EXECUTE: $\sum \tau_A = 0$

$+T(7.0$ m$) \cos\theta - mg(4.5$ m$) \cos\theta = 0$

$T = mg(4.5$ m$/7.0$ m$) = (4.5/7.0)(5700$ N$) = 3700$ N

$\sum F_y = 0$
$n + T - mg = 0$
$n = mg - T = 5700$ N $- 3700$ N $= 2000$ N

The force exerted by the ground is vertical (upward) and has magnitude 2000 N.

EVALUATE: We can verify that $\sum \tau_z = 0$ for an axis at the cg of the pole. $T > n$ since T acts at a point closer to the cg and therefore has a smaller moment arm for this axis than n does.

b) In the $\sum \tau_A = 0$ equation the angle θ divided out. All forces on the pole are vertical and their moment arms are all proportional to $\cos\theta$.

11.83 IDENTIFY: Apply Newton's 2nd law to the mass to find the tension in the wire.

Then apply Eq.(11.10) to the wire to find the elongation this tensile force produces.

a) SET UP: Calculate the tension in the wire as the mass passes through the lowest point. Free-body diagram for the mass:

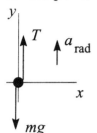

The mass moves in an arc of a circle with radius $R = 0.50$ m. It has acceleration $\vec{a}_{\text{rad}}$ directed in toward the center of the circle, so at this point $\vec{a}_{\text{rad}}$ is upward.

EXECUTE: $\sum F_y = ma_y$

$T - mg = mR\omega^2$ so that $T = m(g + R\omega^2)$.

But ω must be in rad/s:

$\omega = (120 \text{ rev/min})(2\pi \text{ rad/1 rev})(1 \text{ min}/60 \text{ s}) = 12.57 \text{ rad/s}$.

Then $T = (12.0 \text{ kg})(9.80 \text{ m/s}^2 + (0.50 \text{ m})(12.57 \text{ rad/s})^2) = 1066 \text{ N}$.

Now calculate the elongation Δl of the wire that this tensile force produces:

$$Y = \frac{F_\perp l_0}{A \Delta l} \text{ so } \Delta l = \frac{F_\perp l_0}{YA} = \frac{(1066 \text{ N})(0.50 \text{ m})}{(7.0 \times 10^{10} \text{ Pa})(0.014 \times 10^{-4} \text{ m}^2)} = 0.54 \text{ cm}.$$

b) SET UP: The acceleration $\vec{a}_{\text{rad}}$ is directed in towards the center of the circular path, and at this point in the motion this direction is downward.

EXECUTE:
$\sum F_y = ma_y$
$mg + T = mR\omega^2$
$T = m(R\omega^2 - g)$

$T = (12.0 \text{ kg})((0.50 \text{ m})(12.57 \text{ rad/s})^2 - 9.80 \text{ m/s}^2) = 830 \text{ N}$

$$\Delta l = \frac{F_\perp l_0}{YA} = \frac{(830 \text{ N})(0.50 \text{ m})}{(7.0 \times 10^{10} \text{ Pa})(0.014 \times 10^{-4} \text{ m}^2)} = 0.42 \text{ cm}.$$

EVALUATE: At the lowest point T and w are in opposite directions and at the highest point they are in the same direction, so T is greater at the lowest point and the elongation is greatest there. The elongation is at most 1% of the length.

11.85 IDENTIFY: Use the second condition of equilibrium to relate the tension in the two wires to the distance w is from the left end. Use Eqs.(11.8) and (11.10) to relate the tension in each wire to its stress and strain.

a) SET UP: stress $= F_\perp/A$, so equal stress implies T/A same for each wire.

$T_A/2.00$ mm$^2 = T_B/4.00$ mm^2 so $T_B = 2.00T_A$

The question is where along the rod to hang the weight in order to produce this relation between the tensions in the two wires. Let the weight be suspended at point C, a distance x to the right of wire A. The free-body diagram for the rod is then

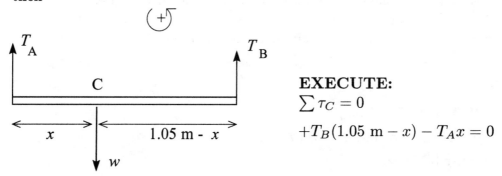

EXECUTE:

$\sum \tau_C = 0$

$+T_B(1.05 \text{ m} - x) - T_A x = 0$

But $T_B = 2.00T_A$ so $2.00T_A(1.05 \text{ m} - x) - T_A x = 0$

$2.10 \text{ m} - 2.00x = x$ and $x = 2.10 \text{ m}/3.00 = 0.70$ m (measured from A).

b) SET UP: $Y = $ stress/strain gives that strain = stress/$Y = F_\perp/AY$.

EXECUTE: Equal strain thus implies

$$\frac{T_A}{(2.00 \text{ mm}^2)(1.80 \times 10^{11} \text{ Pa})} = \frac{T_B}{(4.00 \text{ mm}^2)(1.20 \times 10^{11} \text{ Pa})}$$

$$T_B = \left(\frac{4.00}{2.00}\right)\left(\frac{1.20}{1.80}\right)T_A = 1.333T_A.$$

The $\sum \tau_C = 0$ equation still gives $T_B(1.05 \text{ m} - x) - T_A x = 0$.

But now $T_B = 1.333T_A$ so $(1.333T_A)(1.05 \text{ m} - x) - T_A x = 0$

$1.40 \text{ m} = 2.33x$ and $x = 1.40 \text{ m}/2.33 = 0.60$ m (measured from A).

EVALUATE: Wire B has twice the diameter so it takes twice the tension to produce the same stress. For equal stress the moment arm for T_B (0.35 m) is half that for T_A (0.70 m), since the torques must be equal. The smaller Y for B partially compensates for the larger area in determining the strain and for equal strain the moment arms are closer to being equal.

11.87   IDENTIFY and **SET UP:** The tension is the same at all points along the composite rod. Apply Eqs.(11.8) and (11.10) to relate the elongations, stresses, and strains for each rod in the compound.

EXECUTE: Each piece of the composite rod is subjected to a tensile force of 4.00×10^4 N.

a) $Y = \dfrac{F_\perp l_0}{A \Delta l}$ so $\Delta l = \dfrac{F_\perp l_0}{YA}$

$\Delta l_b = \Delta l_n$ gives that $\dfrac{F_\perp l_{0,b}}{Y_b A_b} = \dfrac{F_\perp l_{0,n}}{Y_n A_n}$ (b for brass and n for nickel); $l_{0,n} = L$

But the $F_\perp$ is the same for both, so

$l_{0,n} = \dfrac{Y_n}{Y_b} \dfrac{A_n}{A_b} l_{0,b}$

$L = \left(\dfrac{21 \times 10^{10} \text{ Pa}}{9.0 \times 10^{10} \text{ Pa}} \right) \left(\dfrac{1.00 \text{ cm}^2}{2.00 \text{ cm}^2} \right) (1.40 \text{ m}) = 1.63 \text{ m}$

b) stress $= F_\perp/A = T/A$

brass: stress $= T/A = (4.00 \times 10^4 \text{ N})/(2.00 \times 10^{-4} \text{ m}^2) = 2.00 \times 10^8 \text{ Pa}$

nickel: stress $= T/A = (4.00 \times 10^4 \text{ N})/(1.00 \times 10^{-4} \text{ m}^2) = 4.00 \times 10^8 \text{ Pa}$

c) $Y =$ stress/strain and strain $=$ stress/Y

brass: strain $= (2.00 \times 10^8 \text{ Pa})/(9.0 \times 10^{10} \text{ Pa}) = 2.22 \times 10^{-3}$

nickel: strain $= (4.00 \times 10^8 \text{ Pa})/(21 \times 10^{10} \text{ Pa}) = 1.90 \times 10^{-3}$

EVALUATE: Larger Y means less Δl and smaller A means greater Δl, so the two effects largely cancel and the lengths don't differ greatly. Equal Δl and nearly equal l means the strains are nearly the same. But equal tensions and A differing by a factor of 2 means the stresses differ by a factor of 2.

11.89 IDENTIFY and **SET UP:** $Y = F_\perp l_0/A\Delta l$ (Eq.11.10 holds since the problem states that the stress is proportional to the strain.) Thus $\Delta l = F_\perp l_0/AY$. Use proportionality to see how changing the wire properties affects Δl.

EXECUTE:

a) Change l_0 but $F_\perp$ (same floodlamp), A (same diameter wire), and Y (same material) all stay the same.

$\dfrac{\Delta l}{l_0} = \dfrac{F_\perp}{AY} =$ constant, so $\dfrac{\Delta l_1}{l_{01}} = \dfrac{\Delta l_2}{l_{02}}$

$\Delta l_2 = \Delta l_1(l_{02}/l_{01}) = 2\Delta l_1 = 2(0.18 \text{ mm}) = 0.36 \text{ mm}$

b) $A = \pi(d/2)^2 = \frac{1}{4}\pi d^2$, so $\Delta l = \dfrac{F_\perp l_0}{\frac{1}{4}\pi d^2 Y}$

$F_\perp$, l_0, Y all stay the same, so $\Delta l(d^2) = F_\perp l_0/(\frac{1}{4}\pi Y) =$ constant
$\Delta l_1(d_1^2) = \Delta l_2(d_2^2)$
$\Delta l_2 = \Delta l_1(d_1/d_2)^2 = (0.18 \text{ mm})(1/2)^2 = 0.045 \text{ mm}$

c) $F_\perp$, l_0, A all stay the same so $\Delta l\, Y = F_\perp l_0/A = \text{constant}$

$\Delta l_1\, Y_1 = \Delta l_2\, Y_2$

$\Delta l_2 = \Delta l_1 (Y_1/Y_2) = (0.18 \text{ mm})(20 \times 10^{10} \text{ Pa}/11 \times 10^{10} \text{ Pa}) = 0.33 \text{ mm}$

EVALUATE: Greater l means greater Δl, greater diameter means less Δl, and smaller Y means greater Δl.

11.91 **IDENTIFY** and **SET UP:** Apply Eqs.(11.8) and (11.15).

The tensile stress depends on the component of $\vec{F}$ perpendicular to the plane and the shear stress depends on the component of $\vec{F}$ parallel to the plane.

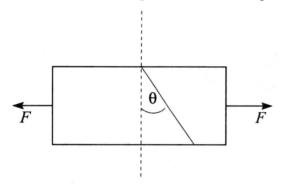

a) EXECUTE:

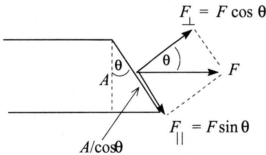

The area of the diagonal face is $A/\cos\theta$.

$$\text{tensile stress} = \frac{F_\perp}{(A/\cos\theta)} = F\cos\theta/(A/\cos\theta) = \frac{F\cos^2\theta}{A}.$$

b) $\text{shear stress} = \dfrac{F_\parallel}{(A/\cos\theta)} = F\sin\theta/(A/\cos\theta) = \dfrac{F\sin\theta\cos\theta}{A} = \dfrac{F\sin 2\theta}{2A}$ (using a trig identity).

EVALUATE:

c) From the result of part (a) the tensile stress is a maximum for $\cos\theta = 1$, so $\theta = 0°$.

d) From the result of part (b) the shear stress is a maximum for $\sin 2\theta = 1$, so for $2\theta = 90°$ and thus $\theta = 45°$

CHAPTER 12
GRAVITATION

Exercises 1, 3, 5, 13, 17, 21, 23, 27, 29, 37, 39, 41, 45
Problems 47, 53, 57, 59, 61, 63, 65, 71, 73, 75, 77, 79, 81, 83

Exercises

12.1 **IDENTIFY** and **SET UP:** Use the law of gravitation, Eq.(12.1), to determine F_g.
EXECUTE:

$$F_{\text{S on m}} = G\frac{m_S m_M}{r_{\text{SM}}^2} \ (\text{S} = \text{sun, M} = \text{moon}); \quad F_{\text{E on m}} = G\frac{m_E m_M}{r_{\text{EM}}^2} \ (\text{E} = \text{earth})$$

$$\frac{F_{\text{S on M}}}{F_{\text{E on M}}} = \left(G\frac{m_S m_M}{r_{\text{SM}}^2}\right)\left(\frac{r_{\text{EM}}^2}{Gm_E m_M}\right) = \frac{m_S}{m_E}\left(\frac{r_{\text{EM}}}{r_{\text{SM}}}\right)^2$$

r_{EM}, the radius of the moon's orbit around the earth is given in Appendix F as 3.84×10^8 m. The moon is much closer to the earth than it is to the sun, so take the distance r_{SM} of the moon from the sun to be r_{SE}, the radius of the earth's orbit around the sun.

$$\frac{F_{\text{S on M}}}{F_{\text{E on M}}} = \left(\frac{1.99 \times 10^{30} \text{ kg}}{5.98 \times 10^{24} \text{ kg}}\right)\left(\frac{3.84 \times 10^8 \text{ m}}{1.50 \times 10^{11} \text{ m}}\right)^2 = 2.18.$$

EVALUATE: The force exerted by the sun is larger than the force exerted by the earth. The moon's motion is a combination of orbiting the sun and orbiting the earth.

12.3 **IDENTIFY** and **SET UP:** Use Eq.(12.1) to calculate F_g.
EXECUTE:

$F_{12} = G\dfrac{m_1 m_2}{r_{12}^2}$ is the gravitational force between the first set of spheres.

For the second set of spheres $F'_{12} = G\dfrac{(nm_1)(nm_2)}{(nr_{12})^2} = \dfrac{n^2}{n^2}G\dfrac{m_1 m_2}{r_{12}^2} = G\dfrac{m_1 m_2}{r_{12}^2}.$

The force is the same as for the first set of spheres.

EVALUATE: The larger masses increase the force by n^2 and the larger separation decreases the force by $1/n^2$.

12.5 **IDENTIFY:** Use Eq.(12.1) to calculate F_g exerted by the earth and by the sun and add these forces as vectors.

a) SET UP:

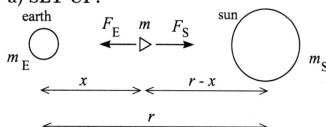

Let $\vec{F}_E$ and $\vec{F}_S$ be the gravitational forces exerted on the spaceship by the earth and by the sun.

EXECUTE: The distance from the earth to the sun is $r = 1.50 \times 10^{11}$ m. Let the ship be a distance x from the earth; it is then a distance $r - x$ from the sun.

$F_E = F_S$ says that $Gmm_E/x^2 = Gmm_S/(r-x)^2$

$m_E/x^2 = m_S/(r-x)^2$ and $(r-x)^2 = x^2(m_S/m_E)$

$r - x = x\sqrt{m_S/m_E}$ and $r = x(1 + \sqrt{m_S/m_E})$

$$x = \frac{r}{1 + \sqrt{m_S/m_E}} = \frac{1.50 \times 10^{11} \text{ m}}{1 + \sqrt{1.99 \times 10^{30} \text{ kg}/5.97 \times 10^{24} \text{ kg}}} = 2.59 \times 10^8 \text{ m (from}$$
center of earth)

b) EVALUATE: At the instant when the spaceship passes through this point its acceleration is zero. Since $m_S \gg m_E$ this equal-force point is much closer to the earth than to the sun.

12.13 IDENTIFY: Use Eq.(12.1) to find the force exerted by each large sphere. Add these forces as vectors to get the net force and then use Newton's 2nd law to calculate the acceleration.

SET UP:

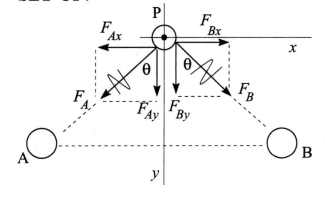

$\sin\theta = 0.80$
$\cos\theta = 0.60$

Take the origin of coordinates at point P.

EXECUTE:

$$F_A = G\frac{m_A m}{r^2} = (6.673 \times 10^{-11} \text{ N} \cdot \text{m}^2/\text{kg}^2)\frac{(0.26 \text{ kg})(0.010 \text{ kg})}{(0.100 \text{ m})^2} = 1.735 \times 10^{-11}$$

N $F_B = G\dfrac{m_B m}{r^2} = 1.735 \times 10^{-11}$ N

$F_{Ax} = -F_A\sin\theta = -(1.735 \times 10^{-11} \text{ N})(0.80) = -1.39 \times 10^{-11}$ N

$F_{Ay} = -F_A\cos\theta = +(1.735 \times 10^{-11} \text{ N})(0.60) = +1.04 \times 10^{-11}$ N

$F_{Bx} = +F_B \sin\theta = +1.39 \times 10^{-11}$ N

$F_{By} = +F_B \sin\theta = +1.04 \times 10^{-11}$ N

$\sum F_x = ma_x$ gives $F_{Ax} + F_{Bx} = ma_x$

$0 = ma_x$ so $a_x = 0$

$\sum F_y = ma_y$ gives $F_{Ay} + F_{By} = ma_y$

$2(1.04 \times 10^{-11}$ N$) = (0.010$ kg$)a_y$

$a_y = 2.1 \times 10^{-9}$ m/s^2, directed downward midway between A and B

EVALUATE: For ordinary size objects the gravitational force is very small, so the initial acceleration is very small. By symmetry there is no x-component of net force and the y-component is in the direction of the two large spheres, since they attract the small sphere.

12.17   **a) IDENTIFY** and **SET UP:** Apply Eq.(12.4) to the earth and to Titania.

The acceleration due to gravity at the surface of Titania is given by $g_T = Gm_T/R_T^2$, where m_T is its mass and R_T is its radius.

For the earth, $g_E = Gm_E/R_E^2$.

EXECUTE: For Titania, $m_T = m_E/1700$ and $R_T = R_E/8$, so

$$g_T = \frac{Gm_T}{R_T^2} = \frac{G(m_E/1700)}{(R_E/8)^2} = \left(\frac{64}{1700}\right)\frac{Gm_E}{R_E^2} = 0.0377g_E.$$

Since $g_E = 9.80$ m/s^2, $g_T = (0.0377)(9.80$ m/s$^2) = 0.37$ m/s^2.

EVALUATE: g on Titania is much smaller than on earth. The smaller mass reduces g and is a greater effect than the smaller radius, which increases g.

b) IDENTIFY and **SET UP:** Use density = mass/volume. Assume Titania is a sphere.

EXECUTE: From Section 12.2 we know that the average density of the earth is 5500 kg/m^3. For Titania,

$$\rho_T = \frac{m_T}{\frac{4}{3}\pi R_T^3} = \frac{m_E/1700}{\frac{4}{3}\pi(R_E/8)^3} = \frac{512}{1700}\rho_E = \frac{512}{1700}(5500 \text{ kg/m}^3) = 1700 \text{ kg/m}^3$$

EVALUATE: The average density of Titania is about a factor of 3 smaller than for earth. We can write Eq.(12.4) for Titania as $g_T = \frac{4}{3}\pi GR_T\rho_T$. $g_T < g_E$ both because $\rho_T < \rho_E$ and $R_T < R_E$.

12.21   **IDENTIFY** and **SET UP:** Use the measured gravitational force to calculate the gravitational constant G, using Eq.(12.1). Then use Eq.12.4 to calculate the mass of the earth:

EXECUTE: $F_g = G\dfrac{m_1 m_2}{r^2}$ so

$$G = \frac{F_g r^2}{m_1 m_2} = \frac{(8.00 \times 10^{-10}\ \mathrm{N})(0.0100\ \mathrm{m})^2}{(0.400\ \mathrm{kg})(3.00 \times 10^{-3}\ \mathrm{kg})} = 6.667 \times 10^{-11}\ \mathrm{N \cdot m^2/kg^2}.$$

$$g = \frac{Gm_E}{R_E^2}\ \text{gives}\ m_E = \frac{R_E^2 g}{G} = \frac{(6.38 \times 10^6\ \mathrm{m})^2(9.80\ \mathrm{m/s^2})}{6.667 \times 10^{-11}\ \mathrm{N \cdot m^2/kg^2}} = 5.98 \times 10^{24}\ \mathrm{kg}.$$

EVALUATE: Our result agrees with the value given in Appendix F.

12.23 IDENTIFY and **SET UP:** Example 12.5 gives the escape speed as $v_1 = \sqrt{2GM/R}$, where M and R are the mass and radius of the astronomical object.
EXECUTE:
$$v_1 = \sqrt{2(6.673 \times 10^{-11}\ \mathrm{N \cdot m^2/kg^2})(3.6 \times 10^{12}\ \mathrm{kg})/700\ \mathrm{m}} = 0.83\ \mathrm{m/s}.$$

EVALUATE: At this speed a person can walk 100 m in 120 s; easily achieved for the average person. We can write the escape speed as $v_1 = \sqrt{\frac{4}{3}\pi \rho G R^2}$, where ρ is the average density of Dactyl. Its radius is much smaller than earth's and its density is about the same, so the escape speed is much less on Dactyl than on earth.

12.27 IDENTIFY and **SET UP:** Apply Newton's 2nd law to the satellite to get an equation for r in terms of v. Then use Eq.(12.13) for the period and $a_{\mathrm{rad}} = v^2/r$ for the radial acceleration.
EXECUTE:
a) Find the orbit radius r: $\frac{Gm m_E}{r^2} = m\frac{v^2}{r}$.
$$r = Gm_E/v^2 = (6.673 \times 10^{-11}\ \mathrm{N \cdot m^2/kg^2})(5.97 \times 10^{24}\ \mathrm{kg})/(6200\ \mathrm{m/s})^2$$
$$= 1.036 \times 10^7\ \mathrm{m}$$
The period (time for one revolution) is then given by
$$T = 2\pi r/v = 2\pi(1.036 \times 10^7\ \mathrm{m})/(6200\ \mathrm{m/s}) = 1.05 \times 10^4\ \mathrm{s} = 175\ \mathrm{min}$$

b) $a_{\mathrm{rad}} = v^2/r = (6200\ \mathrm{m/s})^2/1.036 \times 10^7\ \mathrm{m} = 3.71\ \mathrm{m/s^2}.$
EVALUATE: The radial acceleration of the satellite is due to the gravity force. Gravity becomes weaker with increasing distance above the surface of the earth, so a_{rad} is less than g at the surface of the earth.

12.29 IDENTIFY: Apply Newton's 2nd law to the motion of the satellite and obtain an equation that relates the orbital speed v to the orbital radius r.
SET UP:

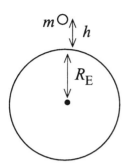

The radius of the orbit is $r = h + R_E$.

$r = 7.80 \times 10^5$ m $+ 6.38 \times 10^6$ m $= 7.16 \times 10^6$ m.

Free-body diagram for the satellite:

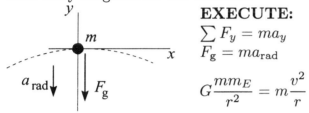

EXECUTE:

$\sum F_y = ma_y$

$F_g = ma_{rad}$

$G\dfrac{mm_E}{r^2} = m\dfrac{v^2}{r}$

$$v = \sqrt{\frac{Gm_E}{r}} = \sqrt{\frac{(6.673 \times 10^{-11} \text{ N} \cdot \text{m}^2/\text{kg}^2)(5.97 \times 10^{24} \text{ kg})}{7.16 \times 10^6 \text{ m}}} = 7.46 \times 10^3 \text{ m/s}$$

EVALUATE: Note that $r = h + R_E$ is the radius of the orbit, measured from the center of the earth. For this satellite r is greater than for the satellite in Example 12.6, so its orbital speed is less.

12.37 a) IDENTIFY: If the orbit is circular, Newton's 2nd law requires a particular relation between its orbit radius and orbital speed.

SET UP: The gravitational force exerted on the spacecraft by the sun is $F_g = Gm_S m_H/r^2$, where m_S is the mass of the sun and m_H is the mass of the Helios B spacecraft.

For a circular orbit, $a_{rad} = v^2/r$ and $\sum F = m_H v^2/r$. If we neglect all forces on the spacecraft except for the force exerted by the sun,

$F_g = \sum F = m_H v^2/r$, so $Gm_S m_H/r^2 = m_H v^2/r$

EXECUTE:

$v = \sqrt{Gm_S/r} = \sqrt{(6.673 \times 10^{-11} \text{ N} \cdot \text{m}^2/\text{kg}^2)(1.99 \times 10^{30} \text{ kg})/43 \times 10^9 \text{ m}}$

$= 5.6 \times 10^4$ m/s $= 56$ km/s

EVALUATE: The actual speed is 71 km/s, so the orbit cannot be circular.

b) IDENTIFY and SET UP: The orbit is a circle or an ellispe if it is closed, a parabola or hyperbola if open. The orbit is closed if the total energy (kinetic + potential) is negative, so that the object cannot reach $r \to \infty$.

EXECUTE: For Helios B,

$K = \frac{1}{2}m_H v^2 = \frac{1}{2}m_H(71 \times 10^3 \text{ m/s})^2 = (2.52 \times 10^9 \text{ m}^2/\text{s}^2)m_H$

$U = -Gm_\text{S}m_\text{H}/r = m_\text{H}(-(6.673 \times 10^{-11}\ \text{N} \cdot \text{m}^2/\text{kg}^2)(1.99 \times 10^{30}\ \text{kg})/(43 \times 10^9\ \text{m})) = -(3.09 \times 10^9\ \text{m}^2/\text{s}^2)m_\text{H}$

$E = K + U = (2.52 \times 10^9\ \text{m}^2/\text{s}^2)m_\text{H} - (3.09 \times 10^9\ \text{m}^2/\text{s}^2)m_\text{H}$

$= -(5.7 \times 10^8\ \text{m}^2/\text{s}^2)m_\text{H}$

EVALUATE: The total energy E is negative, so the orbit is closed. We know from part (a) that it is not circular, so it must be elliptical.

12.39 IDENTIFY: Find the potential due to a small segment of the ring and integrate over the entire ring to find the total U.

a) SET UP:

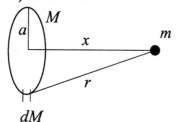

Divide the ring up into small segments dM.

EXECUTE:

The gravitational potential energy of dM and m is $dU = -Gm\,dM/r$.

The total gravitational potential energy of the ring and particle is $U = \int dU = -Gm \int dM/r$.

But $r = \sqrt{x^2 + a^2}$ is the same for all segments of the ring, so

$$U = -\frac{Gm}{r} \int dM = -\frac{GmM}{r} = -\frac{GmM}{\sqrt{x^2 + a^2}}$$

b) EVALUATE: When $x \gg a$, $\sqrt{x^2 + a^2} \to \sqrt{x^2} = x$ and $U = -GmM/x$. This is the gravitational potential energy of two point masses separated by a distance x. This is the expected result.

c) IDENTIFY and **SET UP:** Use $F_x = -dU/dx$ with $U(x)$ from part (a) to calculate F_x.

EXECUTE: $F_x = -\dfrac{dU}{dx} = -\dfrac{d}{dx}\left(-\dfrac{GmM}{\sqrt{x^2 + a^2}}\right)$

$F_x = +GmM \dfrac{d}{dx}(x^2 + a^2)^{-1/2} = GmM(-\tfrac{1}{2}(2x)(x^2 + a^2)^{-3/2})$

$F_x = -GmMx/(x^2 + a^2)^{3/2}$; the minus sign means the force is attractive.

EVALUATE:

d) For $x \gg a$, $(x^2 + a^2)^{3/2} \to (x^2)^{3/2} = x^3$

Then $F_x = -GmMx/x^3 = -GmM/x^2$. This is the force between two point masses separated by a distance x and is the expected result.

e) For $x = 0$, $U = -GMm/a$. Each small segment of the ring is the same distance from the center and the potential is the same as that due to a point charge of mass M located at a distance a.

For $x = 0$, $F_x = 0$. When the particle is at the center of the ring, symmetrically placed segments of the ring exert equal and opposite forces and the total force exerted by the ring is zero.

12.41 **IDENTIFY** and **SET UP:** At the north pole, $F_g = w_0 = mg_0$, where g_0 is given by Eq.(12.4) applied to Neptune. At the equator, the apparent weight is given by Eq.(12.29). The orbital speed v is obtained from the rotational period using Eq.(12.13).

EXECUTE:

a) $g_0 = Gm/R^2 = (6.673 \times 10^{-11} \text{ N} \cdot \text{m}^2/\text{kg}^2)(1.0 \times 10^{26} \text{ kg})/(2.5 \times 10^7 \text{ m})^2$ $= 10.7 \text{ m/s}^2$. This agrees with the value of g given in the problem.

$F = w_0 = mg_0 = (5.0 \text{ kg})(10.7 \text{ m/s}^2) = 53 \text{ N}$; this is the true weight of the object.

b) From Eq.(12.29), $w = w_0 - mv^2/R$

$$T = \frac{2\pi r}{v} \text{ gives } v = \frac{2\pi r}{T} = \frac{2\pi(2.5 \times 10^7 \text{ m})}{(16 \text{ h})(3600 \text{ s}/1 \text{ h})} = 2.727 \times 10^3 \text{ m/s}$$

$v^2/R = (2.727 \times 10^3 \text{ s})^2/2.5 \times 10^7 \text{ m} = 0.297 \text{ m/s}^2$

Then $w = 53 \text{ N} - (5.0 \text{ kg})(0.297 \text{ m/s}^2) = 52 \text{ N}$.

EVALUATE: The apparent weight is less than the true weight. This effect is larger on Neptune than on earth.

12.45 **IDENTIFY** and **SET UP:** A black hole with the earth's mass M has the Schwarzschild radius R_S given by Eq.(12-32).

EXECUTE: $R_S = 2Gm/c^2 = 2(6.673 \times 10^{-11} \text{ N} \cdot \text{m}^2/\text{kg}^2)(5.97 \times 10^{24} \text{ kg})/(2.998 \times 10^8 \text{ m/s})^2 = 8.865 \times 10^{-3} \text{ m}$

The ratio of R_S to the current radius R is $R_S/R = 8.865 \times 10^{-3} \text{ m}/6.38 \times 10^6 \text{ m} = 1.39 \times 10^{-9}$.

EVALUATE: A black hole with the earth's radius is very small.

Problems

12.47 **IDENTIFY:** Use Eq.(12.1) to calculate each gravitational force and add the forces as vectors.

a) **SET UP:**

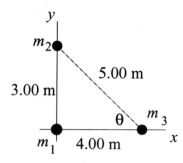

Section 12.6 proves that any two spherically symmetric masses interact as though they were point masses with all the mass concentrated at their centers.

Force diagram for m_3:

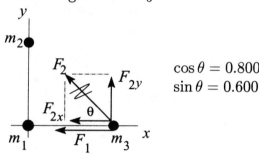

$\cos \theta = 0.800$
$\sin \theta = 0.600$

EXECUTE:

$$F_1 = G \frac{m_1 m_3}{r_{13}^2} = \frac{(6.673 \times 10^{-11} \ \text{N} \cdot \text{m}^2/\text{kg}^2)(60.0 \ \text{kg})(0.500 \ \text{kg})}{(4.00 \ \text{m})^2}$$

$$= 1.251 \times 10^{-10} \ \text{N}$$

$$F_2 = G \frac{m_2 m_3}{r_{23}^2} = \frac{(6.673 \times 10^{-11} \ \text{N} \cdot \text{m}^2/\text{kg}^2)(80.0 \ \text{kg})(0.500 \ \text{kg})}{(5.00 \ \text{m})^2}$$

$$= 1.068 \times 10^{-10} \ \text{N}$$

$F_{1x} = -1.251 \times 10^{-11} \ \text{N}, \quad F_{1y} = 0$

$F_{2x} = -F_2 \cos \theta = -(1.068 \times 10^{-10} \ \text{N})(0.800) = -8.544 \times 10^{-11} \ \text{N}$

$F_{2y} = +F_2 \sin \theta = +(1.068 \times 10^{-10} \ \text{N})(0.600) = +6.408 \times 10^{-11} \ \text{N}$

$F_x = F_{1x} + F_{2x} = -1.251 \times 10^{-10} \ \text{N} - 8.544 \times 10^{-11} \ \text{N} = -2.105 \times 10^{-10} \ \text{N}$

$F_y = F_{1y} + F_{2y} = 0 + 6.408 \times 10^{-11} \ \text{N} = +6.408 \times 10^{-11} \ \text{N}$

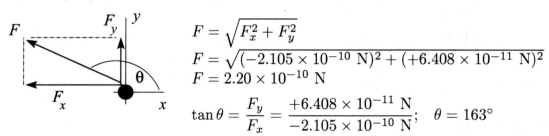

$$F = \sqrt{F_x^2 + F_y^2}$$

$$F = \sqrt{(-2.105 \times 10^{-10} \ \text{N})^2 + (+6.408 \times 10^{-11} \ \text{N})^2}$$

$$F = 2.20 \times 10^{-10} \ \text{N}$$

$$\tan \theta = \frac{F_y}{F_x} = \frac{+6.408 \times 10^{-11} \ \text{N}}{-2.105 \times 10^{-10} \ \text{N}}; \quad \theta = 163°$$

EVALUATE: Both spheres attract the third sphere and the net force is in the second quadrant.

b) SET UP: For the net force to be zero the forces from the two spheres must be equal in magnitude and opposite in direction. For the forces on it to be opposite in direction the third sphere must be on the y-axis and between the other two spheres.

EXECUTE:
$F_\text{net} = 0$ if $F_1 = F_2$

$$G\frac{m_1 m_3}{y^2} = G\frac{m_2 m_3}{(3.00 \text{ m} - y)^2}$$

$$\frac{60.0}{y^2} = \frac{80.0}{(3.00 \text{ m} - y)^2}$$

$\sqrt{80.0}\,y = \sqrt{60.0}(3.00 \text{ m} - y)$

$(\sqrt{80.0} + \sqrt{60.0})y = (3.00 \text{ m})\sqrt{60.0}$ and $y = 1.39$ m

Thus the sphere would have to be placed at the point $x = 0$, $y = 1.39$ m

EVALUATE: For the forces to have the same magnitude the third sphere must be closer to the sphere that has smaller mass.

12.53 IDENTIFY and **SET UP:**

a) To stay above the same point on the surface of the earth the orbital period of the satellite must equal the orbital period of the earth:

$T = 1 \text{ d}(24 \text{ h}/ 1 \text{ d})(3600 \text{ s}/ 1 \text{ h}) = 8.64 \times 10^4$ s

Eq.(12.14) gives the relation between the orbit radius and the period:

EXECUTE: $T = \dfrac{2\pi r^{3/2}}{\sqrt{Gm_\text{E}}}$ and $T^2 = \dfrac{4\pi^2 r^3}{Gm_\text{E}}$

$r = \left(\dfrac{T^2 Gm_\text{E}}{4\pi^2}\right)^{1/3} =$

$\left(\dfrac{(8.64 \times 10^4 \text{ s})^2(6.673 \times 10^{-11} \text{ N}\cdot\text{m}^2/\text{kg}^2)(5.97 \times 10^{24} \text{ kg})}{4\pi^2}\right)^{1/3} = 4.23 \times 10^7$ m

This is the radius of the orbit; it is related to the height h above the earth's surface and the radius R_E of the earth by $r = h + R_\text{E}$. Thus $h = r - R_\text{E} = 4.23 \times 10^7 \text{ m} - 6.38 \times 10^6 \text{ m} = 3.59 \times 10^7$ m.

EVALUATE: The orbital speed of the geosynchronous satellite is $2\pi r/T = 3080$ m/s. The altitude is much larger and the speed is much less than for the satellite in Example 12.6.

b)

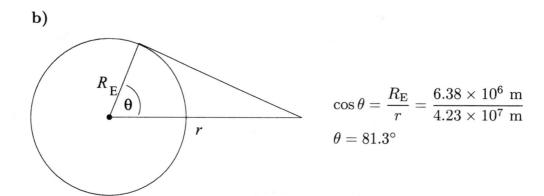

$$\cos\theta = \frac{R_E}{r} = \frac{6.38 \times 10^6 \text{ m}}{4.23 \times 10^7 \text{ m}}$$

$$\theta = 81.3°$$

A line from the satellite is tangent to a point on the earth that is at an angle of 81.3° above the equator. The sketch shows that points at higher latitudes are blocked by the earth from viewing the satellite.

12.57 IDENTIFY and **SET UP:** The observed period allows you to calculate the angular velocity of the satellite relative to you. You know your angular velocity as you rotate with the earth, so you can find the angular velocity of the staellite in a space-fixed reference frame. $v = r\omega$ gives the orbital speed of the satellite and Newton's second law relates this to the orbit radius of the satellite.

EXECUTE:

a) The satellite is revolving west to east, in the same direction the earth is rotating. If the angular speed of the satellite is ω_s and the angular speed of the earth is ω_E, the angular speed ω_{rel} of the satellite relative to you is $\omega_{rel} = \omega_s - \omega_E$.

$\omega_{rel} = (1 \text{ rev})/(12 \text{ h}) = \left(\frac{1}{12}\right) \text{ rev/h}$

$\omega_E = \left(\frac{1}{24}\right) \text{ rev/h}$

$\omega_s = \omega_{rel} + \omega_E = \left(\frac{1}{8}\right) \text{ rev/h} = 2.18 \times 10^{-4} \text{ rad/s}$

$\sum \vec{F} = m\vec{a}$ says $G\dfrac{mm_E}{r^2} = m\dfrac{v^2}{r}$

$v^2 = \dfrac{Gm_E}{r}$ and with $v = r\omega$ this gives $r^3 = \dfrac{Gm_E}{\omega^2}$; $r = 2.03 \times 10^7 \text{ m}$

This is the radius of the satellite's orbit. Its height h above the surface of the earth is $h = r - R_E = 1.39 \times 10^7 \text{ m}$.

EVALUATE: In part (a) the staellite is revolving faster than the earth's rotation and in part (b) it is revolving slower. Slower v and ω means larger orbit radius r.

b) Now the satellite is revolving opposite to the rotation of the earth.

If west to east is positive, then $\omega_{rel} = \left(-\frac{1}{12}\right) \text{ rev/h}$

$\omega_s = \omega_{rel} + \omega_E = \left(-\frac{1}{24}\right) \text{ rev/h} = -7.27 \times 10^{-5} \text{ rad/s}$

$r^3 = \dfrac{Gm_E}{\omega^2}$ gives $r = 4.22 \times 10^7 \text{ m}$ and $h = 3.59 \times 10^7 \text{ m}$

12.59 **IDENTIFY** and **SET UP:** Compare ΔU calculated from Eqs.(7.2) and (12.9). In Eq.(12.9) use the fact that $h << R_E$ to derive an approximate expression.

EXECUTE:

$\underline{U = mgy}$ (Eq.(7.2))

Let $y = 0$ at the earth's surface. Then $U_1 = mgy_1 = 0$ and $U_2 = mgy_2 = mgh$.

$(\Delta U)_{approx} = U_2 - U_1 = mgh$

$\underline{U = -Gm_E m/r}$ (Eq.12.9)

This equation has built into it that $U \to 0$ as $r \to \infty$.

$U_1 = -Gm_E m/R_E$, $U_2 = -Gm_E m/(R_E + h)$

$$\Delta U = -(Gm_E m)\left(\frac{1}{R_E + h} - \frac{1}{R_E}\right) = -\frac{Gm_E m}{R_E}\left(\frac{1}{1 + h/R_E} - 1\right)$$

For only a 1% error in $\Delta U = mgh$ the value of h must be small compared to R_E, so use the binomial theorem to expand $(1 + h/R_E)^{-1}$ in powers of h/R_E:

$(1 + h/R_E)^{-1} = 1 - h/R_E + h^2/R_E^2 - \cdots$

$\Delta U = -(Gm_E m/R_E)(1 - h/R_E + h^2/R_E^2 - \cdots - 1) = (Gm_E m/R_E^2)h(1 - h/R_E)$

$g = Gm_E/R_E^2$, so $\Delta U = mgh(1 - h/R_E)$

Then $\Delta U - (\Delta U)_{approx} = mgh - mgh + mgh(h/R_E) = mgh(h/R_E)$

Want h such that $\dfrac{\Delta U - (\Delta U)_{approx}}{(\Delta U)_{approx}} = 0.010$, so $0.010 = \dfrac{mgh(h/R_E)}{mgh} = h/R_E$

$h = 0.010R_E = 6.4 \times 10^4$ m $= 64$ km

EVALUATE: Eq.(7.2) ignores variation of F_g with altitude. It is very accurate for motion of objects near the surface of the earth.

12.61 **IDENTIFY** and **SET UP:** Use Eq.(12.2) to calculate the gravity force at each location. For the top of Mount Everest write $r = h + R_E$ and use the fact that $h << R_E$ to obtain an expression for the difference in the two forces.

EXECUTE:

At Sacramento, the gravity force on you is $F_1 = G\dfrac{mm_E}{R_E^2}$.

At the top of Mount Everest, a height of $h = 8800$ m above sea level, the gravity force on you is

$F_2 = G\dfrac{mm_E}{(R_E + h)^2} = G\dfrac{mm_E}{R_E^2(1 + h/R_E)^2}$

$(1 + h/R_E)^{-2} \approx 1 - \dfrac{2h}{R_E}, \qquad F_2 = F_1\left(1 - \dfrac{2h}{R_E}\right)$

$$\frac{F_1 - F_2}{F_1} = \frac{2h}{r_E} = 0.28\%$$

EVALUATE: The change in the gravitational force is very small, so for objects near the surface of the earth it is a good approximation to treat it as constant.

12.63 **IDENTIFY** and **SET UP:** First use the radius of the orbit to find the initial orbital speed, from Eq.(12.12) applied to the moon.

EXECUTE:

$v = \sqrt{Gm/r}$ and $r = R_M + h = 1.74 \times 10^6$ m $+ 50.0 \times 10^3$ m $= 1.79 \times 10^6$ m

Thus $v = \sqrt{\dfrac{(6.673 \times 10^{-11} \text{ N} \cdot \text{m}^2/\text{kg}^2)(7.35 \times 10^{22} \text{ kg})}{1.79 \times 10^6 \text{ m}}} = 1.655 \times 10^3$ m/s

After the speed decreases by 20.0 m/s it becomes 1.655×10^3 m/s $- 20.0$ m/s $= 1.635 \times 10^3$ m/s.

IDENTIFY and **SET UP:** Use conservation of energy to find the speed when the spacecraft reaches the lunar surface.

$K_1 + U_1 + W_{\text{other}} = K_2 + U_2$

Gravity is the only force that does work so $W_{\text{other}} = 0$ and $K_2 = K_1 + U_1 - U_2$

EXECUTE: $U_1 = -Gm_m m/r$; $U_2 = -Gm_m m/R_m$

$\frac{1}{2}mv_2^2 = \frac{1}{2}mv_1^2 + Gmm_m(1/R_m - 1/r)$

And the mass m divides out to give $v_2 = \sqrt{v_1^2 + 2Gm_m(1/R_m - 1/r)}$

$v_2 = 1.682 \times 10^3$ m/s(1 km/1000 m)(3600 s/1 h) $= 6060$ km/h

EVALUATE: After the thruster fires the spacecraft is moving too slowly to be in a stable orbit; the gravitational force is larger than what is needed to maintain a circular orbit. The spacecraft gains energy as it is accelerated toward the surface.

12.65 **IDENTIFY** and **SET UP:** Apply conservation of energy. Must use Eq.(12.9) for the gravitational potential energy since h is not small compared to R_E.

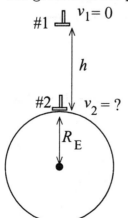

Take point 1 to be where the hammer is released and point 2 to be just above the surface of the earth, so
$r_1 = R_E + h$ and $r_2 = R_E$.

EXECUTE: $K_1 + U_1 + W_{\text{other}} = K_2 + U_2$

Only gravity does work, so $W_{\text{other}} = 0$.

$K_1 = 0$, $K_2 = \frac{1}{2}mv_2^2$

$$U_1 = -G\frac{mm_{\text{E}}}{r_1} = -\frac{Gmm_{\text{E}}}{h + R_{\text{E}}}, \quad U_2 = -G\frac{mm_{\text{E}}}{r_2} = -\frac{Gmm_{\text{E}}}{R_{\text{E}}}$$

Thus $-G\dfrac{mm_{\text{E}}}{h + R_{\text{E}}} = \dfrac{1}{2}mv_2^2 - G\dfrac{mm_{\text{E}}}{R_{\text{E}}}$

$$v_2^2 = 2Gm_{\text{E}}\left(\frac{1}{R_{\text{E}}} - \frac{1}{R_{\text{E}} + h}\right) = \frac{2Gm_{\text{E}}}{R_{\text{E}}(R_{\text{E}} + h)}(R_{\text{E}} + h - R_{\text{E}}) = \frac{2Gm_{\text{E}}h}{R_{\text{E}}(R_{\text{E}} + h)}$$

$$v_2 = \sqrt{\frac{2Gm_{\text{E}}h}{R_{\text{E}}(R_{\text{E}} + h)}}$$

EVALUATE: If $h \to \infty$, $v_2 \to \sqrt{2Gm_{\text{E}}/R_{\text{E}}}$, which equals the escape speed. In this limit this event is the reverse of an object being projected upward from the surface with the escape speed. If $h << R_{\text{E}}$, then $v_2 = \sqrt{2Gm_{\text{E}}/R_{\text{E}}^2} = \sqrt{2gh}$, the same result if used Eq.(7.2) for U.

12.71 IDENTIFY: Use eq.(12.2) to calculate F_g. Apply Newton's 2nd law to the circular motion of each star to find the orbital speed and period. Apply the conservation of energy expression, Eq.(7.13), to calculate the energy input (work) required to separate the two stars to infinity.

a) SET UP: The cm is midway beween the two stars since they have equal masses. Let R be the orbit radius for each star.

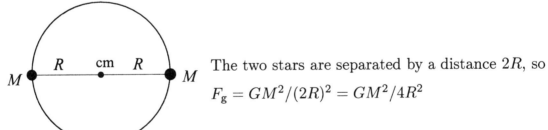

The two stars are separated by a distance $2R$, so

$$F_g = GM^2/(2R)^2 = GM^2/4R^2$$

b) EXECUTE: $F_g = ma_{\text{rad}}$

$GM^2/4R^2 = M(v^2/R)$ so $v = \sqrt{GM/4R}$

And $T = 2\pi R/v = 2\pi R\sqrt{4R/GM} = 4\pi\sqrt{R^3/GM}$

c) SET UP: Apply $K_1 + U_1 + W_{\text{other}} = K_2 + U_2$ to the system of the two stars. Separate to infinity implies $K_2 = 0$ and $U_2 = 0$.

EXECUTE: $K_1 = \frac{1}{2}Mv^2 + \frac{1}{2}Mv^2 = 2(\frac{1}{2}M)(GM/4R) = GM^2/4R$

$U_1 = -GM^2/2R$

Thus the energy required is $W_{\text{other}} = -(K_1 + U_1) = -(GM^2/4R - GM^2/2R) = GM^2/4R$.

EVALUATE: The closer the stars are and the greater their mass, the larger their orbital speed, the shorter their orbital period and the greater the enegy required to separate them.

12.73 **IDENTIFY** and **SET UP:** Use conservation of energy, $K_1 + U_1 + W_{\text{other}} = K_2 + U_2$. The gravity force exerted by the sun is the only force that does work on the comet, so $W_{\text{other}} = 0$.

EXECUTE:

$K_1 = \frac{1}{2}mv_1^2, \quad v_1 = 2.0 \times 10^4$ m/s

$U_1 = -Gm_Sm/r_1, \quad r_1 = 2.5 \times 10^{11}$ m

$K_2 = \frac{1}{2}mv_2^2$

$U_2 = -Gm_Sm/r_2, \quad r_2 = 5.0 \times 10^{10}$ m

$\frac{1}{2}mv_1^2 - Gm_Sm/r_1 = \frac{1}{2}mv_2^2 - Gm_Sm/r_2$

$$v_2^2 = v_1^2 + 2Gm_S\left(\frac{1}{r_2} - \frac{1}{r_1}\right) = v_1^2 + 2Gm_S\left(\frac{r_1 - r_2}{r_1 r_2}\right)$$

$v_2 = 6.8 \times 10^4$ m/s

EVALUATE: The comet has greater speed when it is closer to the sun.

12.75 **a) IDENTIFY** and **SET UP:** Use Eq.(12.19), applied to satellites orbiting the earth rather than the sun.

EXECUTE: Find the value of a for the elliptical orbit:

$2a = r_a + r_p = R_E + h_a + R_E + h_p$, where h_a and h_p are the heights at apogee and perigee, respectively.

$a = R_E + (h_a + h_p)/2$

$a = 6.38 \times 10^6$ m $+ (400 \times 10^3 + 4000 \times 10^3 \text{ m})/2 = 8.58 \times 10^6$ m

$$T = \frac{2\pi a^{3/2}}{\sqrt{GM_E}} = \frac{2\pi(8.58 \times 10^6 \text{ m})^{3/2}}{\sqrt{(6.673 \times 10^{-11} \text{ N} \cdot \text{m}^2/\text{kg}^2)(5.97 \times 10^{24} \text{ kg})}} = 7.91 \times 10^3 \text{ s}$$

b) Conservation of angular momentum gives $r_a v_a = r_p v_p$

$$\frac{v_p}{v_a} = \frac{r_a}{r_p} = \frac{6.38 \times 10^6 \text{ m} + 4.00 \times 10^6 \text{ m}}{6.38 \times 10^6 \text{ m} + 4.00 \times 10^5 \text{ m}} = 1.53$$

c) Conservation of energy applied to apogee and perigee gives $K_a + U_a = K_p + U_p$

$\frac{1}{2}mv_a^2 - Gm_Em/r_a = \frac{1}{2}mv_p^2 - Gm_Em/r_p$

$v_p^2 - v_a^2 = 2Gm_E(1/r_p - 1/r_a) = 2Gm_E(r_a - r_p)/r_a r_p$

But $v_p = 1.532 v_a$, so $1.347 v_a^2 = 2Gm_E(r_a - r_p)/r_a r_p$

$v_a = 5.51 \times 10^3$ m/s, $v_p = 8.43 \times 10^3$ m/s

d) Need v so that $E = 0$, where $E = K + U$.

at perigee: $\frac{1}{2}mv^2 - Gm_E m/r_p = 0$

$v_p = \sqrt{2Gm_E/r_p} =$

$\sqrt{2(6.673 \times 10^{-11}\ \text{N} \cdot \text{m}^2/\text{kg}^2)(5.97 \times 10^{24}\ \text{kg})/6.78 \times 10^6\ \text{m}} = 1.084 \times 10^4$ m/s

This means an increase of 1.084×10^4 m/s $- 8.43 \times 10^3$ m/s $= 2.41 \times 10^3$ m/s.

at apogee: $v_a = \sqrt{2Gm_E/r_a} =$

$\sqrt{2(6.673 \times 10^{-11}\ \text{N} \cdot \text{m}^2/\text{kg}^2)(5.97 \times 10^{24}\ \text{kg})/1.038 \times 10^7\ \text{m}} = 8.761 \times 10^3$ m/s

This means an increase of 8.761×10^3 m/s $- 5.51 \times 10^3$ m/s $= 3.25 \times 10^3$ m/s.

EVALUATE: Perigee is more efficient. At this point r is smaller so v is larger and the satellite has more kinetic energy and more total energy.

12.77 **IDENTIFY** and **SET UP:** Apply conservation of energy (Eq.7.13) and solve for W_{other}. Only $r = h + R_E$ is given, so use Eq.(12.12) to relate r and v.

EXECUTE: $K_1 + U_1 + W_{\text{other}} = K_2 + U_2$

$U_1 = -Gm_M m/r_1$, where m_M is the mass of Mars and $r_1 = R_M + h$, where R_M is the radius of Mars and $h = 2000 \times 10^3$ m

$U_1 = -(6.673 \times 10^{-11}\ \text{N} \cdot \text{m}^2/\text{kg}^2)\dfrac{(6.42 \times 10^{23}\ \text{kg})(3000\ \text{kg})}{3.40 \times 10^6\ \text{m} + 2000 \times 10^3\ \text{m}} = -2.380 \times 10^{10}$ J

$U_2 = -Gm_M m/r_2$, where r_2 is the new orbit radius.

$U_2 = -(6.673 \times 10^{-11}\ \text{N} \cdot \text{m}^2/\text{kg}^2)\dfrac{(6.42 \times 10^{23}\ \text{kg})(3000\ \text{kg})}{3.40 \times 10^6\ \text{m} + 4000 \times 10^3\ \text{m}} = -1.737 \times 10^{10}$ J

For a circular orbit $v = \sqrt{Gm_M/r}$ (Eq.12-12, with the mass of mars rather than the mass of the earth).

Using this gives $K = \frac{1}{2}mv^2 = \frac{1}{2}m(Gm_M/r) = \frac{1}{2}Gm_M m/r$, so $K = -\frac{1}{2}U$

$K_1 = -\frac{1}{2}U_1 = +1.190 \times 10^{10}$ J and $K_2 = -\frac{1}{2}U_2 = +8.685 \times 10^9$ J

Then $K_1 + U_1 + W_{\text{other}} = K_2 + U_2$ gives

$W_{\text{other}} = (K_2 - K_1) + (U_2 - U_1) = (8.685 \times 10^9\ \text{J} - 1.190 \times 10^{10}\ \text{J}) + (-2.380 \times 10^{10}\ \text{J} + 1.737 \times 10^{10}\ \text{J})$

$W_{\text{other}} = -3.215 \times 10^9\ \text{J} + 6.430 \times 10^9\ \text{J} = 3.22 \times 10^9$ J

EVALUATE: When the orbit radius increases the kinetic energy decreases and the gravitational potential energy increases. $K = -U/2$ so $E = K + U = -U/2$ and the total energy also increases (becomes less negative). Positive work must be done to increase the total energy of the satellite.

12.79 **IDENTIFY** and **SET UP:** Use Eq.(12.19) to calculate a.

$T = 30,000 \text{ y}(3.156 \times 10^7 \text{ s/1 y}) = 9.468 \times 10^{11} \text{ s}$

EXECUTE:

Eq.(12.19): $T = \dfrac{2\pi a^{3/2}}{\sqrt{Gm_S}}, \quad T^2 = \dfrac{4\pi^2 a^3}{Gm_S}$

$a = \left(\dfrac{Gm_S T^2}{4\pi^2}\right)^{1/3} = 1.4 \times 10^{14} \text{ m}.$

EVALUATE: The average orbit radius of Pluto is 5.9×10^{12} m (Appendix F); the semi-major axis for this comet is larger by a factor of 24.

4.3 light years = 4.3 light years(9.461×10^{15} m/1 light year) $= 4.1 \times 10^{16}$ m
The distance of Alpha Centauri is larger by a factor of 300.

The orbit of the comet extends well past Pluto but is well within the distance to Alpha Centuri.

12.81 **IDENTIFY** and **SET UP:** Use Eq.(12.1) to calculate the force between the point mass and a small segment of the semicircle.

EXECUTE: The radius of the semicircle is $R = L/\pi$

Divide the semicircle up into small segments of length $R\,d\theta$

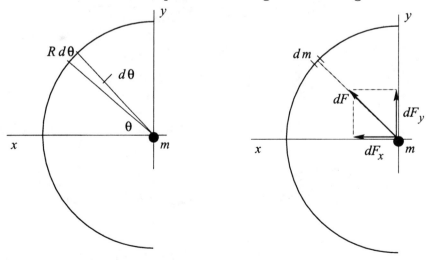

$dM = (M/L)R\,d\theta = (M/\pi)\,d\theta$

$d\vec{F}$ is the gravity force on m exerted by dM

$\int dF_y = 0$; the y-components from the upper half of the semicircle cancel the y-components from the lower half.

The x-components are all in the $+x$-direction and all add.

$$dF = G\frac{m\,dM}{R^2}$$

$$dF_x = G\frac{m\,dM}{R^2}\cos\theta = \frac{Gm\pi M}{L^2}\cos\theta\,d\theta$$

$$F_x = \int_{-\pi/2}^{\pi/2} dF_x = \frac{Gm\pi M}{L^2}\int_{-\pi/2}^{\pi/2}\cos\theta\,d\theta = \frac{Gm\pi M}{L^2}\,(2)$$

$$F = \frac{2\pi GmM}{L^2}$$

EVALUATE: If the semicircle were replaced by a point mass M at $x = R$, the gravity force would be $GmM/R^2 = \pi^2 GmM/L^2$. This is $\pi/2$ times larger than the force exerted by the semicircular wire. For the semicircle it is the x-components that add, and the sum is less than if the force magnitudes were added.

12.83 IDENTIFY: Use Eq.(12.1) for the force between a small segment of the rod and the particle. Integrate over the length of the rod to find the total force.

SET UP: Use a coordinate system with the origin at the left-hand end of the rod and the x'-axis along the rod. Divide the rod into small segments of length dx'. (Use x' for the coordinate so not to confuse with the distance x from the end of the rod to the particle.)

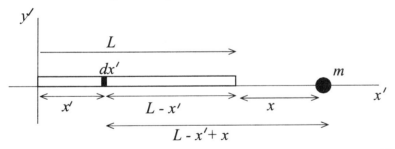

EXECUTE: The mass of each segment is $dM = dx'(M/L)$. Each segment is a distance $L - x' + x$ from mass m, so the force on the particle due to a segment is

$$dF = \frac{Gm\,dM}{(L - x' + x)^2} = \frac{GMm}{L}\frac{dx'}{(L - x' + x)^2}.$$

$$F = \int_L^0 dF = \frac{GMm}{L}\int_L^0\frac{dx'}{(L - x' + x)^2} = \frac{GMm}{L}\left(-\frac{1}{L - x' + x}\,\Big|_L^0\right)$$

$$F = \frac{GMm}{L}\left(\frac{1}{x} - \frac{1}{L + x}\right) = \frac{GMm}{L}\frac{(L + x - x)}{x(L + x)} = \frac{GMm}{x(L + x)}$$

EVALUATE: For $x \gg L$ this result becomes $F = GMm/x^2$, the same as for a pair of point masses.

CHAPTER 13
PERIODIC MOTION

Exercises 1, 7, 11, 13, 15, 17, 23, 25, 27, 29, 33, 35, 39, 41, 43, 49, 51, 55
Problems 59, 61, 65, 69, 71, 73, 75, 83, 85, 89, 91, 93

Exercises

13.1 **IDENTIFY** and **SET UP:** The target variables are the period T and angular frequency ω. We are given the frequency f, so we can find these using Eqs.(13.1) and (13.2)

EXECUTE:

a) $f = 220$ Hz

$T = 1/f = 1/220$ Hz $= 4.54 \times 10^{-3}$ s

$\omega = 2\pi f = 2\pi(220$ Hz$) = 1380$ rad/s

b) $f = 4(220$ Hz$) = 880$ Hz

$T = 1/f = 1/880$ Hz $= 1.14 \times 10^{-3}$ s (smaller by a factor of 4)

$\omega = 2\pi f = 2\pi(880$ Hz$) = 5530$ rad/s (factor of 4 larger)

EVALUATE: The angular frequency is directly proportional to the frequency and the period is inversely proportional to the frequency.

13.7 **IDENTIFY** and **SET UP:** Use Eq.(13.1) to calculate T, Eq.(13.2) to calculate ω, and Eq.(13.10) for m.

EXECUTE:

a) $T = 1/f = 1/6.00$ Hz $= 0.167$ s

b) $\omega = 2\pi f = 2\pi(6.00$ Hz$) = 37.7$ rad/s

c) $\omega = \sqrt{k/m}$ implies $m = k/\omega^2 = (120$ N/m$)/(37.7$ rad/s$)^2 = 0.0844$ kg

EVALUATE: We can verify that k/ω^2 has units of mass.

13.11 **IDENTIFY** and **SET UP:** Use Eqs. (13.13), (13.15), and (13.16).

EXECUTE:

$f = 440$ Hz, $A = 3.0$ mm, $\phi = 0$

a) $x = A\cos(\omega t + \phi)$

$\omega = 2\pi f = 2\pi(440$ Hz$) = 2.76 \times 10^3$ rad/s

$x = (3.0 \times 10^{-3}$ m$)\cos((2.76 \times 10^3$ rad/s$)t)$

b) $v_x = -\omega A\sin(\omega t + \phi)$

$v_{max} = \omega A = (2.76 \times 10^3 \text{ rad/s})(3.0 \times 10^{-3} \text{ m}) = 8.3 \text{ m/s}$ (maximum magnitude of velocity)

$a_x = -\omega^2 A \cos(\omega t + \phi)$

$a_{max} = \omega^2 A = (2.76 \times 10^3 \text{ rad/s})^2(3.0 \times 10^{-3} \text{ m}) = 2.3 \times 10^4 \text{ m/s}^2$ (maximum magnitude of acceleration)

c) $a_x = -\omega^2 A \cos \omega t$

$da_x/dt = +\omega^3 A \sin \omega t = [2\pi(440 \text{ Hz})]^3(3.0 \times 10^{-3} \text{ s}) \sin([2.76 \times 10^3 \text{ rad/s}]t)$

$= (6.3 \times 10^7 \text{ m/s}^3) \sin([2.76 \times 10^3 \text{ rad/s}]t)$

Maximum magnitude of the jerk is $\omega^3 A = 6.3 \times 10^7 \text{ m/s}^3$

EVALUATE: The period of the motion is small, so the maximum acceleration and jerk are large.

13.13 **IDENTIFY** and **SET UP:** We are given k, m, x_0, and v_0. Use Eqs.(13.19), (13.18), and (13.13).

EXECUTE:

a) Eq.(13.19): $A = \sqrt{x_0^2 + v_{0x}^2/\omega^2} = \sqrt{x_0^2 + mv_{0x}^2/k}$

$A = \sqrt{(0.200 \text{ m})^2 + (2.00 \text{ kg})(-4.00 \text{ m/s})^2/(300 \text{ N/m})} = 0.383 \text{ m}$

b) Eq.(13.18): $\phi = \arctan(-v_{0x}/\omega x_0)$

$\omega = \sqrt{k/m} = \sqrt{(300 \text{ N/m})/2.00 \text{ kg}} = 12.25 \text{ rad/s}$

$\phi = \arctan\left(-\dfrac{(-4.00 \text{ m/s})}{12.25 \text{ rad/s})(0.200 \text{ m})}\right) = \arctan(+1.633) = 58.5° \text{ (or 1.02 rad)}$

c) $x = A\cos(\omega t + \phi)$ gives $x = (0.383 \text{ m}) \cos([12.2 \text{ rad/s}]t + 1.02 \text{ rad})$

EVALUATE: At $t = 0$ the block is displaced 0.200 m from equilibrium but is moving, so $A > 0.200$ m. According to Eq.(13.15), a phase angle ϕ in the range $0 < \phi < 90°$ gives $v_{0x} < 0$.

13.15 **IDENTIFY** and **SET UP:** Calculate x using Eq.(13.13).
Use T to calculate ω and x_0 to calculate ϕ.

EXECUTE: $x = 0$ at $t = 0$ implies that $\phi = \pm\pi/2$ rad

Thus $x = A\cos(\omega t \pm \pi/2)$.

$T = 2\pi/\omega$ so $\omega = 2\pi/T = 2\pi/1.20 \text{ s} = 5.236 \text{ rad/s}$

$x = (0.600 \text{ m}) \cos([5.236 \text{ rad/s}][0.480 \text{ s}] \pm \pi/2) = \mp0.353 \text{ m}.$

The distance of the object from the equilibrium position is 0.353 m.

EVALUATE: The problem doesn't specify whether the object is moving in the $+x$ or $-x$ direction at $t = 0$.

13.17 IDENTIFY and **SET UP:** Use eq.(13.12) for T and Eq.(13.4) to relate a_x and k.
EXECUTE: $T = 2\pi\sqrt{m/k}$, $m = 0.400$ kg

Use $a_x = -2.70$ m/s^2 to calculate k:

$$-kx = ma_x \text{ gives } k = -\frac{ma_x}{x} = -\frac{(0.400\text{ kg})(-2.70\text{ m/s}^2)}{0.300\text{ m}} = +3.60\text{ N/m}$$

$T = 2\pi\sqrt{m/k} = 2.09$ s

EVALUATE: a_x is negative when x is positive. ma_x/x has units of N/m and $\sqrt{m/k}$ has units of s.

13.23 IDENTIFY and **SET UP:** Use Eq.(13.21) to relate K and U.
U depends on x and K depends on v_x.
EXECUTE:

a) $U + K = E$ $U = K$ says that $2U = E$
$2(\frac{1}{2}kx^2) = \frac{1}{2}kA^2$ and $x = \pm A/\sqrt{2}$; magnitude is $A/\sqrt{2}$

But can also say $U = K$ implies that $2K = E$
$2(\frac{1}{2}mv_x^2) = \frac{1}{2}kA^2$ and $v_x = \pm\sqrt{k/m}A/\sqrt{2} = \pm\omega A/\sqrt{2}$; magnitude is $\omega A/\sqrt{2}$.

b) In one cycle x goes from A to 0 to $-A$ to 0 to $+A$. Thus $x = +A\sqrt{2}$ twice and $x = -A/\sqrt{2}$ twice in each cycle. Therefore, $U = K$ four times each cycle.

The time between $U = K$ occurrences is the time Δt_a for $x_1 = +A/\sqrt{2}$ to $x_2 = -A\sqrt{2}$, time Δt_b for $x_1 = -A/\sqrt{2}$ to $x_2 = +A\sqrt{2}$, time Δt_c for $x_1 = +A/\sqrt{2}$ to $x_2 = +A\sqrt{2}$, or the time Δt_d for $x_1 = -A/\sqrt{2}$ to $x_2 = -A/\sqrt{2}$,

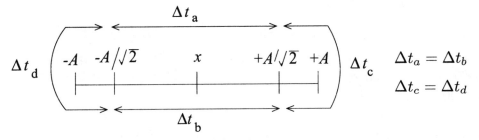

Calculation of Δt_a:
Specify x in $x = A\cos\omega t$ (choose $\phi = 0$ so $x = A$ at $t = 0$) and solve for t.
$x_1 = +A/\sqrt{2}$ implies $A/\sqrt{2} = A\cos(\omega t_1)$
$\cos\omega t_1 = 1\sqrt{2}$ so $\omega t_1 = \arccos(1/\sqrt{2}) = \pi/4$ rad
$t_1 = \pi/4\omega$

$x_2 = -A/\sqrt{2}$ implies $-A/\sqrt{2} = A\cos(\omega t_2)$
$\cos\omega t_2 = -1\sqrt{2}$ so $\omega t_1 = 3\pi/4$ rad
$t_2 = 3\pi/4\omega$

$\Delta t_a = t_2 - t_1 = 3\pi/4\omega - \pi/4\omega = \pi/2\omega$ (Note that this is $T/4$, one fourth period.)

Calculation of Δt_d:

$x_1 = -A/\sqrt{2}$ implies $t_1 = 3\pi/4\omega$

$x_2 = -A/\sqrt{2}$, t_2 is the <u>next</u> time after t_1 that gives $\cos \omega t_2 = -1/\sqrt{2}$

Thus $\omega t_2 = \omega t_1 + \pi/2 = 5\pi/4$ and $t_2 = 5\pi/4\omega$.

$\Delta t_d = t_2 - t_1 = 5\pi/4\omega - 3\pi/4\omega = \pi/2\omega$, so is the same as Δt_a.

Therefore the occurrences of $K = U$ are equally spaced in time, with a time interval between them of $\pi/2\omega$.

EVALUATE: This is one-fourth T, as it must be if there are 4 equally spaced occurrences each period.

c) EXECUTE: $x = A/2$ and $U + K = E$

$K = E - U = \frac{1}{2}kA^2 - \frac{1}{2}kx^2 = \frac{1}{2}kA^2 - \frac{1}{2}k(A/2)^2 = \frac{1}{2}kA^2 - \frac{1}{8}kA^2 = 3kA^2/8$

Then $\dfrac{K}{E} = \dfrac{3kA^2/8}{\frac{1}{2}kA^2} = \dfrac{3}{4}$ and $\dfrac{U}{E} = \dfrac{\frac{1}{8}kA^2}{\frac{1}{2}kA^2} = \dfrac{1}{4}$

EVALUATE: At $x = 0$ all the energy is kinetic and at $x = \pm A$ all the energy is potential. But $K = U$ does not occur at $x = \pm A/2$, since U is not linear in x.

13.25 IDENTIFY and **SET UP:** a_x is related to x by Eq.(13.4) and v_x is related to x by Eq.(13.21). a_x is a maximum when $x = \pm A$ and v_x is a maximum when $x = 0$. t is related to x by Eq.(13.13).

EXECUTE:

a) $-kx = ma_x$ so $a_x = -(k/m)x$ (Eq.13.4) But the maximum $\mid x \mid$ is A, so $a_{max} = (k/m)A = \omega^2 A$.

$f = 0.850$ Hz implies $\omega = \sqrt{k/m} = 2\pi f = 2\pi(0.850$ Hz$) = 5.34$ rad/s

$a_{max} = \omega^2 A = (5.34$ rad/s$)^2(0.180$ m$) = 5.13$ m/s^2.

$\frac{1}{2}mv_x^2 + \frac{1}{2}kx^2 = \frac{1}{2}kA^2$

$v_x = v_{max}$ when $x = 0$ so $\frac{1}{2}mv_{max}^2 = \frac{1}{2}kA^2$

$v_{max} = \sqrt{k/m}A = \omega A = (5.34$ rad/s$)(0.180$ m$) = 0.961$ m/s

b) $a_x = -(k/m)x = -\omega^2 x = -(5.34$ rad/s$)^2(0.090$ m$) = -2.57$ m/s^2

$\frac{1}{2}mv_x^2 + \frac{1}{2}kx^2 = \frac{1}{2}kA^2$ says that $v_x = \pm\sqrt{k/m}\sqrt{A^2 - x^2} = \pm\omega\sqrt{A^2 - x^2}$

$v_x = \pm(5.34$ rad/s$)\sqrt{(0.180 \text{ m})^2 - (0.090 \text{ m})^2} = \pm 0.832$ m/s

The speed is 0.832 m/s.

c) $x = A\cos(\omega t + \phi)$

Let $\phi = -\pi/2$ so that $x = 0$ at $t = 0$.

Then $x = A\cos(\omega t - \pi/2) = A\sin(\omega t)$ [Using the trig identity $\cos(a - \pi/2) = \sin a$]

Find the time t that gives $x = 0.120$ m.

0.120 m $= (0.180$ m$)\sin(\omega t)$

$\sin \omega t = 0.6667$

$t = \arcsin(0.6667)/\omega = 0.7297$ rad$/(5.34$ rad$/$s$) = 0.137$ s

EVALUATE: It takes one-fourth of a period for the object to go from $x = 0$ to $x = A = 0.180$ m. So the time we have calculated should be less than $T/4$. $T = 1/f = 1/0.850$ Hz $= 1.18$ s, $T/4 = 0.295$ s, and the time we calculated is less than this.

Note that the a_x and v_x we calculated in part (b) are smaller in magnitude that the maximum values we calculated in part (b).

d) The conservation of energy equation relates v and x and $F = ma$ relates a and x. So the speed and acceleration can be found by energy methods but the time cannot.

Specifying x uniquely determines a_x but determines only the magnitude of v_x; at a given x the object could be moving either in the $+x$ or $-x$ direction.

13.27 **IDENTIFY** and **SET UP:** Use Eq.(13.21).

$x = \pm A\omega$ when $v_x = 0$ and $v_x = \pm v_{\max}$ when $x = 0$.

EXECUTE:

a) $E = \frac{1}{2}mv^2 + \frac{1}{2}kx^2$ $E = \frac{1}{2}(0.150$ kg$)(0.300$ m$/$s$)^2 + \frac{1}{2}(300$ N$/$m$)(0.012$ m$)^2 = 0.0284$ J

b) $E = \frac{1}{2}kA^2$ so $A = \sqrt{2E/k} = \sqrt{2(0.0284 \text{ J})/300 \text{ N/m}} = 0.014$ m

c) $E = \frac{1}{2}mv_{\max}^2$ so $v_{\max} = \sqrt{2E/m} = \sqrt{2(0.0284 \text{ J})/0.150 \text{ kg}} = 0.615$ m$/$s

EVALUATE: The total energy E is constant but is transferred between kinetic and potential energy during the motion.

13.29 **IDENTIFY** and **SET UP:** The maximum speed is given by Eq.(13.23). Use Eq.(13.12) to relate T and k/m. Use Eq.(13.4) to relate $a_{\max}$ and A.

EXECUTE: $v_{\max} = A\sqrt{k/m}$

Use T to find k/m:

$T = 2\pi\sqrt{m/k}$ so $k/m = (2\pi/T)^2 = 158$ s^{-2}

Use $a_{\max}$ to find A:

$a_{\max} = kA/m$ so $A = a_{\max}/(k/m) = 0.0405$ m

Then $v_{\max} = A\sqrt{k/m} = 0.509$ m$/$s

EVALUATE: We can calculate only k/m, not k or m individually, but the ratio k/m is all that is needed in order to calculate v_{max}.

13.33 **IDENTIFY:** Apply Newton's 2nd law to the cheese to find the amount the spring is stretched when the cheese hangs at rest. The period T and Eq.(13.12) allows us to calculate k for the spring.

SET UP:

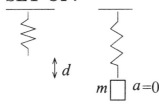

Let d be the distance the spring stretches when the cheese hangs at rest.

Free-body diagram for the cheese:

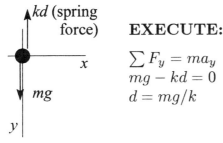

EXECUTE:

$\sum F_y = ma_y$

$mg - kd = 0$

$d = mg/k$

$T = 2\pi\sqrt{m/k}$ says that $m/k = (T/2\pi)^2$

Thus $d = (T/2\pi)^2 g = (0.400 \text{ s}/2\pi)^2 (9.80 \text{ m/s}^2) = 0.0397$ m.

EVALUATE: d depends on m/k and T determines m/k; we didn't use the value of m that was given and our answer is independent of m,

13.35 **IDENTIFY: and SET UP:** The number of ticks per second tells us the period and therefore the frequency. We can use a formula from Table 9.2 to calculate I. Then Eq.(13.24) allows us to calculate the torsion constant κ.

EXECUTE: Ticks four times each second implies 0.25 s per tick. Each tick is half a period, so $T = 0.50$ s and $f = 1/T = 1/0.50$ s $= 2.00$ Hz

a) Thin rim implies $I = MR^2$ (from Table 9-2)

$I = (0.900 \times 10^{-3} \text{ kg})(0.55 \times 10^{-2} \text{ m})^2 = 2.7 \times 10^{-8}$ kg $\cdot$ m^2

b) $T = 2\pi\sqrt{I/\kappa}$ so $\kappa = I(2\pi/T)^2 = (2.7 \times 10^{-8} \text{ kg} \cdot \text{m}^2)(2\pi/0.50 \text{ s})^2$
$= 4.3 \times 10^{-6}$ N$\cdot$m/rad

EVALUATE: Both I and κ are small numbers.

13.39 **IDENTIFY: and SET UP:** Follow the procedure outlined in the problem.
EXECUTE: Eq.(13.25): $U = U_0[(R_0/r)^{12} - 2(R_0/r)^6]$ Let $r = R_0 + x$.

$$U = U_0 \left[\left(\frac{R_0}{R_0 + x} \right)^{12} - 2 \left(\frac{R_0}{R_0 + x} \right)^{6} \right] = U_0 \left(\left(\frac{1}{1 + x/R_0} \right)^{12} - 2 \left(\frac{1}{1 + x/R_0} \right)^{6} \right)$$

$$\left(\frac{1}{1 + x/R_0} \right)^{12} = (1 + x/R_0)^{-12}; \quad | x/R_0 | << 1$$

Apply Eq.(13-28) with $n = -12$ and $u = +x/R_0$:

$$\left(\frac{1}{1 + x/R_0} \right)^{12} = 1 - 12x/R_0 + 66x^2/R_0^2 - \ldots$$

For $\left(\frac{1}{1 + x/R_0} \right)^{6}$ apply Eq.(13-28) with $n = -6$ and $u = +x/R_0$:

$$\left(\frac{1}{1 + x/R_0} \right)^{6} = 1 - 6x/R_0 + 15x^2/R_0^2 - \ldots$$

Thus $U = U_0(1 - 12x/R_0 + 66x^2/R_0^2 - 2 + 12x/R_0 - 30x^2/R_0^2) = -U_0 + 36U_0x^2/R_0^2$.
This is in the form $U = \frac{1}{2}kx^2 - U_0$ with $k = 72U_0/R_0^2$, which is the same as the force constant in Eq.(13.29).

EVALUATE: $F_x = -dU/dx$ so $U(x)$ contains an additive constant that can be set to any value we wish. If $U_0 = 0$ then $U = 0$ when $x = 0$.

13.41   **IDENTIFY:** Use Eq.(13.34) to relate the period to g.

SET UP: Let the period on earth be $T_E = 2\pi \sqrt{L/g_E}$, where $g_E = 9.80 \text{ m/s}^2$, the value on earth.

Let the period on Mars be $T_M = 2\pi \sqrt{L/g_M}$, where $g_M = 3.71 \text{ m/s}^2$, the value on Mars.

We can eliminate L, which we don't know, by taking a ratio:

EXECUTE: $\dfrac{T_M}{T_E} = 2\pi \sqrt{\dfrac{L}{g_M}} \dfrac{1}{2\pi} \sqrt{\dfrac{g_E}{L}} = \sqrt{\dfrac{g_E}{g_M}}.$

$$T_M = T_E \sqrt{\frac{g_E}{g_M}} = (1.60 \text{ s}) \sqrt{\frac{9.80 \text{ m/s}^2}{3.71 \text{ m/s}^2}} = 2.60 \text{ s}.$$

EVALUATE: Gravity is weaker on Mars so the period of the pendulum is longer there.

13.43   **IDENTIFY** and **SET UP:** The bounce frequency is given by Eq.(13.11) and the pendulum frequency by Eq.(13.33). Use the relation between these two frequencies that is specified in the problem to calculate the equilibrium length L of the spring, when the apple hangs at rest on the end of the spring.

EXECUTE:

vertical SHM: $f_b = \dfrac{1}{2\pi}\sqrt{\dfrac{k}{m}}$

pendulum motion (small amplitude): $f_p = \dfrac{1}{2\pi}\sqrt{\dfrac{g}{L}}$

The problem specifies that $f_p = \frac{1}{2}f_b$.

$$\dfrac{1}{2\pi}\sqrt{\dfrac{g}{L}} = \dfrac{1}{2}\dfrac{1}{2\pi}\sqrt{\dfrac{k}{m}}$$

$g/L = k/4m$ so $L = 4gm/k = 4w/k = 4(1.00\text{ N})/1.50\text{ N/m} = 2.67\text{ m}$

EVALUATE: This is the <u>stretched</u> length of the spring, its length when the apple is hanging from it. (Note: Small angle of swing means v is small as the apple passes through the lowest point, so a_rad is small and the component of mg perpendicular to the spring is small. Thus the amount the spring is stretched changes very little as the apple swings back and forth.)

IDENTIFY: Use Newton's second law to calculate the distance the spring is stretched from its unstretched length when the apple hangs from it.

SET UP: Free-body diagram for the apple hanging at rest on the end of the spring:

EXECUTE:
$\sum F_y = ma_y$
$k\,\Delta L - mg = 0$
$\Delta L = mg/k = w/k = 1.00\text{ N}/1.50\text{ N} = 0.667$

Thus the unstretched length of the spring is $2.67\text{ m} - 0.67\text{ m} = 2.00\text{ m}$.

EVALUATE: The spring shortens to its unstretched length when the apple is removed.

13.49 IDENTIFY: The ornament is a physical pendulum: $T = 2\pi\sqrt{I/mgd}$ (Eq.13.39). T is the target variable.

SET UP: $I = 5MR^2/3$, the moment of inertia about an axis at the edge of the sphere.

d is the distance from the axis to the center of gravity, which is at the center of the sphere, so $d = R$.

EXECUTE: $T = 2\pi\sqrt{5/3}\sqrt{R/g} = 2\pi\sqrt{5/3}\sqrt{0.050\text{ m}/(9.80\text{ m/s}^2)} = 0.58\text{ s}$.

EVALUATE: A simple pendulum of length $R = 0.050\text{ m}$ has period 0.45 s; the period of the physical pendulum is longer.

13.51 **IDENTIFY** and **SET UP:** Use Eq.(13.43) to calculate ω', and then $f' = \omega'/2\pi$.

a) EXECUTE:

$$\omega' = \sqrt{(k/m) - (b^2/4m^2)} = \sqrt{\frac{2.50 \text{ N/m}}{0.300 \text{ kg}} - \frac{(0.900 \text{ kg/s})^2}{4(0.300 \text{ kg})^2}} = 2.47 \text{ rad/s}$$

$$f' = \omega'/2\pi = (2.47 \text{ rad/s})/2\pi = 0.393 \text{ Hz}$$

b) IDENTIFY and **SET UP:** The condition for critical damping is $b = 2\sqrt{km}$ (Eq.13.44)

EXECUTE: $b = 2\sqrt{(2.50 \text{ N/m})(0.300 \text{ kg})} = 1.73 \text{ kg/s}$

EVALUATE: The value of b in part (a) is less than the critical dmaping value found in part (b). With no damping, the frequency is $f = 0.459$ Hz; the damping reduces the oscillation frequency.

13.55 **IDENTIFY** and **SET UP:** Apply Eq.(13.46): $A = \dfrac{F_{max}}{\sqrt{(k - m\omega_d^2)^2 + b^2\omega_d^2}}$

EXECUTE:

a) Consider the special case where $k - m\omega_d^2 = 0$, so $A = F_{max}/b\omega_d$ and $b = F_{max}/A\omega_d$.

Units of $\dfrac{F_{max}}{A\omega_d}$ are $\dfrac{\text{kg} \cdot \text{m/s}^2}{(\text{m})(\text{s}^{-1})} = \text{kg/s}$.

For units consistency the units of b must be kg/s.

b) Units of $\sqrt{km}$: $[(\text{N/m})\text{kg}]^{1/2} = (\text{N kg/m})^{1/2} = [(\text{kg}\cdot\text{m/s}^2)(\text{kg})/\text{m}]^{1/2} = (\text{kg}^2/\text{s}^2)^{1/2} = \text{kg/s}$, the same as the units for b.

c) For $\omega_d = \sqrt{k/m}$ (at resonance) $A = (F_{max}/b)\sqrt{m/k}$.

(i) $b = 0.2\sqrt{km}$

$$A = F_{max}\sqrt{\frac{m}{k}}\frac{1}{0.2\sqrt{km}} = \frac{F_{max}}{0.2k} = 5.0\frac{F_{max}}{k}.$$

(ii) $b = 0.4\sqrt{km}$

$$A = F_{max}\sqrt{\frac{m}{k}}\frac{1}{0.4\sqrt{km}} = \frac{F_{max}}{0.4k} = 2.5\frac{F_{max}}{k}.$$

EVALUATE: Both these results agree with what is shown in Fig.13.24. As b increases the maximum amplitude decreases.

Problems

13.59 **IDENTIFY** and **SET UP:** Use Eqs.(13.12), (13.21), and (13.22) to relate the various quantities to the amplitude.

EXECUTE:

a) $T = 2\pi\sqrt{m/k}$; independent of A so period doesn't change $f = 1/T$; doesn't change

$\omega = 2\pi f$; doesn't change

b) $E = \frac{1}{2}kA^2$ when $k = \pm A$. When A is halved E decreases by a factor of 4; $E_2 = E_1/4$.

c) $v_{\max} = \omega A = 2\pi f A$

$v_{\max,1} = 2\pi f A_1 \quad v_{\max,2} = 2\pi f A_2 \quad$ (f doesn't change)

Since $A_2 = \frac{1}{2}A_1$, $v_{\max,2} = 2\pi f(\frac{1}{2}A_1) = \frac{1}{2}2\pi f A_1 = \frac{1}{2}v_{\max,1}$; $v_{\max}$ is one-half as great

d) $v_x = \pm\sqrt{k/m}\sqrt{A^2 - x^2}$

$x = \pm A_1/4$ gives $v_x = \pm\sqrt{k/m}\sqrt{A^2 - A_1^2/16}$

With the original amplitude $v_{1x} = \pm\sqrt{k/m}\sqrt{A_1^2 - A_1^2/16} = \pm\sqrt{15/16}(\sqrt{k/m})A_1$

With the reduced amplitude $v_{2x} = \pm\sqrt{k/m}\sqrt{A_2^2 - A_1^2/16}$

$= \pm\sqrt{k/m}\sqrt{(A_1/2)^2 - A_1^2/16} = \pm\sqrt{3/16}(\sqrt{k/m})A_1$

$v_{1x}/v_{2x} = \sqrt{15/3} = \sqrt{5}$, so $v_2 = v_1/\sqrt{5}$; the speed at this x is $1/\sqrt{5}$ times as great.

e) $U = \frac{1}{2}kx^2$; same x so same U.

$K = \frac{1}{2}mv_x^2; \quad K_1 = \frac{1}{2}mv_{1x}^2$

$K_2 = \frac{1}{2}mv_{2x}^2 = \frac{1}{2}m(v_{1x}/\sqrt{5})^2 = \frac{1}{5}(\frac{1}{2}mv_{1x}^2) = K_1/5; \quad 1/5$ times as great.

EVALUATE: Reducing A reduces the total energy but doesn't affect the period and the frequency.

13.61 **a) IDENTIFY** and **SET UP:** Combine Eqs.(13.12) and (13.21) to relate v_x and x to T.

EXECUTE: $T = 2\pi\sqrt{m/k}$

We are given information about v_x at a particular x. The expression relating these two quantities comes from conservation of energy:

$\frac{1}{2}mv_x^2 + \frac{1}{2}kx^2 = \frac{1}{2}kA^2$

We can solve this equation for $\sqrt{m/k}$, and then use that result to calculate T.

$mv_x^2 = k(A^2 - x^2)$

$$\sqrt{\frac{m}{k}} = \frac{\sqrt{A^2 - x^2}}{v_x} = \frac{\sqrt{(0.100 \text{ m})^2 - (0.060 \text{ m})^2}}{0.300 \text{ m/s}} = 0.267 \text{ s}$$

Then $T = 2\pi\sqrt{m/k} = 2\pi(0.267 \text{ s}) = 1.68 \text{ s}$.

b) IDENTIFY and **SET UP:** We are asked to relate x and v_x, so use conservation of energy equation:

$\frac{1}{2}mv_x^2 + \frac{1}{2}kx^2 = \frac{1}{2}kA^2$

$kx^2 = kA^2 - mv_x^2$

$x = \sqrt{A^2 - (m/k)v_x^2} = \sqrt{(0.100 \text{ m})^2 - (0.267 \text{ s})^2(0.160 \text{ m/s})^2} = 0.090 \text{ m}$

EVALUATE: Smaller $|v_x|$ means larger x.

c) IDENTIFY: If the slice doesn't slip the maximum acceleration of the plate (Eq.13.4) equals the maximum acceleration of the slice, which is determined by applying Newton's 2nd law to the slice.

SET UP: For the plate, $-kx = ma_x$ and $a_x = -(k/m)x$.

The maximum $|x|$ is A, so $a_{\text{max}} = (k/m)A$.

If the carrot slice doesn't slip then the static friction force must be able to give it this much acceleration.

Free-body diagram for the carrot slice (mass m'):

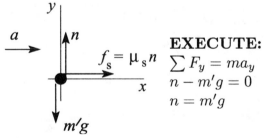

EXECUTE:

$\sum F_y = ma_y$

$n - m'g = 0$

$n = m'g$

$\sum F_x = ma_x$

$\mu_s n = m'a$

$\mu_s m'g = m'a$ and $a = \mu_s g$

But we require that $a = a_{\text{max}} = (k/m)A$, so $(k/m)A = \mu_s g$ and

$$\mu_s = \frac{k}{m}\frac{A}{g} = \left(\frac{1}{0.267 \text{ s}}\right)^2 \left(\frac{0.100 \text{ m}}{9.80 \text{ m/s}^2}\right) = 0.143$$

EVALUATE: We can write this as $\mu_s = \omega^2 A/g$. More friction is required if the frequency or the amplitude are increased.

13.65 IDENTIFY and **SET UP:** Analyze the amplitude using energy considerations, Eq.(13.21). The period is given by Eq.(13.12) and depends only on m and k.

EXECUTE:

a) $m \rightarrow m/2$

Splits at $x = 0$ where energy is all kinetic energy, $E = \frac{1}{2}mv^2$, so $E \rightarrow E/2$

k stays same

$E = \frac{1}{2}kA^2$ so $A = \sqrt{2E/k}$

Then $E \rightarrow E/2$ means $A \rightarrow A/\sqrt{2}$

$T = 2\pi\sqrt{m/k}$ so $m \rightarrow m/2$ means $T \rightarrow T/\sqrt{2}$

b) $m \rightarrow m/2$

Splits at $x = A$ where all the energy is potential energy in the spring, so E doesn't change.

$E = \frac{1}{2}kA^2$ so A stays the same.

$T = 2\pi\sqrt{m/k}$ so $T \rightarrow T/\sqrt{2}$, as in part (a).

c) EVALUATE: In Example 13.5, the mass increased. This means that T increases rather than decreases. When the mass is added at $x = 0$, the energy and amplitude change. When the mass is added at $x = \pm A$, the energy and amplitude remian the same. This is the same as in this problem.

13.69 IDENTIFY and **SET UP:** Measure x from the equilibrium position of the object, where the gravity and spring forces balance. Let $+x$ be downward.

a) Use conservation of energy (Eq.13.21) to relate v_x and x. Use Eq.(13.12) to relate T to k/m.

EXECUTE: $\frac{1}{2}mv_x^2 + \frac{1}{2}kx^2 = \frac{1}{2}kA^2$

For $x = 0$, $\frac{1}{2}mv_x^2 = \frac{1}{2}kA^2$ and $v = A\sqrt{k/m}$, just as for horizontal SHM.

We can use the period to calculate $\sqrt{k/m}$: $T = 2\pi\sqrt{m/k}$ implies $\sqrt{k/m} = 2\pi/T$.

Thus $v = 2\pi A/T = 2\pi(0.100 \text{ m})/4.20 \text{ s} = 0.150 \text{ m/s}$.

b) IDENTIFY and **SET UP:** Use Eq.(13.4) to relate a_x and x.

EXECUTE: $ma_x = -kx$ so $a_x = -(k/m)x$

$+x$-direction is downward, so here $x = -0.050$ m

$a_x = -(2\pi/T)^2(-0.050 \text{ m}) = +(2\pi/4.20 \text{ s})^2(0.050 \text{ m}) = 0.112 \text{ m/s}^2$ (positive, so direction is downward)

c) IDENTIFY and **SET UP:** Use Eq.(13.13) to relate x and t. The time asked for is twice the time it takes to go from $x = 0$ to $x = +0.050$ m.

EXECUTE: $x(t) = A\cos(\omega t + \phi)$

Let $\phi = -\pi/2$, so $x = 0$ at $t = 0$.

Then $x = A\cos(\omega t - \pi/2) = A\sin\omega t = A\sin(2\pi t/T)$.

Find the time t that gives $x = +0.050$ m:

0.050 m $= (0.100$ m$)\sin(2\pi t/T)$

$2\pi t/T = \arcsin(0.50) = \pi/6$ and $t = T/12 = 4.20$ s$/12 = 0.350$ s

The time asked for in the problem is twice this, 0.700 s.

d) IDENTIFY: The problem is asking for the distance d that the spring stretches when the object hangs at rest from it. Apply Newton's 2nd law to the object.

SET UP: Free-body diagram for the object

$a = 0$

EXECUTE:

$\sum F_x = ma_x$

$mg - kd = 0$

$d = (m/k)g$

But $\sqrt{k/m} = 2\pi/T$ (part (a)) and $m/k = (T/2\pi)^2$

$$d = \left(\frac{T}{2\pi}\right)g = \left(\frac{4.20}{2\pi}\right)^2(9.80 \text{ m/s}^2) = 4.38 \text{ m.}$$

EVALUATE: When the displacement is upward (part (b)), the acceleration is downward. The mass of the partridge never entered into the calculation. We used just the ratio k/m, that is determined from T.

13.71 IDENTIFY: Apply conservation of linear momentum to the collision between the steak and the pan. Then apply conservation of energy to the motion after the collision to find the amplitude of the subsequent SHM. Use Eq.(13.12) to calculate the period.

a) SET UP: First find the speed of the steak just before it strikes the pan. Use a coordinate system with $+y$ downward.

$v_{0y} = 0$ (released from rest); $y - y_0 = 0.40$ m; $a_y = +9.80$ m/s^2; $v_y = ?$

$v_y^2 = v_{0y}^2 + 2a_y(y - y_0)$

EXECUTE: $v_y = +\sqrt{2a_y(y - y_0)} = +\sqrt{2(9.80 \text{ m/s}^2)(0.40 \text{ m})} = +2.80$ m/s

SET UP: Apply conservation of momentum to the collision between the steak and the pan. After the collision the steak and the pan are moving together with common velocity v_2. Let A be the steak and B be the pan.

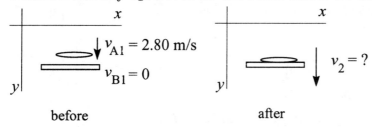

before   after

EXECUTE: P_y conserved: $m_A v_{A1y} + m_B v_{B1y} = (m_A + m_B)v_{2y}$

$m_A v_{A1} = (m_A + m_B)v_2$

$$v_2 = \left(\frac{m_A}{m_A + m_B}\right) v_{A1} = \left(\frac{2.2 \text{ kg}}{2.2 \text{ kg} + 0.20 \text{ kg}}\right)(2.80 \text{ m/s}) = 2.57 \text{ m/s}$$

b) SET UP: Conservation of energy applied to the SHM gives:
$\frac{1}{2}mv_0^2 + \frac{1}{2}kx_0^2 = \frac{1}{2}kA^2$ where v_0 and x_0 are the initial speed and displacement of the object and where the displacement is measured from the equilibrium position of the object.

EXECUTE: The weight of the steak will stretch the spring an additional distance d given by

$$kd = mg \text{ so } d = \frac{mg}{k} = \frac{(2.2 \text{ kg})(9.80 \text{ m/s}^2)}{400 \text{ N/m}} = 0.0539 \text{ m. So just after the steak}$$

hits the pan, before the pan has had time to move, the steak plus pan is 0.0539 m above the equilibrium position of the combined object. Thus $x_0 = 0.0539$ m.

From part (a) $v_0 = 2.57$ m/s, the speed of the combined object just after the collision.

Then $\frac{1}{2}mv_0^2 + \frac{1}{2}kx_0^2 = \frac{1}{2}kA^2$ gives

$$A = \sqrt{\frac{mv_0^2 + kx_0^2}{k}} = \sqrt{\frac{2.4 \text{ kg}(2.57 \text{ m/s})^2 + (400 \text{ N/m})(0.0539 \text{ m})^2}{400 \text{ N/m}}} = 0.21 \text{ m}$$

c) $T = 2\pi\sqrt{m/k} = 2\pi\sqrt{\dfrac{2.4 \text{ kg}}{400 \text{ N/m}}} = 0.49$ s

EVALUATE: The amplitude is less than the initial height of the steak above the pan because mechanical energy is lost in the inelastic collision.

13.73 IDENTIFY and **SET UP:** Use Eq.(13.12) to calculate g and use Eq.(12.4) applied to Newtonia to relate g to the mass of the planet.

EXECUTE: The pendulum swings through $\frac{1}{2}$ cycle in 1.42 s, so $T = 2.84$ s. $L = 1.85$ m.

Use T to find g:
$T = 2\pi\sqrt{L/g}$ so $g = L(2\pi/T)^2 = 9.055$ m/s^2

Use g to find the mass M_{p} of Newtonia:
$g = GM_{\text{p}}/R_{\text{p}}^2$
$2\pi R_{\text{p}} = 5.14 \times 10^7$ m, so $R_p = 8.18 \times 10^6$ m

$$m_{\text{p}} = \frac{gR_p^2}{G} = 9.08 \times 10^{24} \text{ kg}$$

EVALUATE: g is similar to that at the surface of the earth. The radius of

Newtonia is a little less than earth's radius and its mass is a little more.

13.75 **IDENTIFY:** Use Eq.(13.13) to relate x and t. $T = 3.5$ s.
SET UP:

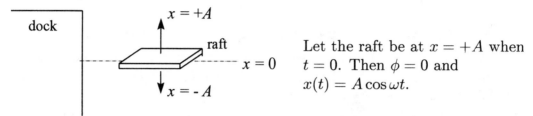

Let the raft be at $x = +A$ when $t = 0$. Then $\phi = 0$ and $x(t) = A\cos\omega t$.

EXECUTE: Calculate the time it takes the raft to move from $x = +A = +0.200$ m to $x = A - 0.100$ m $= 0.100$ m.

Write the equation for $x(t)$ in terms of T rather than ω:

$\omega = 2\pi/T$ gives that $x(t) = A\cos(2\pi t/T)$

$x = A$ at $t = 0$

$x = 0.100$ m implies 0.100 m $= (0.200$ m$)\cos(2\pi t/T)$

$\cos(2\pi t/T) = 0.500$ so $2\pi t/T = \arccos(0.500) = 1.047$ rad

$t = (T/2\pi)(1.047 \text{ rad}) = (3.5 \text{ s}/2\pi)(1.047 \text{ rad}) = 0.583$ s

This is the time for the raft to move down from $x = 0.200$ m to $x = 0.100$ m. But people can also get off while the raft is moving up from $x = 0.100$ m to $x = 0.200$ m, so during each period of the motion the time the people have to get off is $2t = 2(0.583 \text{ s}) = 1.17$ s.

EVALUATE: The time to go from $x = 0$ to $x = A$ and return is $T/2 = 1.75$ s. The time to go from $x = A/2$ to A and return is less than this.

13.83 **IDENTIFY** and **SET UP:** Use $F_r = -dU/dr$ to relate F_r and $U(r)$. At equilibrium, $F_r = 0$. Use Eq.(13.11) to calculate the frequency.
EXECUTE:

a) $U = A\left[\dfrac{1}{r} - \dfrac{1}{(r - 2R_0)}\right]$

$F_r = -\dfrac{dU}{dr} = -A\left[-\dfrac{1}{r^2} + \dfrac{1}{(r - 2R_0)^2}\right]$, as was to be shown.

b) At equilibirum $F_r = 0$ so $1/r^2 - 1/(r - 2R_0)^2 = 0$

$r^2 = (r - 2R_0)^2$ and $r = \pm(r - 2R_0)$

$r = +(r - 2R_0)$; no solution unless $R_0 = 0$

or $r = -(r - 2R_0)$, which says $r = R_0$, as was to be shown.

c) Write $r = R_0 + x$, where x/R_0 is small.

Note that $(r - 2R_0)^2 = (R_0 + x - 2R_0)^2 = (x - R_0)^2 = (R_0 - x)^2$, so

$$F_r = A \left[\frac{1}{(R_0 + x)^2} - \frac{1}{(R_0 - x)^2} \right] = \frac{A}{R_0^2} \left[\frac{1}{(1 + x/R_0)^2} - \frac{1}{(1 - x/R_o)^2} \right]$$

For $(1 + x/R_0)^{-2}$ apply Eq.(13-28) with $n = -2$ and $u = x/R_0$.
This gives $(1 + x/R_0)^{-2} \sim 1 - 2(x/R_0)$.
For $(1 - x/R_0)^{-2}$ apply Eq.(13-28) with $n = -2$ and $u = -x/R_0$. This gives $(1 - x/R_0)^{-2} \sim 1 + 2(x/R_0)$.
Then $F_r \sim (A/R_0^2)[1 - 2x/R_0 - (1 + 2x/R_0)] = -4Ax/R_0^3$.
$F_r = -kx$ gives $k = 4A/R_0^3$.

d) $f = (1/2\pi)\sqrt{k/m}$.

Using $k = \dfrac{4A}{R_0^3}$ from part (c) gives $f = \dfrac{1}{2\pi} \sqrt{\dfrac{4A}{mR_0^3}} = \dfrac{1}{\pi} \sqrt{\dfrac{A}{mR_0^3}}$.

EVALUATE: The minus sign in $F_r(x)$ says that when the center particle is displaced in either direction the net force on it is directed to return it to equilibrium.

13.85 **IDENTIFY:** Apply $\sum \tau_z = I_{\mathrm{cm}}\alpha_z$ and $\sum F_x = Ma_{\mathrm{cm}-x}$ to the cylinders. Solve for $a_{\mathrm{cm}-x}$. Compare to Eq.(13.8) to find the angular frequency and period, $T = 2\pi\omega$.

SET UP:

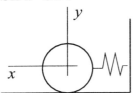

Let the origin of coordinates be at the center of the cylinders when they are at their equilibrium position.

Free-body diagram for the cylinders when they are displaced a distance x to the left:

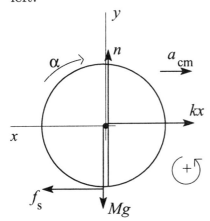

EXECUTE:
$\sum \tau_z = I_{\mathrm{cm}}\alpha_z$
$f_s R = (\frac{1}{2}MR^2)\alpha$
$f_s = \frac{1}{2}MR\alpha$

But $R\alpha = a_{\mathrm{cm}}$ so
$f_s = \frac{1}{2}Ma_{\mathrm{cm}}$

$$\sum F_x = ma_x$$

$$f_s - kx = -Ma_{cm}$$

$$\tfrac{1}{2}Ma_{cm} - kx = -Ma_{cm}$$

$$kx = \tfrac{3}{2}Ma_{cm}$$

$$(2k/3M)x = a_{cm}$$

Eq.(13.8): $a = -\omega^2 x$ (The minus sign says that x and a have opposite directions, as our diagram shows.)

Our result for a_{cm} is of this form, with $\omega^2 = 2k/3M$ and $\omega = \sqrt{2k/3M}$.

Thus $T = 2\pi/\omega = 2\pi\sqrt{3M/2k}$.

EVALUATE: If there were no friction and the cylinder didn't roll, the period would be $2\pi\sqrt{M/k}$. The period when there is rolling without slipping is larger than this.

13.89 **IDENTIFY** and **SET UP:** Eq.(13.39) gives the period for the bell and Eq.(13.34) gives the period for the clapper.

EXECUTE:

The bell swings as a physical pendulum so its period of oscillation is given by
$$T = 2\pi\sqrt{I/mgd} = 2\pi\sqrt{18.0\ \text{kg}\cdot\text{m}^2/(34.0\ \text{kg})(9.80\ \text{m/s}^2)(0.60\ \text{m})} = 1.885\ \text{s}$$

The clapper is a simple pendulum so its period is given by $T = 2\pi\sqrt{L/g}$.

Thus $L = g(T/2\pi)^2 = (9.80\ \text{m/s}^2)(1.885\ \text{s}/2\pi)^2 = 0.88\ \text{m}$.

EVALUATE: If the cm of the bell were at the geometrical center of the bell, the bell would extend 1.20 m from the pivot, so the clapper is well inside the bell.

13.91 **IDENTIFY** and **SET UP:** Use Eq.(13.34) for the simple pendulum.
Use a physical pendulum (Eq.13.39) for the pendulum in the case.
EXECUTE:

a) $T = 2\pi\sqrt{L/g}$ $L = g(T/2\pi)^2 = (9.80\ \text{m/s}^2)(4.00\ \text{s}/2\pi)^2 = 3.97\ \text{m}$

b) Use a uniform slender rod of mass M and length $L = 0.50$ m. Pivot the rod about an axis that is a distance d abover the center of the rod. The rod will oscillate as a physical pendulum with period $T = 2\pi\sqrt{I/Mgd}$.

Choose d so that $T = 4.00$ s.

$$I = I_{cm} + Md^2 = \tfrac{1}{12}ML^2 + Md^2 = M(\tfrac{1}{12}L^2 + d^2)$$

$$T = 2\pi\sqrt{\frac{I}{Mgd}} = 2\pi\sqrt{\frac{M(\tfrac{1}{12}L^2 + d^2)}{Mgd}} = 2\pi\sqrt{\frac{\tfrac{1}{12}L^2 + d^2}{gd}}.$$

Solve for d and set $L = 0.50$ m and $T = 4.00$ s:

$$gd(T/2\pi)^2 = \tfrac{1}{12}L^2 + d^2$$

$$d^2 - (T/2\pi)^2 gd + L^2/12 = 0$$

$$d^2 - (4.00 \text{ s}/2\pi)^2(9.80 \text{ m/s}^2)d + (0.50 \text{ m})^2/12 = 0$$

$$d^2 - 3.9718d + 0.020833 = 0$$

The quadratic formula gives

$$d = \tfrac{1}{2}[3.9718 \pm \sqrt{(3.9718)^2 - 4(0.020833)}] \text{ m}$$

$$d = (1.9859 \pm 1.9806) \text{ m so } d = 3.97 \text{ m or } d = 0.0053 \text{ m}.$$

The maximum value d can have is $L/2 = 0.25$ m, so the answer we want is $d = 0.0053$ m $= 0.53$ cm.

Therefore, take a slender rod of length 0.50 m and pivot it about an axis that is 0.53 cm above its center.

EVALUATE: Note that $T \to \infty$ as $d \to 0$ (pivot at center of rod) and that if the pivot is at the top of rod then $d = L/2$ and

$$T = 2\pi\sqrt{\frac{\tfrac{1}{12}L^2 + \tfrac{1}{4}L^2}{Lg/2}} = 2\pi\sqrt{\frac{L}{g}\frac{4}{6}} = 2\pi\sqrt{\frac{2L}{3g}} = 2\pi\sqrt{\frac{2(0.50 \text{ m})}{3(9.80 \text{ m/s}^2)}} = 1.16 \text{ s, which}$$

is less than the desired 4.00 s. Thus it is reasonable to expect that there is a value of d between 0 and $L/2$ for which $T = 4.00$ s.

13.93 IDENTIFY: The angular frequency is given by Eq.(13.38). Use the parallel-axis theorem to calculate I in terms of x.

a) SET UP:

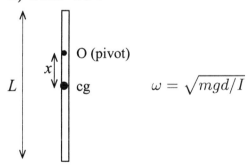

$\omega = \sqrt{mgd/I}$

$d = x$, the distance from the cg of the object (which is at its geometrical center) from the pivot

EXECUTE:

I is the moment of inertia about the axis of rotation through O. By the parallel axis theorem $I_0 = md^2 + I_{cm}$. $I_{cm} = \tfrac{1}{12}mL^2$ (Table 9.2), so $I_0 = mx^2 + \tfrac{1}{12}mL^2$.

$$\omega = \sqrt{\frac{mgx}{mx^2 + \tfrac{1}{12}mL^2}} = \sqrt{\frac{gx}{x^2 + L^2/12}}$$

b) The maximum ω as vary x occurs when $d\omega/dx = 0$.

$$\frac{d\omega}{dx} = 0 \text{ gives } \sqrt{g}\frac{d}{dx}\left(\frac{x^{1/2}}{(x^2 + L^2/12)^{1/2}}\right) = 0.$$

$$\frac{\frac{1}{2}x^{-1/2}}{(x^2 + L^2/12)^{1/2}} - \frac{1}{2}\frac{2x}{(x^2 + L^2/12)^{3/2}}(x^{1/2}) = 0$$

$$x^{-1/2} - \frac{2x^{3/2}}{x^2 + L^2/12} = 0$$

$x^2 + L^2/12 = 2x^2$ so $x = L/\sqrt{12}$. Get maximum ω when the pivot is a distance $L/\sqrt{12}$ above the center of the rod.

c) To answer this question we need an expression for ω_{max}:

In $\omega = \sqrt{\dfrac{gx}{x^2 + L^2/12}}$ substitute $x = L/\sqrt{12}$.

$$\omega_{max} = \sqrt{\frac{g(L/\sqrt{12})}{L^2/12 + L^2/12}} = \frac{g^{1/2}(12)^{-1/4}}{(L/6)^{1/2}} = \sqrt{g/L}(12)^{-1/4}(6)^{1/2} = \sqrt{g/L}(3)^{1/4}$$

$\omega_{max}^2 = (g/L)\sqrt{3}$ and $L = g\sqrt{3}/\omega_{max}^2$

$\omega_{max} = 2\pi$ rad/s gives $L = \dfrac{(9.80 \text{ m/s}^2)\sqrt{3}}{(2\pi \text{ rad/s})^2} = 0.430$ m.

EVALUATE: $\omega \to 0$ as $x \to 0$ and $\omega \to \sqrt{3g/(2L)} = 1.115\sqrt{g/L}$ when $x \to L/2$. ω_{max} is greater than the $x = L/2$ value. A simple pendulum has $\omega = \sqrt{g/L}$; ω_{max} is greater than this.

CHAPTER 14
FLUID MECHANICS

Exercises 1, 11, 13, 15, 19, 21, 25, 31, 33, 35, 39, 41
Problems 45, 47, 51, 53, 55, 59, 63, 65, 67, 69, 71, 73, 79, 81, 83, 87

Exercises

14.1 **IDENTIFY:** Use Eq.(14.1) to calculate the mass and then use $w = mg$ to calculate the weight.

SET UP: $\rho = m/V$ so $m = \rho V$ From Table 14.1, $\rho = 7.8 \times 10^3$ kg/m^3.

EXECUTE: For a cylinder of length L and radius R,

$V = (\pi R^2)L = \pi(0.01425 \text{ m})^2(0.858 \text{ m}) = 5.474 \times 10^{-4}$ m^3.

Then $m = \rho V = (7.8 \times 10^3 \text{ kg/m}^3)(5.474 \times 10^{-4} \text{ m}^3) = 4.27$ kg, and

$w = mg = (4.27 \text{ kg})(9.80 \text{ m/s}^2) = 41.8$ N (about 9.4 lbs). A cart is not needed.

EVALUATE: The rod is less than 1m long and less than 3 cm in diameter, so a weight of around 10 lbs seems reasonable.

14.11 **IDENTIFY and SET UP:** Use Eq.(14.8) to calculate the gauge pressure at this depth. Use Eq.(14.3) to calculate the force the inside and outside pressures exert on the window, and combine the forces as vectors to find the net force.

EXECUTE:

a) gauge pressure $= p - p_0 = \rho g h$ From Table 14.1 the density of seawater is 1.03×10^3 kg/m^3, so

$p - p_0 - \rho g h = (1.03 \times 10^3 \text{ kg/m}^3)(9.80 \text{ m/s}^2)(250 \text{ m}) = 2.52 \times 10^6$ Pa

b) The force on each side of the window is $F = pA$. Inside the pressure is p_0 and outside in the water the pressure is $p = p_0 + \rho g h$.

inside bell $\quad$ outside bell

$F_1 = p_0 A \qquad F_2 = (p_0 + \rho g h)A$

The net force is
$F_2 - F_1 = (p_0 + \rho g h)A - p_0 A = (\rho g h)A$
$F_2 - F_1 = (2.52 \times 10^6 \text{ Pa})\pi(0.150 \text{ m})^2$
$F_2 - F_1 = 1.78 \times 10^5$ N

EVALUATE: The pressure at this depth is very large, over 20 times normal air pressure, and the net force on the window is huge. Diving bells used at such depths must be constructed to withstand these large forces.

14.13 **IDENTIFY and SET UP:** Use Eq.(14.6) to calculate the pressure at the specified depths in the open tube. The pressure is the same at all points the same distance from the bottom of the tubes, so the pressure calculated in part (b) is the pressure in the tank. Gauge pressure is the difference between the absolute pressure and air pressure.

EXECUTE: $p_a = 980$ millibar $= 9.80 \times 10^4$ Pa

a) Apply $p = p_0 + \rho g h$ to the right-hand tube. The top of this tube is open to the air so $p_0 = p_a$. The density of the liquid (mercury) is 13.6×10^3 kg/m^3.

Thus $p = 9.80 \times 10^4$ Pa $+ (13.6 \times 10^3$ kg/m$^3)(9.80$ m/s$^2)(0.0700$ m$) = 1.07 \times 10^5$ Pa.

b) $p = p_0 + \rho g h = 9.80 \times 10^4$ Pa $+ (13.6 \times 10^3$ kg/m$^3)(9.80$ m/s$^2)(0.0400$ m$) = 1.03 \times 10^5$ Pa.

c) Since $y_2 - y_1 = 4.00$ cm the pressure at the mercury surface in the left-hand end tube equals that calculated in part (b). Thus the absolute pressure of gas in the tank is 1.03×10^5 Pa.

d) $p - p_0 = \rho g h = (13.6 \times 10^3$ kg/m$^3)(9.80$ m/s$^2)(0.0400$ m$) = 5.33 \times 10^3$ Pa.

EVALUATE: If Eq.(14.8) is evaluated with the density of mercury and $p - p_a = 1$ atm $= 1.01 \times 10^5$ Pa, then $h = 76$ cm. The mercury columns here are much shorter than 76 cm, so the gauge pressures are much less than 1.0×10^5 Pa.

14.15 **IDENTIFY** and **SET UP:** Apply Eq.(14.6) to the water and mercury columns. The pressure at the bottom of the water column is the pressure at the top of the mercury column.

EXECUTE: With just the mercury, the gauge pressure at the bottom of the cylinder is $p = p_0 + \rho_m g h_m$. With the water to a depth h_w, the gauge pressure at the bottom of the cylinder is $p = p_0 + \rho_m g h_m + \rho_w g h_w$. If this is to be double the first value, then $\rho_w g h_w = \rho_m g h_m$.

$h_w = h_m(\rho_m/\rho_w) = (0.0500$ m$)(13.6 \times 10^3/1.00 \times 10^3) = 0.680$ m

The volume of water is $V = hA = (0.680$ m$)(12.0 \times 10^{-4}$ m$^2) = 8.16 \times 10^{-4}$ m$^3 = 816$ cm^3

EVALUATE: The density of mercury is 13.6 times the density of water and $(13.6)(5$ cm$) = 68$ cm, so the pressure increase from the top to the bottom of a 68-cm tall column of water is the same as the pressure increase from top to bottom for a 5-cm tall column of mercury.

14.19 **IDENTIFY** and **SET UP:** Use Eq.(14.3) to calculate the pressure at the bottom of the kerosene, use Eq.(14.1) to calculate the mass of the kerosene, and then apply Eq.(14.6) and solve for h.

EXECUTE:

The pressure at the bottom of the tank is $p = \dfrac{F_\perp}{A} = \dfrac{16.4 \times 10^3 \text{ N}}{0.0700 \text{ m}^2} = 2.343 \times 10^5 \text{ Pa}$

The density of the kerosene is $\rho = \dfrac{m}{V} = \dfrac{205 \text{ kg}}{0.250 \text{ m}^3} = 820 \text{ kg/m}^3$.

$p = p_0 + \rho g h$ gives that $h = \dfrac{p - p_0}{\rho g} = \dfrac{2.343 \times 10^5 \text{ Pa} - 2.01 \times 10^5 \text{ Pa}}{(820 \text{ kg/m}^3)(9.80 \text{ m/s}^2)} = 4.14 \text{ m}$

EVALUATE: If the tank weren't tapered we could calculate the depth of kerosene from its volume: $h = V/A = 3.57$ m. A depth of 4.14 m for the tapered tank is reasonable.

14.21 **IDENTIFY:** Apply Newton's 2nd law to the woman plus slab. The buoyancy force exerted by the water is upward and given by $B = \rho_{\text{water}} V_{\text{displ}} g$, where V_{displ} is the volume of water displaced.

SET UP: The floating object is the slab of ice plus the woman; the buoyant force must support both. The volume of water displaced equals the volume V_{ice} of the ice.

EXECUTE:

$\sum F_y = ma_y$

$B - m_{\text{tot}} g = 0$

$\rho_{\text{water}} V_{\text{ice}} g = (45.0 \text{ kg} + m_{\text{ice}}) g$

But $\rho = m/V$ so $m_{\text{ice}} = \rho_{\text{ice}} V_{\text{ice}}$

$V_{\text{ice}} = \dfrac{45.0 \text{ kg}}{\rho_{\text{water}} - \rho_{\text{ice}}} = \dfrac{45.0 \text{ kg}}{1000 \text{ kg/m}^3 - 920 \text{ kg/m}^3} = 0.562 \text{ m}^3$.

EVALUATE: The mass of ice is $m_{\text{ice}} = \rho_{\text{ice}} V_{\text{ice}} = 562$ kg.

14.25 **IDENTIFY** and **SET UP:** Use Eq.(14.8) to calculate the gauge pressure at the two depths.

a)

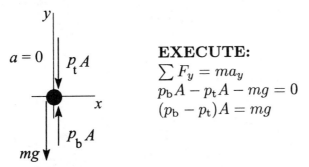

EXECUTE:

$$p - p_0 = \rho g h$$

The upper face is 1.50 cm below the top of the oil, so

$$p - p_0 = (790 \ \text{kg/m}^3)(9.80 \ \text{m/s}^2)(0.0150 \ \text{m})$$

$$p - p_0 = 116 \ \text{Pa}$$

b) The pressure at the interface is $p_{\text{interface}} = p_a + \rho_{\text{oil}}g(0.100 \ \text{m})$. The lower face of the block is 1.50 cm below the interface, so the pressure there is $p = p_{\text{interface}} + \rho_{\text{water}}g(0.0150 \ \text{m})$. Combining these two equations gives

$$p - p_a = \rho_{\text{oil}}g(0.100 \ \text{m}) + \rho_{\text{water}}g(0.0150 \ \text{m})$$

$$p - p_a = [(790 \ \text{kg/m}^3)(0.100 \ \text{m}) + (1000 \ \text{kg/m}^3)(0.0150 \ \text{m})](9.80 \ \text{m/s}^2)$$

$$p - p_a = 921 \ \text{Pa}$$

c) IDENTIFY and **SET UP:** Consider the forces on the block. The area of each face of the block is $A = (0.100 \ \text{m})^2 = 0.0100 \ \text{m}^2$. Let the absolute pressure at the top face be p_t and the pressure at the bottom face be p_b. In Eq.(14.3) use these pressures to calculate the force exerted by the fluids at the top and bottom of the block. Then the free-body diagram for the block is:

y

$a = 0$ $p_t A$

x

$p_b A$

mg

EXECUTE:
$$\sum F_y = ma_y$$
$$p_b A - p_t A - mg = 0$$
$$(p_b - p_t)A = mg$$

Note that $(p_b - p_t) = (p_b - p_a) - (p_t - p_a) = 921 \ \text{Pa} - 116 \ \text{Pa} = 805 \ \text{Pa}$; the difference in absolute pressures equals the difference in gauge pressures.

$$m = \frac{(p_b - p_t)A}{g} = \frac{(805 \ \text{Pa})(0.0100 \ \text{m}^2)}{9.80 \ \text{m/s}^2} = 0.821 \ \text{kg}.$$

And then $\rho = m/V = 0.821 \ \text{kg}/(0.100 \ \text{m})^3 = 821 \ \text{kg/m}^3$.

EVALUATE: We can calculate the buoyant force as $B = (\rho_{\text{oil}}V_{\text{oil}} + \rho_{\text{water}}V_{\text{water}})g$ where $V_{\text{oil}} = (0.0100 \ \text{m}^2)(0.0850 \ \text{m}) = 8.50 \times 10^{-4} \ \text{m}^3$ is the volume of oil displaced by the block and $V_{\text{water}} = (0.0100 \ \text{m}^2)(0.0150 \ \text{m}) = 1.50 \times 10^{-4} \ \text{m}^3$ is the volume of water displaced by the block. This gives $B = (0.821 \ \text{kg})g$. The mass of water displaced equals the mass of the block.

14.31 IDENTIFY and **SET UP:** Apply Eq.(14.10). In part (a) the target variable is V. In part (b) solve for A and then from that get the radius of the pipe.

EXECUTE:

a) $vA = 1.20 \text{ m}^3/\text{s}$

$$v = \frac{1.20 \text{ m}^3/\text{s}}{A} = \frac{1.20 \text{ m}^3/\text{s}}{\pi r^2} = \frac{1.20 \text{ m}^3/\text{s}}{\pi (0.150 \text{ m})^2} = 17.0 \text{ m/s}$$

b) $vA = 1.20 \text{ m}^3/\text{s}$

$v\pi r^2 = 1.20 \text{ m}^3/\text{s}$

$$r = \sqrt{\frac{1.20 \text{ m}^3/\text{s}}{v\pi}} = \sqrt{\frac{1.20 \text{ m}^3/\text{s}}{(3.80 \text{ m/s})\pi}} = 0.317 \text{ m}$$

EVALUATE: The speed is greater where the area and radius are smaller.

14.33 IDENTIFY and **SET UP:**

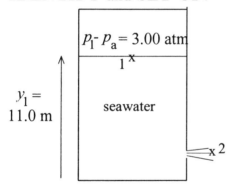

Apply Bernoulli's equation with points 1 and 2 chosen as shown in the sketch. Let $y = 0$ at the bottom of the tank so $y_1 = 11.0$ m and $y_2 = 0$. The target variable is v_2.

$$p_1 + \rho g y_1 + \tfrac{1}{2}\rho v_1^2 = p_2 + \rho g y_2 + \tfrac{1}{2}\rho v_2^2$$

$A_1 v_1 = A_2 v_2$, so $v_1 = (A_2/A_1)v_2$. But the cross-section area of the tank (A_1) is much larger than the cross-section area of the hole (A_2), so $v_1 << v_2$ and the $\tfrac{1}{2}\rho v_1^2$ term can be neglected.

EXECUTE: This gives $\tfrac{1}{2}\rho v_2^2 = (p_1 - p_2) + \rho g y_1$.

Use $p_2 = p_a$ and solve for v_2:

$$v_2 = \sqrt{2(p_1 - p_a)/\rho + 2g y_1} = \sqrt{\frac{2(3.039 \times 10^5 \text{ Pa})}{1030 \text{ kg/m}^3} + 2(9.80 \text{ m/s}^2)(11.0 \text{ m})}$$

$v_2 = 28.4 \text{ m/s}$

EVALUATE: If the pressure at the top surface of the water were air pressure, then Torricelli's theorem (Example 14.8) gives $v_2 = \sqrt{2g(y_1 - y_2)} = 14.7$ m/s. The actual efflux speed is much larger than this due to the excess pressure at the top of the tank.

14.35 IDENTIFY and SET UP:

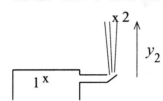

Apply Bernoulli's equation to points 1 and 2 as shown in the sketch. Point 1 is in the mains and point 2 is at the maximum height reached by the stream, so $v_2 = 0$.

Solve for p_1 and then convert this absolute pressure to gauge pressure.

EXECUTE: $p_1 + \rho g y_1 + \frac{1}{2}\rho v_1^2 = p_2 + \rho g y_2 + \frac{1}{2}\rho v_2^2$

Let $y_1 = 0$, $y_2 = 15.0$ m. The mains have large diameter so $v_1 \approx 0$.

Thus $p_1 = p_2 + \rho g y_2$.

But $p_2 = p_a$, so $p_1 - p_a = \rho g y_2 = (1000 \text{ kg/m}^3)(9.80 \text{ m/s}^2)(15.0 \text{ m}) = 1.47 \times 10^5$ Pa.

EVALUATE: This is the gauge pressure at the bottom of a column of water 15.0 m high.

14.39 IDENTIFY: Apply Bernoulli's equation to two points in the pipe: point 1 is where the radius of the pipe is 2.05 cm and point 2 is in the constriction.

SET UP:

$p_1 + \rho g y_1 + \frac{1}{2}\rho v_1^2 = p_2 + \rho g y_2 + \frac{1}{2}\rho v_2^2$

EXECUTE: Horizontal pipe implies $y_1 = y_2$ so $p_1 + \frac{1}{2}\rho v_1^2 = p_2 + \frac{1}{2}\rho v_2^2$

If we solve for v_2 then we can use the discharge rate to calculate A_2 and hence r_2.

Also, $v_1 A_1 = 4.65 \times 10^{-4}$ m³/s so $v_1 = \dfrac{4.65 \times 10^{-4} \text{ m}^3/\text{s}}{1.32 \times 10^{-3} \text{ m}^2} = 0.3523$ m/s.

Then $\frac{1}{2}\rho v_2^2 = \frac{1}{2}\rho v_1^2 + (p_1 - p_2)$ gives

$$v_2 = \sqrt{v_1^2 + \frac{2(p_1 - p_2)}{\rho}} = \sqrt{(0.3523 \text{ m/s})^2 + \frac{2(1.60 \times 10^5 \text{ Pa} - 1.20 \times 10^5 \text{ Pa})}{1000 \text{ kg/m}^3}}$$

$v_2 = 8.951$ m/s

Then $v_2 A_2 = v_2 \pi r_2^2 = 4.65 \times 10^{-4}$ m³/s gives

$$r_2 = \sqrt{\frac{4.65 \times 10^{-4} \text{ m}^3/\text{s}}{v_2 \pi}} = \sqrt{\frac{4.65 \times 10^{-4} \text{ m}^3/\text{s}}{(8.951 \text{ m/s})\pi}} = 4.1 \times 10^{-3} \text{ m} = 0.41 \text{ cm}.$$

EVALUATE: The radius is smaller at the constriction, where the water speed is greater. When y is constant, as it is here, the pressure is less where the speed is greater.

14.41 IDENTIFY and SET UP: Let point 1 be where $r_1 = 4.00$ cm and point 2 be where $r_2 = 2.00$ cm. The volume flow rate vA has the value 7200 cm^3/s at all points in the pipe. Apply Eq.(14.10) to find the fluid speed at points 1 and 2 and then use Bernoulli's equation for these two points to find p_2.

EXECUTE:

$v_1 A_1 = v_1 \pi r_1^2 = 7200$ cm^3, so $v_1 = 1.43$ m/s

$v_2 A_2 = v_2 \pi r_2^2 = 7200$ cm^3, so $v_2 = 5.73$ m/s

$p_1 + \rho g y_1 + \frac{1}{2}\rho v_1^2 = p_2 + \rho g y_2 + \frac{1}{2}\rho v_2^2$

$y_1 = y_2$ and $p_2 = 2.40 \times 10^5$ Pa, so $p_2 = p_1 + \frac{1}{2}\rho(v_1^2 - v_2^2) = 2.25 \times 10^5$ Pa

EVALUATE: Where the area decreases the speed increases and the pressure decreases.

Problems

14.45 IDENTIFY: Use Eq.(14.8) to find the gauge pressure versus depth, use Eq.(14.3) to relate the pressure to the force on a strip of the gate, calculate the torque as force times moment arm, and follow the procedure outlined in the hint to calculate the total torque.

SET UP:

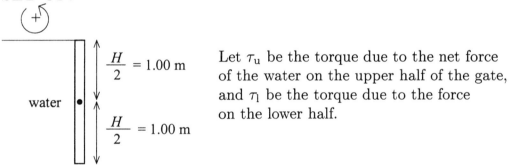

Let τ_u be the torque due to the net force of the water on the upper half of the gate, and τ_l be the torque due to the force on the lower half.

$\dfrac{H}{2} = 1.00$ m

water

$\dfrac{H}{2} = 1.00$ m

With the indicated sign convention, τ_l is positive and τ_u is negative, so the net torque about the hinge is $\tau = \tau_l - \tau_u$. Let H be the height of the gate.

Upper-half of gate:

Calculate the torque due to the force on a narrow strip of height dy located a distance y below the top of the gate. Then integrate to get the total torque.

dF

$\left(\dfrac{H}{2} - y\right)$

y

dy

axis

The net force on the strip is $dF = p(y)\, dA$, where $p(y) = \rho g y$ is the pressure at this depth and $dA = W\, dy$ with $W = 4.00$ m

$dF = \rho g y W\, dy$

The moment arm is $(H/2 - y)$, so $d\tau = \rho g W (H/2 - y) y\, dy$.

$\tau_u = \int_0^{H/2} d\tau = \rho g W \int_0^{H/2}(H/2 - y)y\, dy = \rho g W\left((H/4)y^2 - y^3/3\right)\Big|_0^{H/2}$

$\tau_u = \rho g W(H^3/16 - H^3/24) = \rho g W(H^3/48)$

$\tau_u = (1000 \text{ kg/m}^3)(9.80 \text{ m/s}^2)(4.00 \text{ m})(2.00 \text{ m})^3/48 = 6.533 \times 10^3 \text{ N} \cdot \text{m}$

Lower-half of gate:

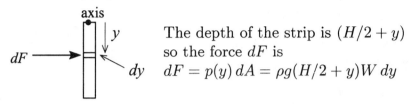

The depth of the strip is $(H/2 + y)$ so the force dF is
$dF = p(y) \, dA = \rho g(H/2 + y)W \, dy$

The moment arm is y, so $d\tau = \rho g W (H/2 + y)y \, dy$.

$\tau_l = \int_0^{H/2} d\tau = \rho g W \int_0^{H/2} (H/2 + y)y \, dy = \rho g W \left((H/4)y^2 + y^3/3\right) \Big|_0^{H/2}$

$\tau_l = \rho g W (H^3/16 + H^3/24) = \rho g W (5H^3/48)$

$\tau_l = (1000 \text{ kg/m}^3)(9.80 \text{ m/s}^2)(4.00 \text{ m})5(2.00 \text{ m})^3/48 = 3.267 \times 10^4 \text{ N} \cdot \text{m}$

Then $\tau = \tau_l - \tau_u = 3.267 \times 10^4 \text{ N} \cdot \text{m} - 6.533 \times 10^3 \text{ N} \cdot \text{m} = 2.61 \times 10^4 \text{ N} \cdot \text{m}$.

EVALUATE: The forces and torques on the upper and lower halves of the gate are in opposite directions so find the net value by subtracting the magnitudes. The torque on the lower half is larger than the torque on the upper half since pressure increases with depth.

14.47 **IDENTIFY** and **SET UP:** Apply Eq.(14.6) and solve for g.
Then use Eq.(12.4) to relate g to the mass of the planet.

EXECUTE: $p - p_0 = \rho g d$.

This expression gives that $g = (p - p_0)/\rho d = (p - p_0)V/md$.

But also $g = Gm_p/R^2$ (Eq.(12.4) applied to the planet rather than to earth.)

Setting these two expressions for g equal gives $Gm_p/R^2 = (p - p_0)V/md$

and $m_p = (p - p_0)VR^2/Gmd$.

EVALUATE: The greater p is at a given depth, the greater g is for the planet and greater g means greater m_p.

14.51 **a) IDENTIFY** and **SET UP:**

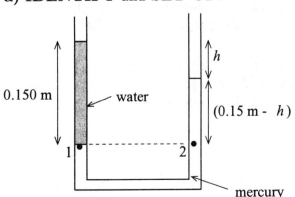

Apply $p = p_0 + \rho g h$ to the water in the left-hand arm of the tube.

EXECUTE: $p_0 = p_a$, so the gauge pressure at the interface (point 1) is

$p - p_a = \rho g h = (1000 \text{ kg/m}^3)(9.80 \text{ m/s}^2)(0.150 \text{ m}) = 1470 \text{ Pa}$

b) IDENTIFY and **SET UP:** The pressure at point 1 equals the pressure at point 2. Apply Eq.(14.6) to the right-hand arm of the tube and solve for h.

EXECUTE: $p_1 = p_a + \rho_w g(0.150 \text{ m})$ and $p_2 = p_a + \rho_{Hg}g(0.150 \text{ m} - h)$

$p_1 = p_2$ implies $\rho_w g(0.150 \text{ m}) = \rho_{Hg}g(0.150 \text{ m} - h)$

$0.150 \text{ m} - h = \dfrac{\rho_w(0.150 \text{ m})}{\rho_{Hg}} = \dfrac{(1000 \text{ kg/m}^3)(0.150 \text{ m})}{13.6 \times 10^3 \text{ kg/m}^3} = 0.011 \text{ m}$

$h = 0.150 \text{ m} - 0.011 \text{ m} = 0.139 \text{ m} = 13.9 \text{ cm}$

EVALUATE: The height of mercury above the bottom level of the water is 1.1 cm. This height of mercury produces the same gauge pressure as a height of 15.0 cm of water.

14.53 IDENTIFY: Apply Newton's 2nd law to the barge plus its contents. Apply Archimedes' principle to express the buoyancy force B in terms of the volume of the barge.

SET UP: Free-body diagram for the barge plus coal.

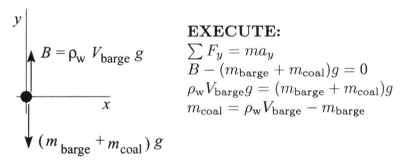

EXECUTE:

$\sum F_y = ma_y$

$B - (m_{barge} + m_{coal})g = 0$

$\rho_w V_{barge}g = (m_{barge} + m_{coal})g$

$m_{coal} = \rho_w V_{barge} - m_{barge}$

$V_{barge} = (22 \text{ m})(12 \text{ m})(40 \text{ m}) = 1.056 \times 10^4 \text{ m}^3$

The mass of the barge is $m_{barge} = \rho_s V_s$, where s refers to steel.

From Table 14.1, $\rho_s = 7800 \text{ kg/m}^3$. The volume V_s is 0.040 m times the total area of the five pieces of steel that make up the barge:

$V_s = (0.040 \text{ m})[2(22 \text{ m})(12 \text{ m}) + 2(40 \text{ m})(12 \text{ m}) + (22 \text{ m})(40 \text{ m})] = 94.7 \text{ m}^3$.

Therefore, $m_{barge} = \rho_s V_s = (7800 \text{ kg/m}^3)(94.7 \text{ m}^3) = 7.39 \times 10^5 \text{ kg}$.

Then $m_{coal} = \rho_w V_{barge} - m_{barge} = (1000 \text{ kg/m}^3)(1.056 \times 10^4 \text{ m}^3) - 7.39 \times 10^5 \text{ kg} = 9.8 \times 10^6 \text{ kg}$.

The volume of this mass of coal is $V_{coal} = m_{coal}/\rho_{coal} = 9.8 \times 10^6 \text{ kg}/1500 \text{ kg/m}^3 = 6500 \text{ m}^3$; this is less that V_{barge} so it will fit into the barge.

EVALUATE: The buoyancy force B must support both the weight of the coal and also the mass of the barge. The weight of the coal is about 13 times the weight of the barge. The buoyancy force increases when more of the barge is submerged, so when it holds the maximum mass of coal the barge is fully submerged.

14.55 **IDENTIFY:** Apply Newton's 2nd law to the car. The buoyancy force is given by Archimedes' principle.

a) SET UP: Free-body diagram for the floating car (V_{sub} is the volume that is submerged.)

EXECUTE:
$$\sum F_y = ma_y$$
$$B - mg = 0$$
$$\rho_w V_{sub} g - mg = 0$$

$V_{sub} = m/\rho_w = (900 \text{ kg})/(1000 \text{ kg/m}^3) = 0.900 \text{ m}^3$

$V_{sub}/V_{obj} = (0.900 \text{ m}^3)/(3.0 \text{ m}^3) = 0.30 = 30\%$

EVALUATE: The average density of the car is $(900 \text{ kg})/(3.0 \text{ m}^3) = 300 \text{ kg/m}^3$. $\rho_{car}/\rho_{water} = 0.30$; this equals V_{sub}/V_{obj}.

b) SET UP: When the car starts to sink it is fully submerged and the buoyant force is equal to the weight of the car plus the water that is inside it.

EXECUTE: When the car is full submerged $V_{sub} = V$, the volume of the car and

$B = \rho_{water} V g = (1000 \text{ kg/m}^3)(3.0 \text{ m}^3)(9.80 \text{ m/s}^2) = 2.94 \times 10^4 \text{ N}$

The weight of the car is $mg = (900 \text{ kg})(9.80 \text{ m/s}^2) = 8820 \text{ N}$.

Thus the weight of the water in the car when it sinks is the buoyant force minus the weight of the car itself:

$m_{water} = (2.94 \times 10^4 \text{ N} - 8820 \text{ N})/(9.80 \text{ m/s}^2) = 2.10 \times 10^3 \text{ kg}$

And $V_{water} = m_{water}/\rho_{water} = (2.10 \times 10^3 \text{ kg})/(1000 \text{ kg/m}^3) = 2.10 \text{ m}^3$

The fraction this is of the total interior volume is $(2.10 \text{ m}^3)/(3.00 \text{ m}^3) = 0.70 = 70\%$

EVALUATE: The average density of the car plus the water inside it is $(900 \text{ kg} + 2100 \text{ kg})/(3.0 \text{ m}^3) = 1000 \text{ kg/m}^3$, so $\rho_{car} = \rho_{water}$ when the car starts to sink.

14.59 **a) IDENTIFY:** Apply Newton's 2nd law to the airship. The buoyancy force is given by Archimedes' principle; the fluid that exerts this force is the air.

SET UP: Free-body diagram for the dirigible. The lift corresponds to a mass $m_{lift} = (120 \times 10^3 \text{ N})/(9.80 \text{ m/s}^2) = 1.224 \times 10^4 \text{ kg}$. The mass m_{tot} is 1.224×10^4 kg plus the mass m_{gas} of the gas that fills the dirigible. B is the buoyant force exerted by the air.

EXECUTE:
$$\sum F_y = ma_y$$
$$B - m_{tot}g = 0$$
$$\rho_{air}Vg = (1.224 \times 10^4 \text{ kg} + m_{gas})g$$

Write m_{gas} in terms of V: $m_{gas} = \rho_{gas}V$

And let g divide out; the equation becomes $\rho_{air}V = 1.224 \times 10^4 \text{ kg} + \rho_{gas}V$

$$V = \frac{1.224 \times 10^4 \text{ kg}}{1.20 \text{ kg/m}^3 - 0.0899 \text{ kg/m}^3} = 1.10 \times 10^4 \text{ m}^3$$

EVALUATE: The density of the airship is less than the density of air and the airship is totally submerged in the air, so the buoyancy force exceeds the weight of the airship.

b) SET UP: Let m_{lift} be the mass that could be lifted.

EXECUTE: From part (a), $m_{lift} = (\rho_{air} - \rho_{gas})V$

$= (1.20 \text{ kg/m}^3 - 0.166 \text{ kg/m}^3)(1.10 \times 10^4 \text{ m}^3) = 1.14 \times 10^4 \text{ kg}.$

The lift force is $m_{lift} = (1.14 \times 10^4 \text{ kg})(9.80 \text{ m/s}^2) = 112 \text{ kN}.$

EVALUATE: The density of helium is less than that of air but greater than that of hydrogen. Helium provides lift, but less lift than hydrogen. Hydrogen is not used because it is highly explosive in air.

14.63 IDENTIFY: Apply the 2nd condition of equilibrium to the balance arm and apply the first condition of equilibrium to the block and to the brass mass. The buoyancy force on the wood is given by Archimedes' principle and the buoyancy force on the brass mass is ignored.

SET UP:

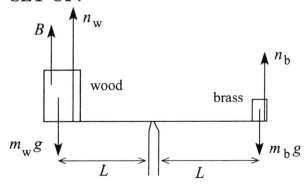

The buoyant force on the brass is neglected, but we include the buoyant force B on the block of wood. n_w and n_b are the normal forces exerted by the balance arm on which the objects sit.

Free-body diagram for the balance arm

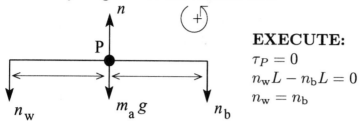

EXECUTE:
$\tau_P = 0$
$n_w L - n_b L = 0$
$n_w = n_b$

SET UP: Free-body diagram for the brass mass

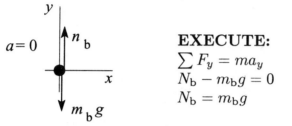

EXECUTE:
$\sum F_y = ma_y$
$N_b - m_b g = 0$
$N_b = m_b g$

Free-body diagram for the block of wood:

$\sum F_y = ma_y$
$n_w + B - m_w g = 0$
$n_w = m_w g - B$

But $n_b = n_w$ implies $m_b g = m_w g - B$.

And $B = \rho_{air} V_w g = \rho_{air}(m_w/\rho_w)g$, so $m_b g = m_w g - \rho_{air}(m_w/\rho_w)g$.

$$m_w = \frac{m_b}{1 - \rho_{air}/\rho_w} = \frac{0.0950 \text{ kg}}{1 - ((1.20 \text{ kg/m}^3)/(150 \text{ kg/m}^3))} = 0.0958 \text{ kg}$$

EVALUATE: The mass of the wood is greater than the mass of the brass; the wood is partially supported by the buoyancy force exerted by the air. The buoyancy in air of the brass can be neglected because the density of brass is much more than the density of air; the buoyancy force exerted on the brass by the air is much less than the weight of the brass. The density of the balsa wood is much less than the density of the brass, so the buoyancy force on the balsa wood is not such a small fraction of its weight.

14.65 IDENTIFY: Apply Newton's 2nd law to the ingot. Use the expression for the buoyancy force given by Archimedes' principle to solve for the volume of the ingot. Then use the facts that the total mass is the mass of the gold plus the mass of the aluminum and that the volume of the ingot is the volume of the gold plus the volume of the aluminum.

SET UP: Free-body diagram for the piece of alloy

EXECUTE:
$$\sum F_y = ma_y$$
$$B + T - m_{tot}g = 0$$
$$B = m_{tot}g - T$$
$$B = 45.0 \text{ N} - 39.0 \text{ N} = 6.0 \text{ N}$$

Also, $m_{tot}g = 45.0$ N so $m_{tot} = 45.0$ N$/(9.80$ m$/$s$^2) = 4.59$ kg.

We can use the known value of the buoyant force to calculate the volume of the object:

$$B = \rho_w V_{obj} g = 6.0 \text{ N}$$

$$V_{obj} = \frac{6.0 \text{ N}}{\rho_w g} = \frac{6.0 \text{ N}}{(1000 \text{ kg/m}^3)(9.80 \text{ m/s}^2)} = 6.122 \times 10^{-4} \text{ m}^3$$

We know two things:

(1) The mass m_g of the gold plus the mass m_a of the aluminum must add to m_{tot}:

$$m_g + m_a = m_{tot}$$

We write this in terms of the volumes V_g and V_a of the gold and aluminum:

$$\rho_g V_g + \rho_a V_a = m_{tot}$$

(2) The volumes V_a and V_g must add to give V_{obj}:

$$V_a + V_g = V_{obj} \text{ so that } V_a = V_{obj} - V_g$$

Use this in the equation in (1) to eliminate V_a:

$$\rho_g V_g + \rho_a (V_{obj} - V_g) = m_{tot}$$

$$V_g = \frac{m_{tot} - \rho_a V_{obj}}{\rho_g - \rho_a} = \frac{4.59 \text{ kg} - (2.7 \times 10^3 \text{ kg/m}^3)(6.122 \times 10^{-4} \text{ m}^3)}{19.3 \times 10^3 \text{ kg/m}^3 - 2.7 \times 10^3 \text{ kg/m}^3} =$$
$$1.769 \times 10^{-4} \text{ m}^3.$$

Then $m_g = \rho_g V_g = (19.3 \times 10^3 \text{ kg/m}^3)(1.769 \times 10^{-4} \text{ m}^3) = 3.41$ kg and the weight of gold is $w_g = m_g g = 33.4$ N.

EVALUATE: The gold is 29% of the volume but 74% of the mass, since the density of gold is much greater than the density of aluminum.

14.67 a) IDENTIFY: Apply Newton's 2nd law to the crown. The buoyancy force is given by Archimedes' principle. The target variable is the ratio ρ_c/ρ_w (c = crown, w = water).

SET UP: Free-body diagram for the crown

EXECUTE:
$$\sum F_y = ma_y$$
$$T + B - w = 0$$

$$T = fw$$
$B = \rho_w V_c g$, where ρ_w = density of water, V_c = volume of crown

Then $fw + \rho_w V_c g - w = 0$.

$(1 - f)w = \rho_w V_c g$

Use $w = \rho_c V_c g$, where ρ_c = density of crown.

$(1 - f)\rho_c V_c g = \rho_w V_c g$

$$\frac{\rho_c}{\rho_w} = \frac{1}{1 - f}, \text{ as was to be shown.}$$

$f \to 0$ gives $\rho_c/\rho_w = 1$ and $T = 0$. These values are consistent. If the density of the crown equals the density of the water, the crown just floats, fully submerged, and the tension should be zero.

When $f \to 1$, $\rho_c \gg \rho_w$ and $T = w$. If $\rho_c \gg \rho_w$ then B is negligible relative to the weight w of the crown and T should equal w.

b) "apparent weight" equals T in rope when the crown is immersed in water. $T = fw$, so need to compute f.

$\rho_c = 19.3 \times 10^3$ kg/m^3; $\rho_w = 1.00 \times 10^3$ kg/m^3

$$\frac{\rho_c}{\rho_w} = \frac{1}{1 - f} \text{ gives } \frac{19.3 \times 10^3 \text{ kg/m}^3}{1.00 \times 10^3 \text{ kg/m}^3} = \frac{1}{1 - f}$$

$19.3 = 1/(1 - f)$ and $f = 0.9482$

Then $T = fw = (0.9482)(12.9 \text{ N}) = 12.2$ N.

c) Now the density of the crown is very nearly the density of lead;

$\rho_c = 11.3 \times 10^3$ kg/m^3.

$$\frac{\rho_c}{\rho_w} = \frac{1}{1 - f} \text{ gives } \frac{11.3 \times 10^3 \text{ kg/m}^3}{1.00 \times 10^3 \text{ kg/m}^3} = \frac{1}{1 - f}$$

$11.3 = 1/(1 - f)$ and $f = 0.9115$

Then $T = fw = (0.9115)(12.9 \text{ N}) = 11.8$ N.

EVALUATE: In part (c) the average density of the crown is less than in part (b), so the volume is greater. B is greater and T is less. These measurements can be used to determine if the crown is solid gold, without damaging the crown.

14.69 IDENTIFY and **SET UP:** Use Archimedes' principle for B.

a) $B = \rho_{water} V_{tot} g$, where V_{tot} is the total volume of the object.

$V_{tot} = V_m + V_0$, where V_m is the volume of the metal.

EXECUTE: $V_m = w/g\rho_m$ so $V_{tot} = w/g\rho_m + V_0$

This gives $B = \rho_{water} g(w/g\rho_m + V_0)$

Solving for V_0 gives $V_0 = B/(\rho_{water} g) - w/(\rho_m g)$, as was to be shown.

b) The expression derived in part (a) gives

$$V_0 = \frac{20 \text{ N}}{(1000 \text{ kg/m}^3)(9.80 \text{ m/s}^2)} - \frac{156 \text{ N}}{(8.9 \times 10^3 \text{ kg/m}^3)(9.80 \text{ m/s}^2)} = 2.52 \times 10^{-4} \text{ m}^3$$

$$V_{tot} = \frac{B}{\rho_{water} g} = \frac{20 \text{ N}}{(1000 \text{ kg/m}^3)(9.80 \text{ m/s}^2)} = 2.04 \times 10^{-3} \text{ m}^3 \text{ and}$$

$V_0/V_{tot} = (2.52 \times 10^{-4} \text{ m}^3)/(2.04 \times 10^{-3} \text{ m}^3) = 0.124$.

EVALUATE: When $V_0 \to 0$, the object is solid and $V_{obj} = V_m = w/(\rho_m g)$. For $V_0 = 0$, the result in part (a) gives $B = (w/\rho_m)\rho_w = V_m \rho_w g = V_{obj}\rho_w g$, which agrees with Archimedes' principle. As V_0 increases with the weight kept fixed, the total volume of the object increases and there is an increase in B.

14.71 IDENTIFY and **SET UP:** Apply the first condition of equilibrium to the barge plus the anchor. Use Archimedes' principle to relate the weight of the boat and anchor to the amount of water displaced. In both cases the total buoyant force must equal the weight of the barge plus the weight of the anchor. Thus the total amount of water displaced must be the same when the anchor is in the boat as when it is over the side. When the anchor is in the water the barge displaces less water, less by the amount the anchor displaces. Thus the barge rises in the water.

EXECUTE: The volume of the anchor is

$V_{anchor} = m/\rho = (35.0 \text{ kg})/(7860 \text{ kg/m}^3) = 4.453 \times 10^{-3} \text{ m}^3$.

The barge rises in the water a vertical distance h given by $hA = 4.453 \times 19^{-3} \text{ m}^3$, where A is the area of the bottom of the barge.

$h = (4.453 \times 10^{-3} \text{ m}^3)/(8.00 \text{ m}^2) = 5.57 \times 10^{-4} \text{ m}$.

EVALUATE: The barge rises a very small amount. The buoyancy force on the barge plus the buoyancy force on the anchor must equal the weight of the barge plus the weight of the anchor. When the anchor is in the water, the buoyancy force on it is less than its weight (the anchor doesn't float on its own), so part of the buoyancy force on the barge is used to help support the anchor. If the rope is cut, the buoyancy force on the barge must equal only the weight of the barge and the barge rises still farther.

14.73 IDENTIFY: Apply Newton's 2nd law to the block. In part (a), use Archimedes' principle for the buoyancy force. In part (b), use Eq.(14.6) to find the pressure at the lower face of the block and then use Eq.(14.3) to calculate the force the fluid exerts.

a) SET UP: Free-body diagram for the block

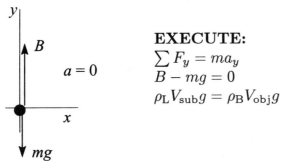

EXECUTE:
$$\sum F_y = ma_y$$
$$B - mg = 0$$
$$\rho_L V_{sub} g = \rho_B V_{obj} g$$

The fraction of the volume that is submerged is $V_{sub}/V_{obj} = \rho_B/\rho_L$.

Thus the fraction that is *above* the surface is $V_{above}/V_{obj} = 1 - \rho_B/\rho_L$.

EVALUATE: If $\rho_B = \rho_L$ the block is totally submerged as it floats.

b) SET UP: Let the water layer have depth d.

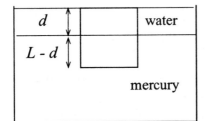

EXECUTE:
$$p = p_0 + \rho_w g d + \rho_L g(L - d)$$

Applying $\sum F_y = ma_y$ to the block gives
$(p - p_0)A - mg = 0$.

$$[\rho_w g d + \rho_L g(L - d)]A = \rho_B LAg$$

A and g divide out and $\rho_w d + \rho_L(L - d) = \rho_B L$

$$d(\rho_w - \rho_L) = (\rho_B - \rho_L)L$$

$$d = \left(\frac{\rho_L - \rho_B}{\rho_L - \rho_w}\right) L$$

c) $d = \left(\dfrac{13.6 \times 10^3 \text{ kg/m}^3 - 7.8 \times 10^3 \text{ kg/m}^3}{13.6 \times 10^3 \text{ kg/m}^3 - 1000 \text{ kg/m}^3}\right)(0.100 \text{ m}) = 0.0460 \text{ m} = 4.60 \text{ cm}$

EVALUATE: In the expression derived in part (b), if $\rho_B = \rho_L$ the block floats in the liquid totally submerged and no water needs to be added. If $\rho_L \to \rho_w$ the block continues to float with a fraction $1 - \rho_B/\rho_w$ above the water as water is added, and the water never reaches the top of the block ($d \to \infty$).

14.79 IDENTIFY and SET UP:

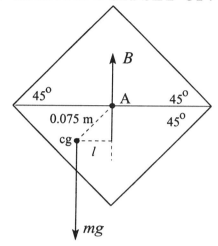

The resultant buoyant force acts at the geometrical center of the submerged portion of the object. The weight of the object acts at the center of gravity of the object. These two points are displaced from each other as in Fig.14.39b, and this gives rise to a restoring torque about point A.

EXECUTE: For an axis at point A

$\tau_B = 0$ and $\tau_{mg} = mgl = mg(0.075 \text{ m})\cos 45° = (0.053 \text{ m})mg$

The block is floating, so $B = mg$. Calculate B in order to find the weight mg of the block. Half of the volume of the block is submerged, so

$V_{\text{sub}} = \frac{1}{2}(0.30 \text{ m})^3 = 0.0135 \text{ m}^3$.

Then $B = \rho_w V_{\text{sub}} g = (1000 \text{ kg/m}^3)(0.0135 \text{ m}^3)(9.80 \text{ m/s}^2) = 132.3 \text{ N}$.

Therefore, $\sum \tau_A = \tau_{\text{mg}} = (0.053 \text{ m})(132.3 \text{ N}) = 7.0 \text{ N} \cdot \text{m}$.

EVALUATE: The net torque is a restoring torque; it is directed to bring the block back to its equilibrium position, the position shown in Fig.14.39a. In the position shown in Fig.14.39a, both the weight and B have lines of action that pass through the center of the block and there is no torque about this point.

14.81 IDENTIFY: Use Bernoulli's equation to find the velocity with which the water flows out the hole.

SET UP: The water level in the vessel will rise until the volume flow rate into the vessel, 2.40×10^{-4} m^3/s, equals the volume flow rate out the hole in the bottom.

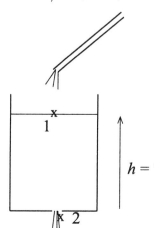

Let points 1 and 2 be chosen as in the sketch.

EXECUTE:

Bernoulli's equation: $p_1 + \rho g y_1 + \frac{1}{2}\rho v_1^2 = p_2 + \rho g y_2 + \frac{1}{2}\rho v_2^2$

Volume flow rate out of hole equals volume flow rate from tube gives that

$$v_2 A_2 = 2.40 \times 10^{-4} \text{ m}^3/\text{s} \text{ and } v_2 = \frac{2.40 \times 10^{-4} \text{ m}^3/\text{s}}{1.50 \times 10^{-4} \text{ m}^2} = 1.60 \text{ m/s}$$

$A_1 >> A_2$ and $v_1 A_1 = v_2 A_2$ says that $\frac{1}{2}\rho v_1^2 << \frac{1}{2}\rho v_2^2$; neglect the $\frac{1}{2}\rho v_1^2$ term.

Measure y from the botom of the bucket, so $y_2 = 0$ and $y_1 = h$.

$p_1 = p_2 = p_a$ (air pressure)

Then $p_a + \rho g h = p_a + \frac{1}{2}\rho v_2^2$ and

$h = v_2^2/2g = (1.60 \text{ m/s})^2/2(9.80 \text{ m/s}^2) = 0.131 \text{ m} = 13.1 \text{ cm}$

EVALUATE: The greater the flow rate into the bucket, the larger v_2 will be at equilibrium and the higher the water will rise in the bucket.

14.83 IDENTIFY and SET UP:

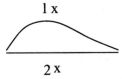

1 x

2 x

Apply Bernoulli's equation to points 1 and 2, where point 1 is just above the wing and point 2 is just below the wing.

The target variable is v_1.

$p_1 + \rho g y_1 + \frac{1}{2}\rho v_1^2 = p_2 + \rho g y_2 + \frac{1}{2}\rho v_2^2$

A "lift" of 2000 N/m² means that $p_2 - p_1 = 2000$ Pa.

EXECUTE: Solving for v_1 gives $v_1 = \sqrt{2(p_2 - p_1)/\rho + v_2^2 - 2g(y_1 - y_2)}$

Note that $2(p_2 - p_1)/\rho = 2(2000 \text{ Pa})/(1.20 \text{ kg/m}^3) = 3333 \text{ m}^2/\text{s}^2$. We aren't given a value for $y_1 - y_2$, but it must be 1 m or so. For $y_1 - y_2 = 1$ m, $2g(y_1 - y_2) = 19.6$ m²/s².

So, $2g(y_1 - y_2) << 2(p_2 - p_1)/\rho$ and can be neglected.

Then $v_1 = \sqrt{2(p_2 - p_1)/\rho + v_2^2} = \sqrt{3333 \text{ m}^2/\text{s}^2 + (120 \text{ m/s})^2} = 133 \text{ m/s}$.

EVALUATE: The larger v_1 is the greater $p_2 - p_1$ and the larger is the lift. If v_1 is too large the flow is no longer streamline but becomes turbulent and Bernoulli's equation no longer applies.

14.87 a) IDENTIFY: Apply constant acceleration equations to the falling liquid to find its speed as a function of the distance below the outlet. Then apply Eq.(14.10) to relate the speed to the radius of the stream.

SET UP:

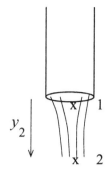

Let point 1 be at the end of the pipe and let point 2 be in the stream of liquid at a distance y_2 below the end of the tube.

Consider the free-fall of the liquid. Take $+y$ to be downward.

Free-fall implies $a_y = g$. v_y is positive, so replace it by the speed v.

EXECUTE: $v_2^2 = v_1^2 + 2a(y - y_0)$ gives $v_2^2 = v_1^2 + 2gy_2$ and $v_2 = \sqrt{v_1^2 + 2gy_2}$.

Equation of continuity says $v_1 A_1 = v_2 A_2$

And since $A = \pi r^2$ this becomes $v_1 \pi r_1^2 = v_2 \pi r_2^2$ and $v_2 = v_1 (r_1/r_2)^2$.

Use this in the above to eliminate v_2: $v_1 (r_1^2/r_2^2) = \sqrt{v_1^2 + 2gy_2}$

$r_2 = r_1 \sqrt{v_1}/(v_1^2 + 2gy_2)^{1/4}$

To correspond to the notation in the problem, let $v_1 = v_0$ and $r_1 = r_0$, since point 1 is where the liquid first leaves the pipe, and let r_2 be r and y_2 be y. The equation we have derived then becomes $r = r_0 \sqrt{v_0}/(v_0^2 + 2gy)^{1/4}$

b) $v_0 = 1.20$ m/s

We want the value of y that gives $r = \frac{1}{2}r_0$, or $r_0 = 2r$

The result obtained in part (a) says $r^4(v_0^2 + 2gy) = r_0^4 v_0^2$

Solving for y gives $y = \dfrac{[(r_0/r)^4 - 1]v_0^2}{2g} = \dfrac{(16 - 1)(1.20 \text{ m/s})^2}{2(9.80 \text{ m/s}^2)} = 1.10$ m

EVALUATE: The equation derived in part (a) says that r decreases with distance below the end of the pipe.

CHAPTER 15

MECHANICAL WAVES

Exercises

15.1 **IDENTIFY:** Determine λ and f from the information given and then use Eq.(15.1) to calculate v. The total vertical distance traveled is twice the amplitude.

SET UP and **EXECUTE:**

a) The distance between crests is the wavelength; $\lambda = 6.0$ m.

Time between largest upward displacement to largest downward displacement is one-half the period; $T/2 = 2.5$ s so $T = 5.0$ s.

$f = 1/T = 1/5.0$ s $= 0.20$ Hz

Then $v = f\lambda = (0.20$ Hz$)(6.0$ m$) = 1.2$ m/s.

b) The distance between the highest and lowest points is twice the amplitude, so $2A = 0.62$ m and $A = 0.31$ m.

c) v unchanged but A becomes 0.15 m.

d) EVALUATE: No. Expect longitudinal as well as transverse surface waves.

v is unchanged in part (c) because the wave speed is independent of the amplitude.

15.5 **IDENTIFY** and **SET UP:** Use Eq.(15.1) to calculate the wavelength λ.

EXECUTE:

a) $v = 344$ m/s; $v = f\lambda$ so $\lambda = v/f$

$f = 20.0$ Hz: $\lambda = (344$ m/s$)/(20.0$ Hz$) = 17.2$ m

$f = 20,000$ Hz: $\lambda = (344$ m/s$)/(20,000$ Hz$) = 0.0172$ m

b) $v = 1480$ m/s

$f = 20.0$ Hz: $\lambda = (1480$ m/s$)/(20.0$ Hz$) = 74.0$ m

$f = 20,000$ Hz: $\lambda = (1480$ m/s$)/(20,000$ Hz$) = 0.0740$ m

EVALUATE: For a given v, λ decreases when f increases. For a given f, λ increases when v increases.

15.7 **IDENTIFY:** Use Eq.(15.1) to calculate v. $T = 1/f$ and k is defined by Eq.(15.5). The general form of the wave function is given by Eq.(15.8), which is the equation for the transverse displacement.

SET UP: $v = 8.00$ m/s, $A = 0.0700$ m, $\lambda = 0.320$ m

EXECUTE:

a) $v = f\lambda$ so $f = v/\lambda = (8.00 \text{ m/s})/(0.320 \text{ m}) = 25.0$ Hz

$T = 1/f = 1/25.0$ Hz $= 0.0400$ s

$k = 2\pi/\lambda = 2\pi$ rad$/0.320$ m $= 19.6$ rad/m

b) For a wave traveling in the $-x$-direction,

$y(x, t) = A \cos 2\pi(x/\lambda + t/T)$ (Eq.(15.8).

At $x = 0$, $y(0, t) = A \cos 2\pi(t/T)$, so $y = 0$ at $t = 0$ and y is positive for t slightly greater than zero. This equation describes the wave specified in the problem.

Substitute in numerical values:

$y(x, t) = (0.0700 \text{ m}) \cos(2\pi(x/0.320 \text{ m} + t/0.0400 \text{ s}))$.

Or, $y(x, t) = (0.0700 \text{ m}) \cos((19.6 \text{ m}^{-1})x + (157 \text{ rad/s})t)$.

c) From part (b), $y = (0.0700 \text{ m}) \cos(2\pi(x/0.320 \text{ m} + t/0.0400 \text{ s}))$.

Plug in $x = 0.360$ m and $t = 0.150$ s:

$y = (0.0700 \text{ m}) \cos(2\pi(0.360 \text{ m}/0.320 \text{ m} + 0.150 \text{ s}/0.0400 \text{ s}))$

$y = (0.0700 \text{ m}) \cos[2\pi(4.875 \text{ rad})] = +0.0495$ m $= +4.95$ cm

d) In part (c) $t = 0.150$ s.

$y = A$ means $\cos(2\pi(x/\lambda + t/T)) = A$

$\cos\theta = 1$ for $\theta = 0, 2\pi, 4\pi, \ldots = n(2\pi)$ for $n = 0, 1, 2, \ldots$

So $y = A$ when $2\pi(x/\lambda + t/T) = n(2\pi)$ or $x/\lambda + t/T = n$

$t = T(n - x/\lambda) = (0.0400 \text{ s})(n - 0.360 \text{ m}/0.320 \text{ m}) = (0.0400 \text{ s})(n - 1.125)$

For $n = 4$, $t = 0.1115$ s (before the instant in part (c))

For $n = 5$, $t = 0.1550$ s (the first occurrence of $y = A$ after the instant in part (c))

Thus the elapsed time is 0.1550 s $- 0.1500$ s $= 0.0050$ s.

EVALUATE: Part (d) says $y = A$ at 0.115 s and next at 0.155 s; the difference between these two times is 0.040 s, which is the period. At $t = 0.150$ s the particle at $x = 0.360$ m is at $y = 4.95$ cm and traveling upward. It takes $T/4 = 0.0100$ s for it to travel from $y = 0$ to $y = A$, so our answer of 0.0050 s is reasonable.

15.11 **IDENTIFY** and **SET UP:** Read A and T from the graph. Apply Eq.(15.4) to determine λ and then use Eq.(15.1) to calculate v.

EXECUTE:

a) The maximum y is 4 mm (read from graph).

b) For either x the time for one full cycle is 0.040 s; this is the period.

c) Since $y = 0$ for $x = 0$ and $t = 0$ and since the wave is traveling in the $+x$-direction then $y(x,t) = A\sin[2\pi(t/T - x/\lambda)]$. (The phase is different from the wave described by Eq.(15.4); for that wave $y = A$ for $x = 0$, $t = 0$.)

From the graph, if the wave is traveling in the $+x$-direction and if $x = 0$ and $x = 0.090$ m are within one wavelength the peak at $t = 0.01$ s for $x = 0$ moves so that it occurs at $t = 0.035$ s (read from graph so is approximate) for $x = 0.090$ m. The peak for $x = 0$ is the first peak past $t = 0$ so corresponds to the first maximum in $\sin[2\pi(t/T - x/\lambda)]$ and hence occurs at $2\pi(t/T - x/\lambda) = \pi/2$. If this same peak moves to $t_1 = 0.035$ s at $x_1 = 0.090$ m, then

$2\pi(t_1/T - x_1/\lambda) = \pi/2$

Solve for λ: $t_1/T - x_1/\lambda = 1/4$

$x_1/\lambda = t_1/T - 1/4 = 0.035$ s$/0.040$ s $- 0.25 = 0.625$

$\lambda = x_1/0.625 = 0.090$ m$/0.625 = 0.14$ m.

Then $v = f\lambda = \lambda/T = 0.14$ m$/0.040$ s $= 3.5$ m/s.

d) If the wave is traveling in the $-x$-direction, then $y(x,t) = A\sin(2\pi(t/T + x/\lambda)$ and the peak at $t = 0.050$ s for $x = 0$ corresponds to the peak at $t_1 = 0.035$ s for $x_1 = 0.090$ m. This peak at $x = 0$ is the second peak past the origin so corresponds to $2\pi(t/T + x/\lambda) = 5\pi/2$. If this same peak moves to $t_1 = 0.035$ s for

$x_1 = 0.090$ m, then $2\pi(t_1/T + x_1/\lambda) = 5\pi/2$.

$t_1/T + x_1/\lambda = 5/4$

$x_1/\lambda = 5/4 - t_1/T = 5/4 - 0.035$ s$/0.040$ s $= 0.375$

$\lambda = x_1/0.375 = 0.090$ m$/0.375 = 0.24$ m.

Then $v = f\lambda = \lambda/T = 0.24$ m$/0.040$ s $= 6.0$ m/s.

e) EVALUATE: No. Wouldn't know which point in the wave at $x = 0$ moved to which point at $x = 0.090$ m.

15.15 IDENTIFY and **SET UP:** Use Eq.(15.13) to calculate the wave speed. Then use Eq.(15.1) to calculate the wavelength.

EXECUTE:

a) The tension F in the rope is the weight of the hanging mass:

$F = mg = (1.50$ kg$)(9.80$ m/s$^2) = 14.7$ N

$v = \sqrt{F/\mu} = \sqrt{14.7 \text{ N}/(0.0550 \text{ kg/m})} = 16.3$ m/s

b) $v = f\lambda$ so $\lambda = v/f = (16.3$ m/s$)/120$ Hz $= 0.136$ m.

c) EVALUATE: $v = \sqrt{F/\mu}$, where $F = mg$. Doubling m increases v by a factor of $\sqrt{2}$.

$\lambda = v/f$. f remains 120 Hz and v increases by a factor of $\sqrt{2}$, so λ increases by a factor of $\sqrt{2}$.

15.17 IDENTIFY and **SET UP:** Use Eq.(15.13) to calculate the wave speed for transverse waves on the rubber tube. Then the time is the distance traveled divided by the wave speed.

EXECUTE: The tension in the tube is equal to the weight of the suspended mass:
$F = mg = (7.50 \text{ kg})(9.80 \text{ m/s}^2) = 73.5 \text{ N}$.
The mass per unit length for the tube is
$\mu = m/L = 0.800 \text{ kg}/14.0 \text{ m} = 0.0571 \text{ kg/m}$.

$$v = \sqrt{\frac{F}{\mu}} = \sqrt{\frac{73.5 \text{ N}}{(0.0571 \text{ kg/m})}} = 35.9 \text{ m/s}$$

Then $t = 14.0 \text{ m}/(35.9 \text{ m/s}) = 0.390 \text{ s}$

EVALUATE: The wave speed we calculated is the same order of magnitude as the wave speeds in Examples 15.3 and 15.4.

15.23 IDENTIFY and **SET UP:** Apply Eq.(15.26) to relate I and r.
Power is related to intensity at a distance r by $P = I(4\pi r^2)$. Energy is power times time.

EXECUTE:
a) $I_1 r_1^2 = I_2 r_2^2$
$I_2 = I_1 (r_1/r_2)^2 = (0.026 \text{ W/m}^2)(4.3 \text{ m}/3.1 \text{ m})^2 = 0.050 \text{ W/m}^2$

b) $P = 4\pi r^2 I = 4\pi (4.3 \text{ m})^2 (0.026 \text{ W/m}^2) = 6.04 \text{ W}$
Energy $= Pt = (6.04 \text{ W})(3600 \text{ s}) = 2.2 \times 10^4 \text{ J}$

EVALUATE: We could have used $r = 3.1$ m and $I = 0.050$ W/m^2 in $P = 4\pi r^2 I$ and would have obtained thes same P. Intensity becomes less as r increases because the radiated power spreads over a sphere of larger area.

15.31 IDENTIFY and **SET UP:** Nodes occur where $\sin kx = 0$ and antinodes are where $\sin kx = \pm 1$.
EXECUTE: Eq.(15.28): $y = (A_{\text{SW}} \sin kx) \sin \omega t$
a) At a node $y = 0$ for all t. This requires that $\sin kx = 0$ and this occurs for $kx = n\pi$, $n = 0, 1, 2, \ldots$

$$x = n\pi/k = \frac{n\pi}{0.750\pi \text{ rad/m}} = (1.33 \text{ m})n, \; n = 0, 1, 2, \ldots$$

b) At an antinode $\sin kx = \pm 1$ so y will have maximum amplitude.

This occurs when $kx = (n + \frac{1}{2})\pi$, $n = 0, 1, 2, \ldots$

$$x = (n + \tfrac{1}{2})\pi/k = (n + \tfrac{1}{2})\frac{\pi}{0.750\pi \text{ rad/m}} = (1.33 \text{ m})(n + \frac{1}{2}), \; n = 0, 1, 2, \ldots$$

EVALUATE: $\lambda = 2\pi/k = 2.66$ m. Adjacent nodes are separated by $\lambda/2$, adjacent antinodes are separated by $\lambda/2$, and the node to antinode distance is $\lambda/4$.

15.33 IDENTIFY: Apply Eqs.(15.28) and (15.1).

a) SET UP: $y = (A_{SW} \sin kx) \sin \omega t$ (Eq.15.28) $A_{SW} = 0.850$ cm

$T = 0.0750$ s, $f = 1/T = 1/0.0750$ s $= 13.33$ Hz, $\omega = 2\pi f = 26.6\pi$ rad/s

antinode to antinode distance is $\lambda/2 = 15.0$ cm, so $\lambda = 30.0$ cm

$k = 2\pi/\lambda = 2\pi/30.0$ cm $= \pi/15.0$ cm

EXECUTE: $y(x, t) = (0.850 \text{ cm}) \sin(\pi x/15.0 \text{ cm}) \sin((26.6\pi \text{ s}^{-1})t)$

b) $v = f\lambda = (13.33 \text{ Hz})(0.300 \text{ m}) = 4.00$ m/s

c) The amplitude of the simple harmonic motion at a particular x is

$A_{SW} \sin kx = (0.850 \text{ cm}) \sin(\pi x/15.0 \text{ cm})$, where x is measured from the node at the fixed end of the string.

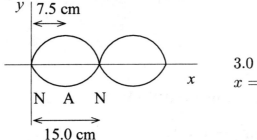

3.0 cm to the right of the first antinode so
$x = 7.50$ cm $+ 3.0$ cm $= 10.5$ cm

$x = 10.5$ cm implies $A_{SW} \sin kx = (0.850 \text{ cm}) \sin(\pi(10.5 \text{ cm})/15.0 \text{ cm}) = 0.688$ cm

EVALUATE: At an antinode the amplitude is 0.850 cm. A node is $\lambda/2 = 15.0$ cm to the right and the amplitude is zero there. So, an amplitude of 0.688 cm at a point 3.0 cm to the right of an antinode is reasonable.

15.37 IDENTIFY: Use Eq.(15.1) for v and Eq.(15.13) for the tension F. $v_y = \partial y/\partial t$ and $a_y = \partial v_y/\partial t$.

a) SET UP: fundamental

← $L = 0.800$ m →

$f = 60.0$ Hz

N   A   N

$\frac{\lambda}{2}$

fundamental

From the sketch,
$\lambda/2 = L$ so
$\lambda = 2L = 1.60$ m

EXECUTE: $v = f\lambda = (60.0 \text{ Hz})(1.60 \text{ m}) = 96.0$ m/s

b) The tension is related to the wave speed by Eq.(15.13):
$v = \sqrt{F/\mu}$ so $F = \mu v^2$.
$\mu = m/L = 0.0400 \text{ kg}/0.800 \text{ m} = 0.0500$ kg/m
$F = \mu v^2 = (0.0500 \text{ kg/m})(96.0 \text{ m/s})^2 = 461$ N.

c) $\omega = 2\pi f = 377$ rad/s and $y(x,t) = A_{SW} \sin kx \sin \omega t$
$v_y = \omega A_{SW} \sin kx \cos \omega t$; $a_y = -\omega^2 A_{SW} \sin kx \sin \omega t$
$(v_y)_{max} = \omega A_{SW} = (377 \text{ rad/s})(0.300 \text{ cm}) = 1.13$ m/s.
$(a_y)_{max} = \omega^2 A_{SW} = (377 \text{ rad/s})^2(0.300 \text{ cm}) = 426$ m/s².

EVALUATE: The transverse velocity is different from the wave velocity. The wave velocity and tension are similar in magnitude to the values in the Examples in the text. Note that the transverse acceleration is quite large.

15.39 IDENTIFY: Compare $y(x,t)$ given in the problem to Eq.(15.28). From the frequency and wavelength for the third harmonic find these values for the eighth harmonic.

a) SET UP:

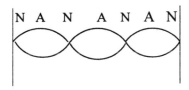

N A N A N A N

EXECUTE: b) Eq. (15.28) gives the general equation for a standing wave on a string:
$y(x,t) = (A_{SW} \sin kx) \sin \omega t$
$A_{SW} = 2A$, so $A = A_{SW}/2 = (5.60 \text{ cm})/2 = 2.80$ cm

c) The sketch in part (a) shows that $L = 3(\lambda/2)$. $k = 2\pi/\lambda$, $\lambda = 2\pi/k$
Comparison of $y(x,t)$ given in the problem to Eq. (15.28) gives $k = 0.0340$ rad/cm.
So, $\lambda = 2\pi/(0.0340 \text{ rad/cm}) = 184.8$ cm

$L = 3(\lambda/3) = 277$ cm

d) $\lambda = 185$ cm, from part (c)

$\omega = 50.0$ rad/s so $f = \omega/2\pi = 7.96$ Hz

period $T = 1/f = 0.126$ s

$v = f\lambda = 1470$ cm/s

e) $v_y = dy/dt = \omega A_{\mathrm{SW}} \sin kx \cos \omega t$

$v_{y,\mathrm{max}} = \omega A_{\mathrm{SW}} = (50.0 \text{ rad/s})(5.60 \text{ cm}) = 280$ cm/s

f) $f_3 = 7.96$ Hz $= 3f_1$, so $f_1 = 2.65$ Hz is the fundamental

$f_8 = 8f_1 = 21.2$ Hz; $\omega_8 = 2\pi f_8 = 133$ rad/s

$\lambda = v/f = (1470 \text{ cm/s})/(21.2 \text{ Hz}) = 69.3$ cm and $k = 2\pi/\lambda = 0.0906$ rad/cm

$y(x,t) = (5.60 \text{ cm}) \sin([0.0906 \text{ rad/cm}]x) \sin([133 \text{ rad/s}]t)$

EVALUATE: The wavelength and frequency of the standing wave equals the wavelength and frequency of the two traveling waves that combine to form the standing wave. In the 8th harmonic the frequency and wave number are larger than in the 3rd harmonic.

15.41 a) IDENTIFY and **SET UP:** Use the angular frequency and wave number for the traveling waves in Eq.(15.28) for the standing wave.

EXECUTE:

The traveling wave is $y(x,t) = (2.30 \text{ m}) \cos([6.98 \text{ rad/m}]x) + [742 \text{ rad/s}]t)$

$A = 2.30$ mm so $A_{\mathrm{SW}} = 4.60$ mm; $k = 6.98$ rad/m and $\omega = 742$ rad/s

The general equation for a standing wave is $y(x,t) = (A_{\mathrm{SW}} \sin kx) \sin \omega t$, so

$y(x,t) = (4.60 \text{ mm}) \sin([6.98 \text{ rad/m}]x) \sin([742 \text{ rad/s}]t)$

b) IDENTIFY and **SET UP:** Compare the wavelength to the length of the rope in order to identify the harmonic.

EXECUTE: $L = 1.35$ m (from Exercise 15.24)

$\lambda = 2\pi/k = 0.900$ m

$L = 3(\lambda/2)$, so this is the 3rd harmonic

c) For this 3rd harmonic, $f = \omega/2\pi = 118$ Hz

$f_3 = 3f_1$ so $f_1 = (118 \text{ Hz})/3 = 39.3$ Hz

EVALUATE: The wavelength and frequency of the standing wave equals the wavelength and frequency of the two traveling waves that combine to form the standing wave. The nth harmonic has n node-to-node segments and the node-to-node distance is $\lambda/2$, so the relation between L and λ for the nth harmonic is $L = n(\lambda/2)$.

15.43 IDENTIFY and **SET UP:** Use the information given about the A_4 note to find the wave speed, that depends on the linear mass density of the string and the tension. The wave speed isn't affected by the placement of the fingers on the bridge. Then find the wavelength for the D_5 note and relate this to the length of the vibrating portion of the string.

EXECUTE:

a) $f = 440$ Hz when a length $L = 0.600$ m vibrates; use this information to calculate the speed v of waves on the string.

For the fundamental $\lambda/2 = L$ so $\lambda = 2L = 2(0.600 \text{ m}) = 1.20$ m.

Then $v = f\lambda = (440 \text{ Hz})(1.20 \text{ m}) = 528$ m/s.

Now find the length $L = x$ of the string that makes $f = 587$ Hz.

$$\lambda = \frac{v}{f} = \frac{528 \text{ m/s}}{587 \text{ Hz}} = 0.900 \text{ m}$$

$L = \lambda/2 = 0.450$ m, so $x = 0.450$ m $= 45.0$ cm.

b) No retuning means same wave speed as in part (a). Find the length of vibrating string needed to produce $f = 392$ Hz.

$$\lambda = \frac{v}{f} = \frac{528 \text{ m/s}}{392 \text{ Hz}} = 1.35 \text{ m}$$

$L = \lambda/2 = 0.675$ m; string is shorter than this. No, not possible.

EVALUATE: Shortening the length of the vibrating string increases the frequency of the fundamental.

Problems

15.47 IDENTIFY and **SET UP:** Calculate v, ω, and k from Eqs.(15.1), (15.5), and (15.6). Then apply Eq.(15.7) to obtain $y(x,t)$.

$A = 2.50 \times 10^{-3}$ m, $\quad \lambda = 1.80$ m, $\quad v = 36.0$ m/s

EXECUTE:

a) $v = f\lambda$ so $f = v/\lambda = (36.0 \text{ m/s})/1.80 \text{ m} = 20.0$ Hz

$\omega = 2\pi f = 2\pi(20.0 \text{ Hz}) = 126$ rad/s

$k = 2\pi/\lambda = 2\pi \text{ rad}/1.80 \text{ m} = 3.49$ rad/m

b) For a wave traveling to the right, $y(x,t) = A\cos(kx - \omega t)$. This equation gives that the $x = 0$ end of the string has maximum upward displacement at $t = 0$.

Put in the numbers: $y(x,t) = (2.50 \times 10^{-3} \text{ m})\cos((3.49 \text{ rad/m})x - (126 \text{ rad/s})t)$.

c) The left hand end is located at $x = 0$. Put this value into the equation of part (b): $y(0,t) = +(2.50 \times 10^{-3} \text{ m})\cos((126 \text{ rad/s})t)$.

d) Put $x = 1.35$ m into the equation of part (b):

$y(1.35 \text{ m}, t) = (2.50 \times 10^{-3} \text{ m}) \cos((3.49 \text{ rad/m})(1.35 \text{ m}) - (126 \text{ rad/s})t)$.

$y(1.35 \text{ m}, t) = (2.50 \times 10^{-3} \text{ m}) \cos(4.71 \text{ rad}) - (125 \text{ rad/s})t$

$4.71 \text{ rad} = 3\pi/2$ and $\cos(\theta) = \cos(-\theta)$, so

$y(1.35 \text{ m}, t) = (2.50 \times 10^{-3} \text{ m}) \cos((125 \text{ rad/s})t - 3\pi/2 \text{ rad})$

e) $y = A \cos(kx - \omega t)$ (part (b))

The transverse velocity is given by

$$v_y = \frac{\partial y}{\partial t} = A \frac{\partial}{\partial t} \cos(kx - \omega t) = +A\omega \sin(kx - \omega t).$$

The maximum v_y is $A\omega = (2.50 \times 10^{-3} \text{ m})(126 \text{ rad/s}) = 0.315$ m/s.

f) $y(x, t) = (2.50 \times 10^{-3} \text{ m}) \cos((3.49 \text{ rad/m})x - (126 \text{ rad/s})t)$

$t = 0.0625$ s and $x = 1.35$ m gives

$y = (2.50 \times 10^{-3} \text{ m}) \cos((3.49 \text{ rad/m})(1.35 \text{ m}) - (126 \text{ rad/s})(0.0625 \text{ s})) = -2.50 \times 10^{-3}$ m.

$v_y = +A\omega \sin(kx - \omega t) = +(0.315 \text{ m/s}) \sin((3.49 \text{ rad/m})x - (126 \text{ rad/s})t)$

$t = 0.0625$ s and $x = 1.35$ m gives

$v_y = (0.315 \text{ m/s}) \sin((3.49 \text{ rad/m})(1.35 \text{ m}) - (125 \text{ rad/s})(0.0625 \text{ s})) = 0.0$

EVALUATE: The results of part (f) illustrate that $v_y = 0$ when $y = \pm A$, as we saw for SHM in Chapter 13.

15.51 **IDENTIFY** and **SET UP:** The transverse speed of a point of the rope is $v_y = \partial y/\partial t$ where $y(x, t)$ is given by Eq.(15.7).

EXECUTE:

a) $y(x, t) = A \cos(kx - \omega t)$

$v_y = dy/dt = +A\omega \sin(kx - \omega t)$

$v_{y,\text{max}} = A\omega = 2\pi f A$

$f = \dfrac{v}{\lambda}$ and $v = \sqrt{\dfrac{F}{(m/L)}}$, so $f = \left(\dfrac{1}{\lambda}\right)\sqrt{\dfrac{FL}{M}}$

$v_{y,\text{max}} = \left(\dfrac{2\pi A}{\lambda}\right)\sqrt{\dfrac{FL}{M}}$

b) To double $v_{y,\text{max}}$ increase F by a factor of 4.

EVALUATE: Increasing the tension increases the wave speed v which in turn increases the osicllation frequency. With the amplitude held fixed, increasing the number of oscillations per second increases the transverse velocity.

15.53 **IDENTIFY** and **SET UP:** Use Eq.(15.1) and $\omega = 2\pi f$ to replace v by ω in Eq.(15.13). Compare this equation to $\omega = \sqrt{k'/m}$ from Chapter 13 to deduce k'.
EXECUTE:

a) $\omega = 2\pi f$, $f = v/\lambda$, and $v = \sqrt{F/\mu}$ These equations combine to give $\omega = 2\pi f = 2\pi (v/\lambda) = (2\pi/\lambda)\sqrt{F/\mu}$.

But also $\omega = \sqrt{k'/m}$. Equating these expressions for ω gives

$k' = m(2\pi/\lambda)^2 (F/\mu)$

But $m = \mu \Delta x$ so $k' = \Delta x (2\pi/\lambda)^2 F$

b) EVALUATE: The "force constant" k' is independent of the amplitude A and mass per unit length μ, just as is the case for a simple harmonic oscillator. The force constant is proportional to the tension in the string F and inversely proportional to the wavelength λ. The tension supplies the restoring force and the $1/\lambda^2$ factor represents the dependence of the restoring force on the curvature of the string.

15.61 **IDENTIFY** and **SET UP:** Use Eq.(15.13) to replace μ, and then Eq.(15.6) to replace v.
EXECUTE:

a) Eq.(19-33): $P_{\text{av}} = \frac{1}{2}\sqrt{\mu F}\omega^2 A^2$

$v = \sqrt{F/\mu}$ says $\sqrt{\mu} = \sqrt{F}/v$ so $P_{\text{av}} = \frac{1}{2}(\sqrt{F}/v)\sqrt{F}\omega^2 A^2 = \frac{1}{2}F\omega^2 A^2/v$

$\omega = 2\pi f$ so $\omega/v = 2\pi f/v = 2\pi/\lambda = k$ and
$P_{\text{av}} = \frac{1}{2}Fk\omega A^2$, as was to be shown.

b) IDENTIFY: For the ω dependence, use Eq.(15.25) since it involves just ω, not k: $P_{\text{av}} = \frac{1}{2}\sqrt{\mu F}\omega^2 A^2$.
SET UP: P_{av}, μ, A all constant so $\sqrt{F}\omega^2$ is constant, and $\sqrt{F_1}\omega_1^2 = \sqrt{F_2}\omega_2^2$.
EXECUTE: $\omega_2 = \omega_1 (F_1/F_2)^{1/4} = \omega_1 (F_1/4F_1)^{1/4} = \omega_1 (4)^{-1/4} = \omega_1/\sqrt{2}$
ω must be changed by a factor of $1/\sqrt{2}$ (decreased)

IDENTIFY:
For the k dependence, use the equation derived in part (a), $P_{\text{av}} = \frac{1}{2}Fk\omega A^2$.
SET UP:
If P_{av} and A are constant then $Fk\omega$ must be constant, and $F_1 k_1 \omega_1 = F_2 k_2 \omega_2$.
EXECUTE:
$k_2 = k_1 \left(\frac{F_1}{F_2}\right)\left(\frac{\omega_1}{\omega_2}\right) = k_1 \left(\frac{F_1}{4F_1}\right)\left(\frac{\omega_1}{\omega_1/\sqrt{2}}\right) = k_1 \frac{\sqrt{2}}{4} = k_1\sqrt{\frac{2}{16}} = k_1/\sqrt{8}$

k must be changed by a factor of $1/\sqrt{8}$ (decreased).

EVALUATE: Increasing F increases the wave speed v. Decreasing k increases λ and decreases f. Decreasing ω and decreases f.

15.63 IDENTIFY and **SET UP:** The average power is given by Eq.(15.25). Rewrite this expression in terms of v and λ in place of F and ω.

EXECUTE:

a) $P_{av} = \frac{1}{2}\sqrt{\mu F}\omega^2 A^2$

$v = \sqrt{F/\mu}$ so $\sqrt{F} = v\sqrt{\mu}$

$\omega = 2\pi f = 2\pi(v/\lambda)$

Using these two expressions to replace $\sqrt{F}$ and ω gives

$P_{av} = 2\mu\pi^2 v^3 A^2/\lambda^2$; $\mu = (6.00 \times 10^{-3} \text{ kg})/(8.00 \text{ m})$

$A = \left(\dfrac{2\lambda^2 P_{av}}{4\pi^2 v^3 \mu}\right)^{1/2} = 7.07 \text{ cm}$

b) EVALUATE: $P_{av} \sim v^3$ so doubling v increases P_{av} by a factor of 8.

$P_{av} = 8(50.0 \text{ W}) = 400.0 \text{ W}$

15.65 IDENTIFY and **SET UP:** $v = \sqrt{F/\mu}$. Eq.(17.12) gives the change in F when the temperature is changed, in terms of the coefficient of linear expansion α. Combine the two equations and solve for α.

EXECUTE: $v_1 = \sqrt{F/\mu}$, $v_1^2 = F/\mu$ and $F = \mu v_1^2$

The length and hence μ stay the same but the tension decreases by

$\Delta F = -Y\alpha A\,\Delta T$.

$v_2 = \sqrt{(F + \Delta F)/\mu} = \sqrt{(F - Y\alpha A\,\Delta T)/\mu}$

$v_2^2 = F/\mu - Y\alpha A\,\Delta T/\mu = v_1^2 - Y\alpha A\,\Delta T/\mu$

And $\mu = m/L$ so $A/\mu = AL/m = V/m = 1/\rho$. ($A$ is the cross-sectional area of the wire, V is the volume of a length L.)

Thus $v_1^2 - v_2^2 = \alpha(Y\,\Delta T/\rho)$ and

$\alpha = \dfrac{v_1^2 - v_2^2}{(Y/\rho)\,\Delta T}$

EVALUATE: When T increases the tension decreases and v decreases.

15.67 IDENTIFY and **SET UP:** There is a node at the post and there must be a node at the clothespin. There could be aditional nodes in between. The distance between adjacent nodes is $\lambda/2$, so the distance between *any* two nodes is $n(\lambda/2)$ for $n = 1, 2, 3, \ldots$ This must equal 45.0 cm, since there are nodes at the post and clothespin. Use this in Eq.(15.1) to get an expression for the possible frequencies f.

EXECUTE: 45.0 cm $= n(\lambda/2)$, $\lambda = v/f$, so

$f = n[v/(90.0 \text{ cm})] = (0.800 \text{ Hz})n, n = 1, 2, 3, \ldots$

EVALUATE: Higher frequencies have smaller wavelengths, so more node-to-node segments fit between the post and clothespin.

15.71 **IDENTIFY:** Carry out the derivation as done in the text for Eq.(15.28). The transverse velocity is $v_y = \partial y / \partial t$ and the transverse acceleration is $a_y = \partial v_y / \partial t$.

a) SET UP: For reflection from a free end of a string the reflected wave is <u>not</u> inverted, so $y(x,t) = y_1(x,t) + y_2(x,t)$, where

$y_1(x,t) = A\cos(kx + \omega t)$ (traveling to the left)

$y_2(x,t) = A\cos(kx - \omega t)$ (traveling to the right)

Thus $y(x,t) = A[\cos(kx + \omega t) + \cos(kx - \omega t)]$.

EXECUTE: Apply the trig identity $\cos(a \pm b) = \cos a \cos b \mp \sin a \sin b$ with $a = kx$ and $b = \omega t$:

$\cos(kx + \omega t) = \cos kx \cos \omega t - \sin kx \sin \omega t$ and

$\cos(kx - \omega t) = \cos kx \cos \omega t + \sin kx \sin \omega t$.

Then $y(x,t) = (2A\cos kx)\cos \omega t$ (the other two terms cancel)

b) For $x = 0$, $\cos kx = 1$ and $y(x,t) = 2A\cos \omega t$. The amplitude of the simple harmonic motion at $x = 0$ is $2A$, which is the maximum for this standing wave, so $x = 0$ is an antinode.

c) $y_{\max} = 2A$ from part (b).

$$v_y = \frac{\partial y}{\partial t} = \frac{\partial}{\partial t}[(2A\cos kx)\cos \omega t] = 2A\cos kx \frac{\partial \cos \omega t}{\partial t} = -2A\omega \cos kx \sin \omega t.$$

At $x = 0$, $v_y = -2A\omega \sin \omega t$ and $(v_y)_{\max} = 2A\omega$

$$a_y = \frac{\partial^2 y}{\partial t^2} = \frac{\partial v_y}{\partial t} = -2A\omega \cos kx \frac{\partial \sin \omega t}{\partial t} = -2A\omega^2 \cos kx \cos \omega t$$

At $x = 0$, $a_y = -2A\omega^2 \cos \omega t$ and $(a_y)_{\max} = 2A\omega^2$.

EVALUATE: The expressions for $(v_y)_{\max}$ and $(a_y)_{\max}$ are the same as at the antinodes for the standing wave of a string fixed at both ends.

15.73 **a) IDENTIFY:** Determine the wavelength and frequency and use Eq.(15.1) to calculate v.

SET UP: Plank oscillates with maximum amplitude at its center says that it is oscillating in its fundamental mode.

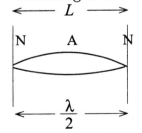

EXECUTE:

$\lambda/2 = L$

$\lambda = 2L = 2(5.00 \text{ m}) = 10.0 \text{ m}$

Student jumps upward two times per second so $f = 2.00$ Hz.

$v = f\lambda = (2.00 \text{ Hz})(10.0 \text{ m}) = 20.0 \text{ m/s}$.

b) IDENTIFY: v is the same as in part (a). Determine λ and use eq.(15.1) to calculate f.

SET UP: There now must be an antinode 1.25 m (a distance of $L/4$) from one end. The nodal structure for the standing wave must be:

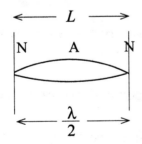

EXECUTE:
Now $\lambda = L = 5.00$ m.

v depends on the properties of the plank, so it is the same as in part (a).

$f = \dfrac{v}{\lambda} = \dfrac{20.0 \text{ m/s}}{5.00 \text{ m}} = 4.00$ Hz

The student now has to jump four times each second.

EVALUATE: To produce an antinode at a location other than the center of the plank, the plank must be vibrating in an overtone. The overtone has a higher frequency than the fundamental and the student must jump more rapidly. There are higher overtones that also have an anitnode at this location.

15.77 IDENTIFY: Compute the wavelength from the length of the string. Use Eq.(15.1) to calculate the wave speed and then apply Eq.(15.13) to relate this to the tension.

a) SET UP: The tension F is related to the wave speed by $v = \sqrt{F/\mu}$ (Eq.(15.13), so use the information given to calculate v.

EXECUTE:
$\lambda/2 = L$
$\lambda = 2L = 2(0.600 \text{ m}) = 1.20$ m

fundamental

$v = f\lambda = (65.4 \text{ Hz})(1.20 \text{ m}) = 78.5$ m/s

$\mu = m/L = 14.4 \times 10^{-3} \text{ kg}/0.600 \text{ m} = 0.024$ kg/m

Then $F = \mu v^2 = (0.024 \text{ kg/m})(78.5 \text{ m/s})^2 = 148$ N.

b) SET UP: $F = \mu v^2$ and $v = f\lambda$ give $F = \mu f^2 \lambda^2$.

μ is a property of the string so is constant.

λ is determined by the length of the string so stays constant.

μ, λ constant implies $F/f^2 = \mu\lambda^2 = \text{constant}$, so $F_1/f_1^2 = F_2/f_2^2$

EXECUTE: $F_2 = F_1 \left(\dfrac{f_2}{f_1} \right)^2 = (148 \text{ N}) \left(\dfrac{73.4 \text{ Hz}}{65.4 \text{ Hz}} \right)^2 = 186 \text{ N}.$

The percent change in F is $\dfrac{F_2 - F_1}{F_1} = \dfrac{186 \text{ N} - 148 \text{ N}}{148 \text{ N}} = 0.26 = 26\%.$

EVALUATE: The wave speed and tension we calculated are similar in magnitude to values in the Examples. Since the frequency is proportional to $\sqrt{F}$, a 26% increase in tension is required to produce a 13% increase in the frequency.

CHAPTER 16
SOUND AND HEARING

Exercises 1, 3, 11, 17, 21, 23, 25, 29, 31, 41, 45
Problems 49, 53, 57, 65, 67, 73, 75

Exercises

16.1 **IDENTIFY** and **SET UP:** Eq.(15.1) gives the wavelength in terms of the frequency. Use Eq.(16.5) to relate the pressure and displacement amplitudes.

EXECUTE:

a) $\lambda = v/f = (344 \text{ m/s})/1000 \text{ Hz} = 0.344 \text{ m}$

b) $p_{max} = BkA$ and Bk is constant gives $p_{max1}/A_1 = p_{max2}/A_2$

$$A_2 = A_1 \left(\frac{p_{max2}}{p_{max1}} \right) = 1.2 \times 10^{-8} \text{ m} \left(\frac{30 \text{ Pa}}{3.0 \times 10^{-2} \text{ Pa}} \right) = 1.2 \times 10^{-5} \text{ m}$$

c) $p_{max} = BkA = 2\pi BA/\lambda$

$p_{max}\lambda = 2\pi BA = $ constant so $p_{max1}\lambda_1 = p_{max2}\lambda_2$ and

$$\lambda_2 = \lambda_1 \left(\frac{p_{max1}}{p_{max2}} \right) = (0.344 \text{ m}) \left(\frac{3.0 \times 10^{-2} \text{ Pa}}{1.5 \times 10^{-3} \text{ Pa}} \right) = 6.9 \text{ m}$$

$f = v/\lambda = (344 \text{ m/s})/6.9 \text{ m} = 50 \text{ Hz}$

EVALUATE: The pressure amplitude and displacement amplitude are directly proportional. For the same displacement amplitude, the pressure amplitude decreases when the frequency decreases and the wavelength increases.

16.3 **IDENTIFY:** Use Eq.(16.5) to relate the pressure and displacement amplitudes.

SET UP: As computed in Example 16.1 the adiabatic bulk modulus for air is $B = 1.42 \times 10^5$ Pa. Use Eq.(15.1) to calculate λ from f, and then $k = 2\pi/\lambda$.

EXECUTE:

a) $f = 150$ Hz

Need to calculate k: $\lambda = v/f$ and $k = 2\pi/\lambda$ so

$k = 2\pi f/v = (2\pi \text{ rad})(150 \text{ Hz})/344 \text{ m/s} = 2.74 \text{ rad/m}$.

Then $p_{max} = BkA = (1.42 \times 10^5 \text{ Pa})(2.74 \text{ rad/m})(0.0200 \times 10^{-3} \text{ m}) = 7.78 \text{ Pa}$.

This is below the pain threshold of 30 Pa.

b) f is larger by a factor of 10 so $k = 2\pi f/v$ is larger by a factor of 10, and $p_{max} = BkA$ is larger by a factor of 10. $p_{max} = 77.8$ Pa, above the pain threshold.

c) There is again an increase in f, k, and p_{max} of a factor of 10, so $p_{max} = 778$ Pa, far above the pain threshold.

EVALUATE: When f increases, λ decreases so k increases and the pressure amplitude increases.

16.11 **IDENTIFY** and **SET UP:** Use $t = $ distance/speed. Calculate the time it takes each sound wave to travel the $L = 80.0$ m length of the pipe. Use Eq.(16.8) to calculate the speed of sound in the brass rod.

EXECUTE:

wave in air: $t = 80.0$ m$/(344$ m/s$) = 0.2326$ s

wave in the metal: $v = \sqrt{\dfrac{Y}{\rho}} = \sqrt{\dfrac{9.0 \times 10^{10} \text{ Pa}}{8600 \text{ kg/m}^3}} = 3235$ m/s

$t = \dfrac{80.0 \text{ m}}{3235 \text{ m/s}} = 0.0247$ s

The time interval between the two sounds is $\Delta t = 0.2326$ s $- 0.0247$ s $= 0.208$ s

EVALUATE: The restoring forces that propagate the sound waves are much greater in solid brass than in air, so v is much larger in brass.

16.17 **IDENTIFY** and **SET UP:** Apply Eqs.(16.5), (16.12)and (16.15).

EXECUTE:

a) $\omega = 2\pi f = (2\pi \text{ rad})(150 \text{ Hz}) = 942.5$ rad/s

$k = \dfrac{2\pi}{\lambda} = \dfrac{2\pi f}{v} = \dfrac{\omega}{v} = \dfrac{942.5 \text{ rad/s}}{344 \text{ m/s}} = 2.74$ rad/m

$B = 1.42 \times 10^5$ Pa (Example 16.1)

Then $p_{max} = BkA = (1.42 \times 10^5 \text{ Pa})(2.74 \text{ rad/m})(5.00 \times 10^{-6} \text{ m}) = 1.95$ Pa.

b) Eq.(16.12): $I = \frac{1}{2}\omega BkA^2$

$I = \frac{1}{2}(942.5 \text{ rad/s})(1.42 \times 10^5 \text{ Pa})(2.74 \text{ rad/m})(5.00 \times 10^{-6} \text{ m})^2 = 4.58 \times 10^{-3}$ W/m^2.

c) Eq.(16.15): $\beta = (10 \text{ dB}) \log(I/I_0)$, with $I_0 = 1 \times 10^{-12}$ W/m^2.

$\beta = (10 \text{ dB}) \log((4.58 \times 10^{-3} \text{ W/m}^2)/(1 \times 10^{-12} \text{ W/m}^2)) = 96.6$ dB.

EVALUATE: Even though the displacement amplitude is very small, this is a very intense sound. Compare the sound intensity level to tHE values in Table 16.2.

16.21 **IDENTIFY** and **SET UP:** Let 1 refer to the mother and 2 to the father. Use the result dervied in Example 16.11 for the difference in sound intensity level for the two sounds. Relate intensity to distance from the source using Eq.(15.26).

EXECUTE: From Example 16.11, $\beta_2 - \beta_1 = (10 \text{ dB}) \log(I_2/I_1)$

Eq.(15.26): $I_1/I_2 = r_2^2/r_1^2$ or $I_2/I_1 = r_1^2/r_2^2$

$\Delta\beta = \beta_2 - \beta_1 = (10 \text{ dB}) \log(I_2/I_1) = (10 \text{ dB}) \log(r_1/r_2)^2 = (20 \text{ dB}) \log(r_1/r_2)$

$\Delta\beta = (20 \text{ dB}) \log(1.50 \text{ m}/0.30 \text{ m}) = 14.0 \text{ dB}$.

EVALUATE: The father is 5 times closer so the intensity at his location is 25 times greater.

16.23 a) IDENTIFY and **SET UP:** From Example 16.11 $\Delta\beta = (10 \text{ dB}) \log(I_2/I_1)$
Set $\Delta\beta = 13.0 \text{ dB}$ and solve for I_2/I_1.

EXECUTE: $13.0 \text{ dB} = 10 \text{ dB} \log(I_2/I_1)$ so $1.3 = \log(I_2/I_1)$ and $I_2/I_1 = 20.0$.

b) EVALUATE: According to the equation in part (a) the difference in two sound intensity levels is determined by the ratio of the sound intensities. So you don't need to know I_1, just the ratio I_2/I_1.

16.25 IDENTIFY and **SET UP:** An open end is a displacement antinode and a closed end is a displacement node. Sketch the standing wave pattern and use the sketch to relate the node-to-antinode distance to the length of the pipe. A displacement node is a pressure antinode and a displacement antinode is a pressure node.

EXECUTE:

a) The placement of the displacement nodes and antinodes along the pipe is as sketched below. The open ends are displacement antinodes.

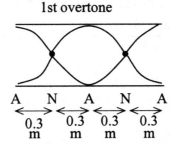 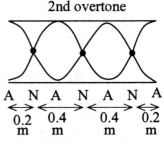

Location of the displacement nodes (N) measured from the left end:
fundamental 0.60 m
1st overtone 0.30 m, 0.90 m
2nd overtone 0.20 m, 0.60 m, 1.00 m

Location of the pressure nodes (displacement antinodes (A)) measured from the left end: fundamental 0, 1.20 m
1st overtone 0, 0.60 m, 1.20 m
2nd overtone 0, 0.40 m, 0.80 m, 1.20 m

b) The open end is a displacement antinode and the closed end is a displacement node.

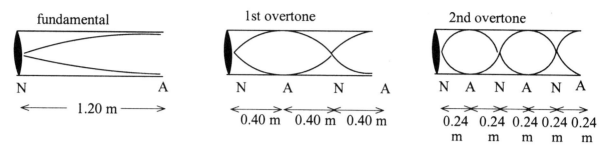

| fundamental | 1st overtone | 2nd overtone |

Location of the displacement nodes (N) measured from the closed end:
fundamental 0
1st overtone 0, 0.80 m
2nd overtone 0, 0.48 m, 0.96 m

Location of the pressure nodes (displacement antinodes (A)) measured from the closed end: fundamental 1.20 m
1st overtone 0.40 m, 1.20 m
2nd overtone 0.24 m, 0.72 m, 1.20 m

EVALUATE: The node-to-node or anitnode-to-antinode distance is $\lambda/2$. For the higher overtones the frequency is higher and the wavelength is smaller.

16.29 **IDENTIFY** and **SET UP:** Use the standing wave pattern to relate the wavelength of the standing wave to the length of the air column and then use Eq.(15.1) to calculate f. There is a displacement antinode at the top (open) end of the air column and a node at the bottom (closed) end.

EXECUTE:

a)

$$\lambda/4 = L$$
$$\lambda = 4L = 4(0.140 \text{ m}) = 0.560 \text{ m}$$

$$f = \frac{v}{\lambda} = \frac{344 \text{ m/s}}{0.560 \text{ m}} = 614 \text{ Hz}$$

b) Now the length L of the air column becomes $\frac{1}{2}(0.140 \text{ m}) = 0.070 \text{ m}$ and $\lambda = 4L = 0.280 \text{ m}$.

$$f = \frac{v}{\lambda} = \frac{344 \text{ m/s}}{0.280 \text{ m}} = 1230 \text{ Hz}$$

EVALUATE: Smaller L means smaller λ which in turn corresponds to larger f.

16.31

A ◁━━━━━━━━━━→ B ◁━━━ Q
        2.00 m   1.00 m

a) IDENTIFY and **SET UP:** Path difference from points A and B to point Q is $3.00 \text{ m} - 1.00 \text{ m} = 2.00 \text{ m}$.

Constructive interference implies path difference $= n\lambda$, $n = 1, 2, 3, \ldots$

EXECUTE: $2.00 \text{ m} = n\lambda$ so $\lambda = 2.00 \text{ m}/n$

$$f = \frac{v}{\lambda} = \frac{nv}{2.00 \text{ m}} = \frac{n(344 \text{ m/s})}{2.00 \text{ m}} = n(172 \text{ Hz}), \quad n = 1, 2, 3, \ldots$$

The lowest frequency for which constructive interference occurs is 172 Hz.

b) IDENTIFY and **SET UP:** Destructive interference implies path difference $= (n/2)\lambda$, $n = 1, 3, 5, \ldots$

EXECUTE: $2.00 \text{ m} = (n/2)\lambda$ so $\lambda = 4.00 \text{ m}/n$

$$f = \frac{v}{\lambda} = \frac{nv}{4.00 \text{ m}} = \frac{n(344 \text{ m/s})}{4.00 \text{ m})} = n(86 \text{ Hz}), \quad n = 1, 3, 5, \ldots$$

The lowest frequency for which destructive interference occurs is 86 Hz.

EVALUATE: As the frequency is slowly increased, the intensity at Q will fluctuate, as the interference changes between destructive and constructive.

16.41 IDENTIFY and **SET UP:** Apply Eqs.(16.27) and (16.28) for the wavelengths in front of and behind the source. Then $f = v/\lambda$.

When the source is at rest $\lambda = \dfrac{v}{f_S} = \dfrac{344 \text{ m/s}}{400 \text{ Hz}} = 0.860 \text{ m}$.

EXECUTE:

a) Eq.(16.27): $\lambda = \dfrac{v - v_S}{f_S} = \dfrac{344 \text{ m/s} - 25.0 \text{ m/s}}{400 \text{ Hz}} = 0.798 \text{ m}$

b) Eq.(16.28): $\lambda = \dfrac{v + v_S}{f_S} = \dfrac{344 \text{ m/s} + 25.0 \text{ m/s}}{400 \text{ Hz}} = 0.922 \text{ m}$

c) $f_L = v/\lambda$ (since $v_L = 0$). so $f_L = (344 \text{ m/s})/0.798 \text{ m} = 431 \text{ Hz}$

d) $f_L = v/\lambda = (344 \text{ m/s})/0.922 \text{ m} = 373 \text{ Hz}$

EVALUATE: In front of the source (source moving toward listener) the wavelength is decreased and the frequency is increased. Behind the source (source moving away from lsitener) the wavelength is increased and the frequency is decreased.

16.45 IDENTIFY: Apply the Doppler effect formula, Eq.(16.29).

a) SET UP: The positive direction is from the listener toward the source.

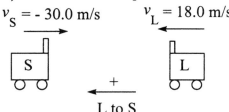

$v_S = -30.0$ m/s $v_L = 18.0$ m/s

L to S

EXECUTE:

$$f_L = \left(\frac{v + v_L}{v + v_S}\right) f_S = \left(\frac{344 \text{ m/s} + 18.0 \text{ m/s}}{344 \text{ m/s} - 30.0 \text{ m/s}}\right) (262 \text{ Hz}) = 302 \text{ Hz}$$

EVALUATE: Listener and source are approaching and $f_L > f_S$.

b) SET UP:

$v_L = -18.0$ m/s $v_S = +30.0$ m/s

L to S

EXECUTE:

$$f_L = \left(\frac{v + v_L}{v + v_S}\right) f_S = \left(\frac{344 \text{ m/s} - 18.0 \text{ m/s}}{344 \text{ m/s} + 30.0 \text{ m/s}}\right) (262 \text{ Hz}) = 228 \text{ Hz}$$

EVALUATE: Listener and source are moving away from each other and $f_L < f_S$.

Problems

16.49 IDENTIFY: Use the equations that relate intensity level and intensity, intensity and pressure amplitude, pressure amplitude and displacement amplitude, and intensity and distance.

a) SET UP: Use the intensity level β to calculate I at this distance. $\beta = (10 \text{ dB}) \log(I/I_0)$

EXECUTE: 52.0 dB $= (10 \text{ dB}) \log(I/(10^{-12} \text{ W/m}^2))$

$\log(I/(10^{-12} \text{ W/m}^2)) = 5.20$ implies $I = 1.585 \times 10^{-7} \text{ W/m}^2$

SET UP: Then use Eq.(16.12) to calculate p_{max}:

$$I = \frac{p_{max}^2}{2\rho v} \text{ so } p_{max} = \sqrt{2\rho v I}$$

From Example 16.6, $\rho = 1.20 \text{ kg/m}^3$ for air at 20°C.

EXECUTE:

$$p_{\text{max}} = \sqrt{2\rho v I} = \sqrt{2(1.20 \text{ kg/m}^3)(344 \text{ m/s})(1.585 \times 10^{-7} \text{ W/m}^2)} = 0.0114 \text{ Pa}$$

b) SET UP: Eq.(16.5): $p_{\text{max}} = BkA$ so $A = \dfrac{p_{\text{max}}}{Bk}$

For air $B = 1.42 \times 10^5$ Pa (Example 16.1).

EXECUTE: $k = \dfrac{2\pi}{\lambda} = \dfrac{2\pi f}{v} = \dfrac{(2\pi \text{ rad})(587 \text{ Hz})}{344 \text{ m/s}} = 10.72 \text{ rad/m}$

$$A = \frac{p_{\text{max}}}{Bk} = \frac{0.0114 \text{ Pa}}{(1.42 \times 10^5 \text{ Pa})(10.72 \text{ rad/m})} = 7.49 \times 10^{-9} \text{ m}$$

c) SET UP: $\beta_2 - \beta_1 = (10 \text{ dB}) \log(I_2/I_1)$ (Example 16.11).

Eq.(15.26): $I_1/I_2 = r_2^2/r_1^2$ so $I_2/I_1 = r_1^2/r_2^2$

EXECUTE: $\beta_2 - \beta_1 = (10 \text{ dB}) \log(r_1/r_2)^2 = (20 \text{ dB}) \log(r_1/r_2)$.

$\beta_2 = 52.0$ dB and $r_2 = 5.00$ m. Then $\beta_1 = 30.0$ dB and we need to calculate r_1.

52.0 dB $-$ 30.0 dB $= (20 \text{ dB}) \log(r_1/r_2)$

22.0 dB $= (20 \text{ dB}) \log(r_1/r_2)$

$\log(r_1/r_2) = 1.10$ so $r_1 = 12.6 r_2 = 63.0$ m.

EVALUATE: The decrease in intensity level corresponds to a decrease in intensity, and this means an increase in distance. The intensity level uses a logarithmic scale, so simple proportionality between r and β doesn't apply.

16.53 IDENTIFY and **SET UP:** The frequency of any harmonic is an integer multiple of the fundamental. For a stopped pipe only odd harmonics are present. For an open pipe, all harmonics are present. See which pattern of harmonics fit to the observed values in order to determine which type of pipe it is. Then solve for the fundamantal frequency and relate that to the length of the pipe.

EXECUTE:

a) For an open pipe the successive harmonics are $f_n = nf_1$, $n = 1, 2, 3, \ldots$. For a stopped pipe the successive harmonics are $f_n = nf_1$, $n = 1, 3, 5, \ldots$.

If the pipe is open and these harmonics are successive, then $f_n = nf_1 = 1372$ Hz and $f_{n+1} = (n+1)f_1 = 1764$ Hz.

Subtract the first equation from the second: $(n+1)f_1 - nf_1 = 1764 \text{ Hz} - 1372 \text{ Hz}$. This gives $f_1 = 392$ Hz.

Then $n = \dfrac{1372 \text{ Hz}}{392 \text{ Hz}} = 3.5$. But n must be an integer, so the pipe can't be open.

If the pipe is stopped and these harmonics are successive, then $f_n = nf_1 = 1372$ Hz and $f_{n+2} = (n+2)f_1 = 1764$ Hz (in this case succesive harmonics differ in n by 2).

Subtracting one equation from the other gives $2f_1 = 392$ Hz and $f_1 = 196$ Hz.

Then $n = 1372 \text{ Hz}/f_1 = 7$ so $1372 \text{ Hz} = 7f_1$ and $1764 \text{ Hz} = 9f_1$.
The solution gives integer n as it should; the pipe is stopped.

b) From part (a) these are the 7th and 9th harmonics.

c) From part (a) $f_1 = 196 \text{ Hz}$.

For a stopped pipe $f_1 = \dfrac{v}{4L}$ and $L = \dfrac{v}{4f_1} = \dfrac{344 \text{ m/s}}{4(196 \text{ Hz})} = 0.439 \text{ m}$.

EVALUATE: It is essential to know that there are successive harmonics and to realize that 1372 Hz is not the fundamental. There are other lower frequency standing waves; these are just two successive ones.

16.57 **IDENTIFY** and **SET UP:** There is a node at the piston, so the distance the piston moves is the node to node distance, $\lambda/2$. Use Eq.(15.1) to calculate v and Eq.(16.10) to calculate γ from v.
EXECUTE:
a) $\lambda/2 = 37.5 \text{ cm}$, so $\lambda = 2(37.5 \text{ cm}) = 75.0 \text{ cm} = 0.750 \text{ m}$.
$v = f\lambda = (500 \text{ Hz})(0.750 \text{ m}) = 375 \text{ m/s}$

b) $v = \sqrt{\gamma RT/M}$ (Eq.16.10)

$$\gamma = \frac{Mv^2}{RT} = \frac{(28.8 \times 10^{-3} \text{ kg/mol})(375 \text{ m/s})^2}{(8.3145 \text{ J/mol} \cdot \text{K})(350 \text{ K})} = 1.39.$$

c) **EVALUATE:** There is a node at the piston so when the piston is 18.0 cm from the open end the node is inside the pipe, 18.0 cm from the open end. The node to antinode distance is $\lambda/4 = 18.8 \text{ cm}$, so the antinode is 0.8 cm beyond the open end of the pipe.
The value of γ we calculated agrees with the value given for air in Example 16.5.

16.65 **IDENTIFY** and **SET UP:** Use Eq.(16.12) for the intensity and Eq.(16.14) to relate the intensity and pressure amplitude.
EXECUTE:
a) The amplitude of the oscillations is ΔR.
$I = \frac{1}{2}\sqrt{\rho B}(2\pi f)^2 A^2 = 2\sqrt{\rho B}\pi^2 f^2 (\Delta R)^2$

b) $P = I(4\pi R^2) = 8\pi^3 \sqrt{\rho B} f^2 R^2 (\Delta R)^2$

c) $I_R/I_d = d^2/R^2$
$I_d = (R/d)^2 I_R = 2\pi^2 \sqrt{\rho B}(Rf/d)^2 (\Delta R)^2$
$I = p_{\text{max}}^2/2\sqrt{\rho B}$ so
$p_{\text{max}} = \sqrt{(2\sqrt{\rho B} I)} = 2\pi \sqrt{\rho B}(Rf/d)\,\Delta R$

$$A = \frac{p_{max}}{Bk} = \frac{p_{max}\lambda}{B2\pi} = \frac{p_{max}v}{B2\pi f} = v\sqrt{\rho/B}\,(R/d)\,\Delta R$$

But $v = \sqrt{B/\rho}$ so $v\sqrt{\rho/B} = 1$ so $A = (R/d)\,\Delta R$.

EVALUATE: The pressure amplitude and displacement amplitude fall off like $1/d$ and the intensity like $1/d^2$.

16.67 a) IDENTIFY and **SET UP:** Use Eq.(15.1) to calculate λ.

EXECUTE: $\lambda = \dfrac{v}{f} = \dfrac{1482 \text{ m/s}}{22.0 \times 10^3 \text{ Hz}} = 0.0674 \text{ m}$

b) IDENTIFY: Apply the Doppler effect equation, Eq.(16.29). The Problem-Solving Strategy in the text (Section 16.8) describes how to do this problem. The frequency of the directly radiated waves is $f_S = 22,000$ Hz. The moving whale first plays the role of a moving listener, receiving waves with frequency f'_L. The whale then acts as a moving source, emitting waves with the same frequency, $f'_S = f'_L$ with which they are received. Let the speed of the whale be v_W.

SET UP: <u>whale receives waves</u>

EXECUTE:
$v_L = +v_W$

$$f'_L = f_S\left(\frac{v + v_L}{v + v_S}\right) = f_S\left(\frac{v + v_W}{v}\right)$$

SET UP: <u>whale re-emits the waves</u>

EXECUTE:
$v_S = -v_W$

$$f_L = f_S\left(\frac{v + v_L}{v + v_S}\right) = f'_S\left(\frac{v}{v - v_W}\right)$$

But $f'_S = f'_L$ so $f_L = f_S\left(\dfrac{v + v_W}{v}\right)\left(\dfrac{v}{v - v_W}\right) = f_S\left(\dfrac{v + v_W}{v - v_W}\right)$.

Then $\Delta f = f_S - f_L = f_S\left(1 - \dfrac{v + v_W}{v - v_W}\right) = f_S\left(\dfrac{v - v_W - v - v_W}{v - v_W}\right) = \dfrac{-2f_S v_W}{v - v_W}$.

$$\Delta f = \frac{-2(2.20 \times 10^4 \text{ Hz})(4.95 \text{ m/s})}{1482 \text{ m/s} - 4.95 \text{ m/s}} = 147 \text{ Hz}.$$

EVALUATE: Listener and source are moving toward each other so frequency is raised.

16.73 **IDENTIFY** and **SET UP:** Use Eq.(16.30) that describes the Doppler effect for electromagnetic waves. $v \ll c$, so the simplified form derived in Problem 16.71b can be used.

a) EXECUTE: From Problem 16.71b, $f_R = f_S(1 - v/c)$.

v is negative since the source is approaching:

$v = -(42.0 \text{ km/h})(1000 \text{ m/1 km})(1 \text{ h/3600 s}) = -11.67 \text{ m/s}$

Approaching means that the frequency is increased.

$$\Delta f = f_S \left(-\frac{v}{c}\right) = 2800 \times 10^6 \text{ Hz} \left(-\frac{-11.67 \text{ m/s}}{3.00 \times 10^8 \text{ m/s}}\right) = 109 \text{ Hz}$$

b) EVALUATE: Approaching, so the frequency is increased. $v \ll c$ and therefore $\Delta f / f_S \ll 1$. The frequency of the waves received and reflected by the water is very close to 2880 MHz, so get an additional shift of 109 Hz and the total shift in frequency is $2(109 \text{ Hz}) = 218 \text{ Hz}$.

16.75 **IDENTIFY** and **SET UP:** Use Fig.(16.38) to relate α and T.
Use this in Eq.(16.31) to eliminate $\sin \alpha$.

EXECUTE: Eq.(16.31): $\sin \alpha = v/v_S$ From Fig.16.38 $\tan \alpha = h/v_S T$. And

$$\tan \alpha = \frac{\sin \alpha}{\cos \alpha} = \frac{\sin \alpha}{\sqrt{1 - \sin^2 \alpha}}.$$

Combining these equations we get $\dfrac{h}{v_S T} = \dfrac{v/v_S}{\sqrt{1 - (v/v_S)^2}}$ and $\dfrac{h}{T} = \dfrac{v}{\sqrt{1 - (v/v_S)^2}}$.

$$1 - (v/v_s)^2 = \frac{v^2 T^2}{h^2} \text{ and } v_S^2 = \frac{v^2}{1 - v^2 T^2/h^2}$$

$v_S = \dfrac{hv}{\sqrt{h^2 - v^2 T^2}}$ as was to be shown.

EVALUATE: For a given h, the faster the speed v_S of the plane, the greater is the delay time T. The maximum delay time is h/v, and T approaches this value as $v_S \to \infty$. $T \to 0$ as $v \to v_S$.

CHAPTER 17

TEMPERATURE AND HEAT

Exercises 15, 21, 23, 27, 31, 33, 35, 39, 41, 45, 49, 51, 53, 57, 59, 61, 65, 67, 71, 73, 75, 77
Problems 79, 83, 87, 91, 93, 95, 97, 101, 103, 105, 107, 111, 113, 115, 117

Exercises

17.15 **IDENTIFY** and **SET UP:** Fit the data to a straight line for $p(T)$ and use
this equation to find T when $p = 0$.

EXECUTE: If the pressure varies linearly with temperature, then $p_2 = p_1 + \gamma(T_2 - T_1)$.

$$\gamma = \frac{p_2 - p_1}{T_2 - T_1} = \frac{6.50 \times 10^4 \text{ Pa} - 4.80 \times 10^4 \text{ Pa}}{100°\text{C} - 0.01°\text{C}} = 170.0 \text{ Pa/C}°$$

Apply $p = p_1 + \gamma(T - T_1)$ with $T_1 = 0.01°\text{C}$ and $p = 0$ to solve for T.
$0 = p_1 + \gamma(T - T_1)$

$$T = T_1 - \frac{p_1}{\gamma} = 0.01°\text{C} - \frac{4.80 \times 10^4 \text{ Pa}}{170.0 \text{ Pa/C}°} = -282°\text{C}.$$

b) Let $T_1 = 100°\text{C}$ and $T_2 = 0.01°\text{C}$; use Eq.(17.4) to calculate p_2. Eq.(17.4) says
$T_2/T_1 = p_2/p_1$, where T is in kelvins.

$$p_2 = p_1 \left(\frac{T_2}{T_1}\right) = 6.50 \times 10^4 \text{ Pa} \left(\frac{0.01 + 273.15}{100 + 273.15}\right) = 4.76 \times 10^4 \text{ Pa; this differs from}$$
the 4.80×10^4 Pa that was measured so Eq.(17.4) is not precisely obeyed.

EVALUATE: The answer to part (a) is in reasonable agreement with the accepted
value of $-273°\text{C}$

17.21 **IDENTIFY:** Linear expansion; apply Eq.(17.6) and solve for α.

SET UP: Let $L_0 = 40.125$ cm; $T_0 = 20.0°\text{C}$.

$\Delta T = 45.0°\text{C} - 20.0°\text{C} = 25.0 \text{ C}°$ gives $\Delta L = 0.023$ cm

EXECUTE: $\Delta L = \alpha L_0 \Delta T$ implies

$$\alpha = \frac{\Delta L}{L_0 \Delta T} = \frac{0.023 \text{ cm}}{(40.125 \text{ cm})(25.0 \text{ C}°)} = 2.3 \times 10^{-5} \text{ (C}°)^{-1}.$$

EVALUATE: The value we calculated is the same order of magnitude as the
values for metals in Table 17.1.

17.23 **IDENTIFY:** Volume expansion; apply Eq.(17.8) to calculate ΔV for the ethanol.

SET UP: From Table 17.2, β for ethanol is 75×10^{-5} K^{-1}

EXECUTE: $\Delta T = 10.0°C - 19.0°C = -9.0$ K.

Then $\Delta V = \beta V_0 \, \Delta T = (75 \times 10^{-5} \text{ K}^{-1})(1700 \text{ L})(-9.0 \text{ K}) = -11$ L.

The volume of the air space will be 11 L = 0.011 m^3.

EVALUATE: The temperature decreases, so the volume of the liquid decreases. The volume change is small, less than 1% of the original volume.

17.27 **IDENTIFY** and **SET UP:** Apply the result of Exercise 17.26a to calculate ΔA for the plate, and then $A = A_0 + \Delta A$.

EXECUTE:

a) $A_0 = \pi r_0^2 = \pi (1.350 \text{ cm}/2)^2 = 1.431$ cm^2

b) Exercise 17.26 says $\Delta A = 2\alpha A_0 \, \Delta T$, so
$\Delta A = 2(1.2 \times 10^{-5} \text{ C}°^{-1})(1.431 \text{ cm})(175°C - 25°C) = 5.15 \times 10^{-3}$ cm^2
$A = A_0 + \Delta A = 1.436$ cm^2

EVALUATE: A hole in a flat metal plate expands when the metal is heated just as a piece of metal the same size as the hole would expnad.

17.31 **IDENTIFY** and **SET UP:** For part (a), apply Eq.(17.6) to the linear expansion of the wire. For part (b), apply Eq.(17.12) and calculate F/A.

EXECUTE:

a) $\Delta L = \alpha L_0 \, \Delta T$

$$\alpha = \frac{\Delta L}{L_0 \, \Delta T} = \frac{1.9 \times 10^{-2} \text{ m}}{(1.50 \text{ m})(420°C - 20°C)} = 3.2 \times 10^{-5} \text{ (C}°)^{-1}$$

b) Eq.(17.12): stress $F/A = -Y\alpha \, \Delta T$

$\Delta T = 20°C - 420°C = -400$ C$°$ (ΔT always means final temperature minus initial temperature)

$F/A = -(2.0 \times 10^{11} \text{ Pa})(3.2 \times 10^{-5}(\text{C}°)^{-1})(-400°C) = +2.6 \times 10^9$ Pa

EVALUATE: F/A is positive means that the stress is a tensile (stretching) stress. The answer to part (a) is consistent with the values of α for metals in Table 17.1. The tensile stress for this modest temperature decrease is huge.

17.33 **IDENTIFY** and **SET UP:** Use Eq.(17.16).

EXECUTE:

a) $Q = mc \, \Delta T$

$m = \frac{1}{2}(1.3 \times 10^{-3} \text{ kg}) = 0.65 \times 10^{-3}$ kg

$$Q = (0.65 \times 10^{-3} \text{ kg})(1020 \text{ J/kg} \cdot \text{K})(37°C - (-20°C)) = 38 \text{ J}$$

b) 20 breaths/min (60 min/1 h) = 1200 breaths/h

So $Q = (1200)(38 \text{ J}) = 4.6 \times 10^4 \text{ J}$.

EVALUATE: The heat loss rate is $Q/t = 13$ W.

17.35 IDENTIFY and **SET UP:** Set the change in gravitational potential energy equal to the quantity of heat added to the water.

EXECUTE: The change in mechanical energy equals the decrease in gravitational potential energy, $\Delta U = -mgh$; $| \Delta U |= mgh$.

$Q =| \Delta U |= mgh$ implies $mc\Delta T = mgh$

$\Delta T = gh/c = (9.80 \text{ m/s}^2)(225 \text{ m})/(4190 \text{ J/kg} \cdot \text{K}) = 0.526 \text{ K} = 0.526 \text{ C}°$

EVALUATE: Note that the answer is independent of the mass of the object. Note also the small change in temperature that corresponds to this large change in height!

17.39 IDENTIFY and **SET UP:** Apply Eq.(17.17) to the kettle and water.

EXECUTE:

kettle

$Q = mc \Delta T$, $c = 910$ J/kg $\cdot$ K (from Table 17.3)

$Q = (1.50 \text{ kg})(910 \text{ J/kg} \cdot \text{K})(85.0°C - 20.0°C) = 8.873 \times 10^4 \text{ J}$

water

$Q = mc \Delta T$, $c = 4190$ J/kg $\cdot$ K (from Table 15-3)

$Q = (1.80 \text{ kg})(4190 \text{ J/kg} \cdot \text{K})(85.0°C - 20.0°C) = 4.902 \times 10^5 \text{ J}$

Total $Q = 8.873 \times 10^4 \text{ J} + 4.902 \times 10^5 \text{ J} = 5.79 \times 10^5 \text{ J}$.

EVALUATE: Water has a much larger specific heat capacity than aluminum, so most of the heat goes into raising the temperature of the water.

17.41 IDENTIFY and **SET UP:** Use the power and time to calculate the heat input Q and then use Eq.(17.13) to calculate c.

a) **EXECUTE:** $P = Q/t$, so the total heat transferred to the liquid is $Q = Pt = (65.0 \text{ W})(120 \text{ s}) = 7800 \text{ J}$

Then $Q = mc \Delta T$ gives

$$c = \frac{Q}{m \Delta T} = \frac{7800 \text{ K}}{0.780 \text{ kg}(22.54°C - 18.55°C)} = 2.51 \times 10^3 \text{ J/kg} \cdot \text{K}$$

b) **EVALUATE:** Then the actual Q transferred to the liquid is less than 7800 J so the actual c is less than our calculated value; our result in part (a) is an overestimate.

17.45 **IDENTIFY** and **SET UP:** Heat comes out of the metal and into the water. The final temperature is in the range $0 < T < 100°C$, so there are no phase changes. $Q_{system} = 0$.

a) EXECUTE $Q_{water} + Q_{metal} = 0$

$m_{water} c_{water} \Delta T_{water} + m_{metal} c_{metal} \Delta T_{metal} = 0$

$(1.00 \text{ kg})(4190 \text{ J/kg} \cdot \text{K})(2.0 \text{ C}°) + (0.500 \text{ kg})(c_{metal})(-78.0 \text{ C}°) = 0$

$c_{metal} = 215 \text{ J/kg} \cdot \text{K}$

b) EVALUATE: Water has a larger specific heat capacity so stores more heat per degree of temperature change.

c) If some heat went into the styrofoam then Q_{metal} should actually be larger than in part (a), so the true c_{metal} is larger than we calculated; the value we calculated would be smaller than the true value.

17.49 **IDENTIFY** and **SET UP:** Use Eq.(17.13) for the temperature changes and Eq.(17.20) for the phase changes.

EXECUTE: Heat must be added to do the following

ice at $-10.0°C \rightarrow$ ice at $0°C$

$Q_{ice} = mc_{ice} \Delta T = (12.0 \times 10^{-3} \text{ kg})(2100 \text{ J/kg} \cdot \text{K})(0°C - (-10.0°C)) = 252 \text{ J}$

phase transition ice $(0°C) \rightarrow$ liquid water $(0°C)$ (melting)

$Q_{melt} = +mL_f = (12.0 \times 10^{-3} \text{ kg})(334 \times 10^3 \text{ J/kg}) = 4.008 \times 10^3 \text{ J}$

water at $0°C$ (from melted ice $\rightarrow$ water at $100°C$

$Q_{water} = mc_{water} \Delta T = (12.0 \times 10^{-3} \text{ kg})(4190 \text{ J/kg} \cdot \text{K})(100°C - 0°C) =$ $5.028 \times 10^3 \text{ J}$

phase transition water $(100°C) \rightarrow$ steam $(100°C)$ (boiling)

$Q_{boil} = +mL_v = (12.0 \times 10^{-3} \text{ kg})(2256 \times 10^3 \text{ J/kg}) = 2.707 \times 10^4 \text{ J}$

The total Q is $Q = 252 \text{ J} + 4.008 \times 10^3 \text{ K} + 5.028 \times 10^3 \text{ J} + 2.707 \times 10^4 \text{ J} =$ $3.64 \times 10^4 \text{ J}$

$(3.64 \times 10^4 \text{ J})(1 \text{ cal}/4.186 \text{ J}) = 8.70 \times 10^3 \text{ cal}$

$(3.64 \times 10^4 \text{ J})(1 \text{ Btu}/1055 \text{ J}) = 34.5 \text{ Btu}$

EVALUATE: Q is positive and heat must be added to the material. Note that more heat is needed for the liquid to gas phase change than for the temperature changes.

17.51 **IDENTIFY** and **SET UP:** Use Eq.(17.20)to calculate Q and then $P = Q/t$. Must convert the quantity of ice from lb to kg.

EXECUTE: "two-ton air conditioner" means 2 tons (4000 lbs) of ice can be frozen from water at $0°C$ in 24 h.

Find the mass m that corresponds to 4000 lb (weight of water):

$m = (4000 \text{ lb})(1 \text{ kg}/2.205 \text{ lb}) = 1814 \text{ kg}$ (The kg to lb equivalence from Appendix E has been used.)

The heat that must be removed from the water to freeze it is

$Q = -mL_f = -(1814 \text{ kg})(334 \times 10^3 \text{ J/kg}) = -6.06 \times 10^8 \text{ J}.$

The power required if this is to be done in 24 hours is

$$P = \frac{|Q|}{t} = \frac{6.06 \times 10^8 \text{ J}}{(24 \text{ h})(3600 \text{ s}/1 \text{ h})} = 7010 \text{ W}$$

or $P = (7010 \text{ W})((1 \text{ Btu/h})/(0.293 \text{ W})) = 2.39 \times 10^4 \text{ Btu/h}.$

EVALUATE: The calculated power, the rate at which heat energy is removed by the unit, is equivalent to seventy 100-W light bulbs.

17.53 **IDENTIFY** and **SET UP:** The heat that must be added to a lead bullet of mass m to melt it is $Q = mc\,\Delta T + mL_f$ ($mc\,\Delta T$ is the heat required to raise the temperature from $25°C$ to the melting point of $327.3°C$; mL_f is the heat required to make the solid $\rightarrow$ liquid phase change.)

The kinetic energy of the bullet if its speed is v is $K = \frac{1}{2}mv^2$.

EXECUTE: $K = Q$ says $\frac{1}{2}mv^2 = mc\,\Delta T + mL_f$.

$v = \sqrt{2(c\Delta T + L_f)}$

$v = \sqrt{2[(130 \text{ J/kg} \cdot \text{K})(327.3°C - 25°C) + 24.5 \times 10^3 \text{ J/kg}]} = 357 \text{ m/s}$

EVALUATE: This is a typical speed for a rifle bullet. A bullet fired into a block of wood does partially melt, but in practice not all of the initial kinetic energy is converted to heat that remains in the bullet.

17.57 **IDENTIFY** and **SET UP:** Heat flows out of the iron block and into the copper pot and water. The net heat flow for the system is zero and all three objects have the same final temperature. There are no phase changes.

EXECUTE: $Q_{system} = 0$. Calculate Q for each component of the system:

copper pot

$Q_{can} = mc\,\Delta T = (0.500 \text{ kg})(390 \text{ J/kg} \cdot \text{K})(T - 20.0°C) = (195 \text{ J/K})T - 3900 \text{ J}$

water

$Q_{\text{water}} = mc\,\Delta T = (0.170\text{ kg})(4190\text{ J/kg} \cdot \text{K})(T - 20.0°\text{C}) = (712.3\text{ J/K})T - 1.425 \times 10^4$ J

iron

$Q_{\text{iron}} = mc\,\Delta T = (0.250\text{ kg})(470\text{ J/kg} \cdot \text{K})(T - 85.0°\text{C}) =$
$(117.5\text{ J/K})T - 9987.5$ J

$Q_{\text{system}} = 0$ gives that $Q_{\text{can}} + Q_{\text{water}} + Q_{\text{iron}} = 0$
$(195\text{ J/K})T - 3900\text{ J} + (712.3\text{ J/K})T - 1.425 \times 10^4\text{ J} + (117.5\text{ J/K})T - 9987.5\text{ J} = 0$
$(1024.8\text{ J/K})T = 2.814 \times 10^4$ J

$$T = \frac{2.814 \times 10^4\text{ J}}{1024.8\text{ J/K}} = 27.5°\text{C}$$

EVALUATE: A final T in the range $0 < T < 100°$C is consistent with no phase changes. The water and pot heat up and the iron block cools down.

17.59 **IDENTIFY** and **SET UP:** Heat flows out of the water and into the ice. The net heat flow for the system is zero. The ice warms to $0°$C, melts, and then the water from the melted ice warms from $0°$C to the final temperature.

EXECUTE: $Q_{\text{system}} = 0$; calculate Q for each component of the system:
(Beaker has small mass says that $Q = mc\,\Delta T$ for beaker can be neglected.)

0.250 kg of water (cools from $75.0°$C to $30.0°$C)
$Q_{\text{water}} = mc\,\Delta T = (0.250\text{ kg})(4190\text{ J/kg} \cdot \text{K})(30.0°\text{C} - 75.0°\text{C}) = -4.714 \times 10^4$ J

ice (warms to $0°$C; melts; water from melted ice warms to $30.0°$C)
$Q_{\text{ice}} = mc_{\text{ice}}\Delta T + mL_{\text{f}} + mc_{\text{water}}\Delta T$
$Q_{\text{ice}} = m[(2100\text{ J/kg} \cdot \text{K})(0°\text{C} - (-20.0°\text{C})) + 334 \times 10^3\text{ J/kg} +$
$(4190\text{ J/kg} \cdot \text{K})(30.0°\text{C} - 0°\text{C})]$
$Q_{\text{ice}} = (5.017 \times 10^5\text{ J/kg})m$

$Q_{\text{system}} = 0$ says $Q_{\text{water}} + Q_{\text{ice}} = 0$
$-4.714 \times 10^4\text{ J} + (5.017 \times 10^5\text{ J/kg})m = 0$

$$m = \frac{4.714 \times 10^4\text{ J}}{5.017 \times 10^5\text{ J/kg}} = 0.0940\text{ kg}$$

EVALUATE: Since the final temperature is $30.0°$C we know that all the ice melts and the final system is all liquid water. The mass of ice added is much less than the mass of the $75°$C water; the ice requires a large heat input for the phase change.

17.61 **IDENTIFY** and **SET UP:** Large block of ice implies that ice is left, so $T_2 = 0°$C (final temperature). Heat comes out of the ingot and into the ice.

The net heat flow is zero. The ingot has a temperature cahnge and the ice has a phase change.

EXECUTE: $Q_{system} = 0$; calculate Q for each component of the system:

ingot

$Q_{ingot} = mc\,\Delta T = (4.00 \text{ kg})(234 \text{ J/kg} \cdot \text{K})(0°C - 750°C) = -7.02 \times 10^5 \text{ J}$

ice

$Q_{ice} = +mL_f$, where m is the mass of the ice that changes phase (melts)

$Q_{system} = 0$ says $Q_{ingot} + Q_{ice} = 0$

$-7.02 \times 10^5 \text{ J} + m(334 \times 10^3 \text{ J/kg}) = 0$

$m = \dfrac{7.02 \times 10^5 \text{ J}}{334 \times 10^3 \text{ J/kg}} = 2.10 \text{ kg}$

EVALUATE: The liquid produced by the phase change remains at 0°C since it is in contact with ice.

17.65 **IDENTIFY** and **SET UP:** The temperature gradient is $(T_H - T_C)/L$ and can be calculated directly. Use Eq.(17.21) to calculate the heat current H. In part (c) use H from part (b) and apply Eq.(17.21) to the 12.0-cm section of the left end of the rod. $T_2 = T_H$ and $T_1 = T$, the target variable.

EXECUTE:

a) temperature gradient $= (T_H - T_C)/L =$

$(100.0°C - 0.0°C)/0.450 \text{ m} = 222 \text{ C}°/\text{m} = 222 \text{ K/m}$

b) $H = kA(T_H - T_C)/L$. From Table 17.5, $k = 385 \text{ W/m} \cdot \text{K}$,

so $H = (385 \text{ W/m} \cdot \text{K})(1.25 \times 10^{-4} \text{ m}^2)(222 \text{ K/m}) = 10.7 \text{ W}$

c) $H = 10.7 \text{ W}$ for all sections of the rod.

Apply $H = kA\,\Delta T/L$ to the 12.0 cm section:

$T_H - T = LH/kA$ and $T = T_H - LH/Ak =$

$100.0°C - \dfrac{(0.120 \text{ m})(10.7 \text{ W})}{(1.25 \times 10^{-4} \text{ m}^2)(385 \text{ W/m} \cdot \text{K})} = 73.3°C$

EVALUATE: H is the same at all points along the rod, so $\Delta T/\Delta x$ is the same for any section of the rod with length Δx. Thus $(T_H - T)/(12.0 \text{ cm}) = (T_H - T_C)/(45.0 \text{ cm})$ gives that $T_H - T = 26.7 \text{ C}°$ and $T = 73.3°C$, as we already calculated.

17.67 **IDENTIFY** and **SET UP:** Call the temperature at the interface between the wood and the styrofoam T. The heat current in each material is given by $H = kA(T_H - T_C)/L$.

Heat current through the wood:
$$H_w = k_w A(T - T_1)/L_w$$

Heat current through the stryofoam:
$$H_s = k_s A(T_2 - T)/L_s$$

In steady-state heat does not accumulate in either material. The same heat has to pass through both materials in succession, so $H_w = H_s$.

EXECUTE: This implies $k_w A(T - T_1)/L_w = k_s A(T_2 - T)/L_s$

$$k_w L_s(T - T_1) = k_s L_w(T_2 - T)$$

$$T = \frac{k_w L_s T_1 + k_s L_w T_2}{k_w L_s + k_s L_w} = \frac{-0.0176 \text{ W} \cdot^\circ \text{C/K} + 0.0057 \text{ W} \cdot^\circ \text{C/K}}{0.00206 \text{ W/K}} = -5.8^\circ\text{C}$$

EVALUATE: The temperature at the junction is much closer in value to T_1 than to T_2. The styrofoam has a very large k, so a larger temperature gradient is required for than for wood to establish the same heat current.

b) IDENTIFY and **SET UP:** Heat flow per square meter is $\dfrac{H}{A} = k\left(\dfrac{T_H - T_C}{L}\right)$. We can calculate this either for the wood or for the styrofoam; the results must be the same.

EXECUTE:

wood
$$\frac{H_w}{A} = k_w \frac{T - T_1}{L_w} = (0.080 \text{ W/m} \cdot \text{K})\frac{(-5.8^\circ\text{C} - (-10.0^\circ\text{C}))}{0.030 \text{ m}} = 11 \text{ W/m}^2.$$

styrofoam
$$\frac{H_s}{A} = k_s \frac{T_2 - T}{L_s} = (0.010 \text{ W/m} \cdot \text{K})\frac{(19.0^\circ\text{C} - (-5.8^\circ\text{C}))}{0.022 \text{ m}} = 11 \text{ W/m}^2.$$

EVALUATE: H must be the same for both materials and our numerical results show this. Both materials are good insulators and the heat flow is very small.

17.71 IDENTIFY and **SET UP:** The heat conducted through the bottom of the pot goes into the water at 100°C to convert it to steam at 100°C. We can calculate the amount of heat flow from the mass of material that changes phase. Then use Eq.(17.21) to calculate T_H, the temperature of the lower surface of the pan.

EXECUTE: $Q = mL_v = (0.390 \text{ kg})(2256 \times 10^3 \text{ J/kg}) = 8.798 \times 10^5 \text{ J}$

$H = Q/t = 8.798 \times 10^5 \text{ J}/180 \text{ s} = 4.888 \times 10^3 \text{ J/s}$

Then $H = kA(T_H - T_C)/L$ says that

$$T_H - T_C = \frac{HL}{kA} = \frac{(4.888 \times 10^3 \text{ J/s})(8.50 \times 10^{-3} \text{ m})}{(50.2 \text{ W/m} \cdot \text{K})(0.150 \text{ m}^2)} = 5.52 \text{ C}°$$

$T_H = T_C + 5.52 \text{ C}° = 100°\text{C} + 5.52 \text{ C}° = 105.5°\text{C}$

EVALUATE: The larger $T_H - T_C$ is the larger H is and the faster the water boils.

17.73 **IDENTIFY** and **SET UP:** Use Eq.(17.21) to express the heat current for each section in terms of the temperature difference between the ends of the section. At steady state these two heat currents are equal. Solve for the temperature T at the junction.

EXECUTE: $H_a = H_b$ (a = aluminum, b = brass)

$$H_a = k_a \frac{A(150.0°\text{C} - T)}{L_a}, \qquad H_b = k_b \frac{A(T - 0°\text{C})}{L_b}$$

(It has been assumed that the two sections have the same cross-sectional area.)

$$k_a \frac{A(150.0°\text{C} - T)}{L_a} = k_b \frac{A(T - 0°\text{C})}{L_b}$$

$$\frac{(205.0 \text{ W/m} \cdot \text{K})(150.0°\text{C} - T)}{0.800 \text{ m}} = \frac{(109.0 \text{ W/m} \cdot \text{K})(T - 0°\text{C})}{0.500 \text{ m}}$$

Solving for T gives $T = 90.2°\text{C}$

EVALUATE: k/L is about the same for the two sections so similar temperature differences across the ends of the sections are required to make the heat currents equal. The temperature at the junction is about mid-way between the temperature at the two ends.

17.75 **IDENTIFY:** Use Eq.(17.26) to calculate H_{net}.

SET UP: $H_{net} = Ae\sigma(T^4 - T_s^4)$ (Eq.17.26); T must be in kelvins)

Example 17.16 gives $A = 1.2 \text{ m}^2$, $e = 1.0$, and $T = 30°\text{C} = 303 \text{ K}$ (body surface temperature)

$T_s = 5.0°\text{C} = 278 \text{ K}$

EXECUTE: $H_{net} = 573.5 \text{ W} - 406.4 \text{ W} = 167 \text{ W}$

EVALUATE: Note that this is larger than H_{net} calculated in Example 17.16. The lower temperature of the surroundings increases the rate of heat loss by radiation.

17.77 **IDENTIFY:** Use Eq.(17.26) to calculate A.

SET UP: $H = Ae\sigma T^4$ so $A = H/e\sigma T^4$

150-W and all electrical energy consumed is radiated says $H = 150 \text{ W}$

EXECUTE: $A = \dfrac{150 \text{ W}}{(0.35)(5.67 \times 10^{-8} \text{ W/m}^2 \cdot \text{K}^4)(2450 \text{ K})^4} =$ $2.1 \times 10^{-4} \text{ m}^2 (1 \times 10^4 \text{ cm}^2/1 \text{ m}^2) = 2.1 \text{ cm}^2$

EVALUATE: Light bulb filaments are often in the shape of a tightly wound coil to increase the surface area; larger A means a larger radiated power H.

Problems

17.79 IDENTIFY and **SET UP:** Use the temperature difference in M° and in C° between the melting and boiling points of mercury to relate M° to C°. Also adjust for the different zero points on the two scales to get an equation for T_M in terms of T_C.

a) EXECUTE: normal melting point of mercury: $-30°\text{C} = 0.0°\text{M}$

normal boiling point of mercury: $357°\text{C} = 100.0°\text{M}$

$100 \text{ M}° = 396 \text{ C}°$ so $1 \text{ M}° = 3.96 \text{ C}°$

Zero on the M scale is -39 on the C scale, so to obtain T_C multiple T_M by 3.96 and then subtract 39°: $T_C = 3.96 T_M - 39°$

Solving for T_M gives $T_M = \frac{1}{3.96}(T_C + 39°)$

The normal boiling point of water is $100°\text{C}$; $T_M = \frac{1}{3.96}(100° + 39°) = 35.1°\text{M}$

b) $10.0 \text{ M}° = 39.6 \text{ C}°$

EVALUATE: A M° is larger than a C° since it takes fewer of them to express the difference between the boiling and melting points for mercury.

17.83 IDENTIFY and **SET UP:** Use Eq.(17.8) for the volume expansion of the oil and of the cup. Both the volume of the cup and the volume of the olive oil increase when the temperature increases, but β is larger for the oil so it expands more. When the oil starts to overflow, $\Delta V_{\text{oil}} = \Delta V_{\text{glass}} + (1.00 \times 10^{-3} \text{ m})A$, where A is the cross-sectional area of the cup.

EXECUTE:

$\Delta V_{\text{oil}} = V_{0,\text{oil}}\beta_{\text{oil}}\Delta T = (9.9 \text{ cm})A\beta_{\text{oil}}\Delta T$

$\Delta V_{\text{glass}} = V_{0,\text{glass}}\beta_{\text{glass}}\Delta T = (10.0 \text{ cm})A\beta_{\text{glass}}\Delta T$

$(9.9 \text{ cm})A\beta_{\text{oil}}\Delta T = (10.0 \text{ cm})A\beta_{\text{glass}}\Delta T + (1.00 \times 10^{-3} \text{ m})A$

The area A divides out. Solving for ΔT gives $\Delta T = 15.5 \text{ C}°$

$T_2 = T_1 + \Delta T = 37.5°\text{C}$

EVALUATE: If the expansion of the cup is neglected, the olive oil will have expanded to fill the cup when $(0.100 \text{ cm})A = (9.9 \text{ cm})A\beta_{|rmoil}\Delta T$, so $\Delta T = 15.0 \text{ C}°$ and $T_2 = 37.0°\text{C}$. Our result is slightly higher than this. The cup also expands, bu not very much since $\beta_{\text{glass}} << \beta_{\text{oil}}$.

17.87 IDENTIFY and **SET UP:** Call the metals A and B. Use the data given to calculate α for each metal.

EXECUTE: $\Delta L = L_0 \alpha \Delta T$ so $\alpha = \Delta L / (L_0 \Delta T)$

metal A: $\alpha_A = \dfrac{\Delta L}{L_0 \Delta T} = \dfrac{0.0650 \text{ cm}}{(30.0 \text{ cm})(100 \text{ C}°)} = 2.167 \times 10^{-5} \text{ (C}°)^{-1}$

metal B: $\alpha_B = \dfrac{\Delta L}{L_0 \Delta T} = \dfrac{0.0350 \text{ cm}}{(30.0 \text{ cm})(100 \text{ C}°)} = 1.167 \times 10^{-5} \text{ (C}°)^{-1}$

EVALUATE: L_0 and ΔT are the same, so the rod that expands the most has the larger α.

IDENTIFY and **SET UP:** Now consider the composite rod. Apply Eq.(17.6). The target variables are L_A and L_B, the lengths of metals A and B in the composite rod.

$\Delta T = 100 \text{ C}°$
$\Delta L = 0.058 \text{ cm}$

EXECUTE: $\Delta L = \Delta L_A + \Delta L_B = (\alpha_A L_A + \alpha_B L_B)\Delta T$

$\Delta L / \Delta T = \alpha_A L_A + \alpha_B (0.300 \text{ m} - L_A)$

$L_A = \dfrac{\Delta L / \Delta T - (0.300 \text{ m})\alpha_B}{\alpha_A - \alpha_B} =$

$\dfrac{(0.058 \times 10^{-2} \text{ m}/100 \text{ C}°) - (0.300 \text{ m})(1.167 \times 10^{-5}(\text{C}°)^{-1})}{1.00 \times 10^{-5} \text{ (C}°)^{-1}}$

$L_A = \left(\dfrac{5.80 \times 10^{-6} - 3.50 \times 10^{-6}}{1.00 \times 10^{-5}} \right) \text{ m} = 0.230 \text{ m} = 23.0 \text{ cm}$

$L_B = 30.0 \text{ cm} - L_A = 30.0 \text{ cm} - 23.0 \text{ m} = 7.0 \text{ cm}$

EVALUATE: The expansion of the composite rod is similar to that of rod A, sothe composite rod is mostly metal A.

17.91 a) IDENTIFY and **SET UP:** The diameter of the ring undergoes linear expansion (increases with T) just like a solid steel disk of the same diameter as the hole in the ring. Heat the ring to make its diameter equal to 2.5020 in.

EXECUTE:

$\Delta L = \alpha L_0 \Delta T$ so $\Delta T = \dfrac{\Delta L}{L_0 \alpha} = \dfrac{0.0020 \text{ in.}}{(2.5000 \text{ in.})(1.2 \times 10^{-5}(\text{C}°)^{-1})} = 66.7 \text{ C}°$

$T = T_0 + \Delta T = 20.0°\text{C} + 66.7 \text{ C}° = 87°\text{C}$

b) IDENTIFY and **SET UP:** Apply the linear expansion equation to the diameter of the brass shaft and to the diameter of the hole in the steel ring.

EXECUTE: $L = L_0(1 + \alpha \Delta T)$

Want L_s (steel) $= L_b$ (brass) for the same ΔT for both materials:

$L_{0s}(1 + \alpha_s \Delta T) = L_{0b}(1 + \alpha_b \Delta T)$ so $L_{0s} + L_{0s}\alpha_s \Delta T = L_{0b} + L_{0b}\alpha_b \Delta T$

$$\Delta T = \frac{L_{0b} - L_{0s}}{L_{0s}\alpha_s - L_{0b}\alpha_b} =$$

$$\frac{2.5020 \text{ in.} - 2.5000 \text{ in.}}{(2.5000 \text{ in.}(1.2 \times 10^{-5}(\text{C}^\circ)^{-1}) - (2.5020 \text{ in.}(2.0 \times 10^{-5}(\text{C}^\circ)^{-1})}$$

$$\Delta T = \frac{0.0020}{3.00 \times 10^{-5} - 5.00 \times 10^{-5}} \text{ C}^\circ = -100 \text{ C}^\circ$$

$T = T_0 + \Delta T = 20.0^\circ\text{C} - 100 \text{ C}^\circ = -80^\circ\text{C}$

EVALUATE: Both diameters decrease when the temperature is lowered but the diameter of the brass shaft decreases more since $\alpha_b > \alpha_s$;
$|\Delta L_b| - |\Delta L_s| = 0.0020$ in.

17.93 IDENTIFY: Apply Eq.(11.14) to the volume increase of the liquid due to the pressure decrease. Eq.(17.8) gives the volume decrease of the cylinder and liquid when they are cooled. Can think of the liquid expanding when the pressure is reduced and then contracting to the new volume of the cylinder when the temperature is reduced.

SET UP: Let β_l and β_m be the coefficients of volume expansion for the liquid and for the metal. Let ΔT be the (negative) change in temperature when the system is cooled to the new temperature.

EXECUTE:

Change in volume of cylinder when cool: $\Delta V_m = \beta_m V_0 \Delta T$ (negative)

Change in volume of liquid when cool: $\Delta V_l = \beta_l V_0 \Delta T$ (negative)

The difference $\Delta V_l - \Delta V_m$ must be equal to the negative volume change due to the increase in pressure, which is $-\Delta p V_0 / B = -k \Delta p V_0$.

Thus $\Delta V_l - \Delta V_m = -k \Delta p V_0$.

$$\Delta T = -\frac{k \Delta p}{\beta_l - \beta_m}$$

$$\Delta T = -\frac{(8.50 \times 10^{-10} \text{ Pa}^{-1})(50.0 \text{ atm})(1.013 \times 10^5 \text{ Pa}/1 \text{ atm})}{4.80 \times 10^{-4} \text{ K}^{-1} - 3.90 \times 10^{-5} \text{ K}^{-1}} = -9.8 \text{ C}^\circ$$

$T = T_0 + \Delta T = 30.0^\circ\text{C} - 9.8 \text{ C}^\circ = 20.2^\circ\text{C}.$

EVALUATE: A modest temperature change produces the same volume change as a large change in pressure; $B >> \beta$ for the liquid.

17.95 a) IDENTIFY: Calculate K/Q. We don't know the mass m of the spacecraft, but it divides out of the ratio.

SET UP: The kinetic energy is $K = \frac{1}{2}mv^2$.

The heat required to raise its temperature by 600 C° (but not to melt it) is $Q = mc\,\Delta T$.

EXECUTE: The ratio is $\dfrac{K}{Q} = \dfrac{\frac{1}{2}mv^2}{mc\,\Delta T} = \dfrac{v^2}{2c\,\Delta T} = \dfrac{(7700 \text{ m/s})^2}{2(910 \text{ J/kg} \cdot \text{K})(600 \text{ C}°)} = 54.3$.

b) EVALUATE: The heat generated when friction work (due to friction force exerted by the air) removes the kinetic energy of the spacecraft during reentry is very large, and could melt the spacecraft. Manned space vehicles must have heat shields made of very high melting temperature materials, and reentry must be made slowly.

17.97 IDENTIFY and **SET UP:** To calculate Q, use Eq.(17.18) in the form $dQ = nC\,dT$ and integrate, using $C(T)$ given in the problem. C_{av} is obtained from Eq.(17.19) using the finite temperature range instead of an infinitesimal dT.

EXECUTE:

a) $dQ = mc\,dT$

$$Q = n\int_{T_1}^{T_2} C\,dT = n\int_{T_1}^{T_2} k(T^3/\Theta^3)\,dT = (nk/\Theta^3)\int_{T_1}^{T_2} T^3\,dT = (nk/\Theta^3)(\tfrac{1}{4}T^4\,|_{T_1}^{T_2})$$

$$Q = \frac{nk}{4\Theta^3}(T_2^4 - T_1^4) = \frac{(1.50 \text{ mol})(1940 \text{ J/mol} \cdot \text{K})}{4(281 \text{ K})^3}((40.0 \text{ K})^4 - (10.0 \text{ K})^4) = 83.6 \text{ J}$$

b) $C_{av} = \dfrac{1}{n}\dfrac{\Delta Q}{\Delta T} = \dfrac{1}{1.50 \text{ mol}}\left(\dfrac{83.6 \text{ J}}{40.0 \text{ K} - 10.0 \text{ K}}\right) = 1.86 \text{ J/mol} \cdot \text{K}$

c) $C = k(t/\Theta)^3 = (1940 \text{ J/mol} \cdot \text{K})(40.0 \text{ K}/281 \text{ K})^3 = 5.60 \text{ J/mol} \cdot \text{K}$

EVALUATE: C is increasing with T, so C at the upper end of the temperature interval is larger than its average value over the interval.

17.101 a) IDENTIFY and **SET UP:** Heat comes out of the 0.0°C water and into the ice cube to warm it to 0.0°C. The heat that comes out of the water causes a phase change liquid → solid. The net heat flow for the system is zero.

EXECUTE:

Heat that goes into the ice cube:

$$Q_{ice} = mc_{ice}\,\Delta T = (0.075 \text{ kg})(2100 \text{ J/kg} \cdot \text{K})(0.0°\text{C} - (-10.0°\text{C})) = 1575 \text{ J}$$

Heat that comes out of the water when mass m freezes:

$$Q_{water} = -mL_f.$$

$Q_{system} = 0$ implies $Q_{ice} + Q_{water} = 0$.

$$m = \frac{1575 \text{ J}}{L_f} = \frac{1575 \text{ J}}{334 \times 10^3 \text{ J/kg}} = 4.7 \times 10^{-3} \text{ kg} = 4.7 \text{ g}$$

b) EVALUATE: Yes, just need for the heat that is absorbed by the ice when it warms to $0.0°C$ to equal the heat that must be removed from the water to cause it to freeze.

17.103 a) IDENTIFY and **SET UP:** Assume that all the ice melts and that all the steam condenses. If we calculate a final temperature T that is outside the range $0°C$ to $100°C$ then we know that this assumption is incorrect. Calculate Q for each piece of the system and then set the total $Q_{system} = 0$.

EXECUTE:

copper can (changes temperature from $0.0°C$ to T; no phase change)
$$Q_{can} = mc\,\Delta T = (0.446 \text{ kg})(390 \text{ J/kg} \cdot \text{K})(T - 0.0°C) = (173.9 \text{ J/K})T$$

ice (melting phase change and then the water produced warms to T)
$$Q_{ice} = +mL_f + mc\,\Delta T =$$
$$(0.0950 \text{ kg})(334 \times 10^3 \text{ J/kg}) + (0.0950 \text{ kg})(4190 \text{ J/kg} \cdot \text{K})(T - 0.0°C)$$
$$Q_{ice} = 3.173 \times 10^4 \text{ J} + (398.0 \text{ J/K})T.$$

steam (condenses to liquid and then water produced cools to T)
$$Q_{steam} = -mL_v + mc\,\Delta T =$$
$$-(0.0350 \text{ kg})(2256 \times 10^3 \text{ J/kg}) + (0.0350 \text{ kg})(4190 \text{ J/kg} \cdot \text{K})(T - 100.0°C)$$
$$Q_{steam} = -7.896 \times 10^4 \text{ J} + (146.6 \text{ J/K})T - 1.466 \times 10^4 \text{ J} = -9.362 \times 10^4 \text{ J} + (146.6 \text{ J/K})T$$

$Q_{system} = 0$ implies $Q_{can} + Q_{ice} + Q_{steam} = 0$.
$$(173.9 \text{ J/K})T + 3.173 \times 10^4 \text{ J} + (398.0 \text{ J/K})T - 9.362 \times 10^4 \text{ J} + (146.6 \text{ J/K})T = 0$$
$$(718.5 \text{ J/K})T = 6.189 \times 10^4 \text{ J}$$

$$T = \frac{6.189 \times 10^4 \text{ J}}{718.5 \text{ J/K}} = 86.1°C.$$

EVALUATE: This is between $0°C$ and $100°C$ so our assumptions about the phase changes being complete were correct.

b) No ice, no steam, $0.0950 \text{ kg} + 0.0350 \text{ kg} = 0.130 \text{ kg}$ of liquid water.

17.105 IDENTIFY and **SET UP:** Heat comes out of the steam when it changes phase and heat goes into the water and causes its temperature to rise. $Q_{system} = 0$. First determine what phases are present after the system has come to a uniform final temperature.

a) EXECUTE: Heat that must be removed from steam if all of it condenses is

$$Q = -mL_v = -(0.0400 \text{ kg})(2256 \times 10^3 \text{ J/kg}) = -9.02 \times 10^4 \text{ J}$$

Heat absorbed by the water if it heats all the way to the boiling point of 100°C:
$$Q = mc\,\Delta T = (0.200 \text{ kg})(4190 \text{ J/kg} \cdot \text{K})(50.0 \text{ C°}) = 4.19 \times 10^4 \text{ J}$$

EVALUATE: The water can't absorb enough heat for all the steam to condense. Steam is left and the final temperature then must be 100°C.

b) EXECUTE: Mass of steam that condenses is
$$m = Q/L_v = 4.19 \times 10^4 \text{ J}/2256 \times 10^3 \text{ J/kg} = 0.0186 \text{ kg}$$

Thus there is 0.0400 kg − 0.0186 kg = 0.0214 kg of steam left.

The amount of liquid water is 0.0186 kg + 0.200 kg = 0.219 kg.

17.107 IDENTIFY and **SET UP:** Heat comes out of the steam and goes into the ice. Heat goes into the water if its final temperature is greater than 50.0°C and comes out of the water if its final temperature is less than 50.0°C. $Q_{\text{system}} = 0$. First determine what phases are present after the system has come to a common final temperature.

EXECUTE:

a) Heat that must go into ice for all of it to melt:
$$Q = +mL_f = +(0.150 \text{ kg})(334 \times 10^3 \text{ J/kg}) = 5.01 \times 10^4 \text{ J}.$$

Heat that must come out of the steam for all of it to condense:
$$Q = -mL_v = -(0.0950 \text{ kg})(2256 \times 10^3 \text{ J/kg}) = -2.143 \times 10^5 \text{ J}$$

Heat that goes into the 0.200 kg of water if it warms from 50.0°C to 100.0°C:
$$Q = mc\,\Delta T = (0.200 \text{ kg})(4190 \text{ J/kg} \cdot \text{K})(50.0 \text{ C°}) = +4.19 \times 10^4 \text{ J}$$

Heat that goes into water produced from melting the ice if it warms from 0.0°C (the temperature at which it is produced) to 100.0°C:
$$Q = mc\,\Delta T = (0.150 \text{ kg})(4190 \text{ J/kg} \cdot \text{K})(100.0 \text{ C°}) = +6.285 \times 10^4 \text{ J}$$

Thus the total heat that the rest of the system can absorb from the steam is
$$Q = -(5.01 \times 10^4 \text{ J} + 4.19 \times 10^4 \text{ J} + 6.285 \times 10^4 \text{ J}) = -1.548 \times 10^5 \text{ J}$$

This is less than the 2.143×10^5 J that must be removed from the steam for all of it to condense, so some steam is left and this means the final temperature is 100°C

b) Mass of steam that condenses is $m = (1.548 \times 10^5 \text{ J})/(2256 \times 10^3 \text{ J/kg}) = 0.0686$ kg

Mass of steam left: 0.0950 kg − 0.0686 kg = 0.0264 kg

Total mass of liquid water at the end is the mass of liquid water that started with plus the mass of the ice (all melts) plus the mass of steam that condenses: 0.200 kg + 0.150 kg + 0.0684 kg = 0.418 kg

There is no ice left.

c) Heat that must go into ice for all of it to melt:
$Q = +mL_f = +(0.350 \text{ kg})(334 \times 10^3 \text{ J/kg}) = 1.169 \times 10^5 \text{ J}.$

Heat that must come out of the steam for all of it to condense:
$Q = -mL_v = -(0.0120 \text{ kg})(2256 \times 10^3 \text{ J/kg}) = -2.707 \times 10^4 \text{ J}$

Heat that comes out of the 0.200 kg of water if it cools from 40.0°C to 0.0°C:
$Q = mc\,\Delta T = (0.200 \text{ kg})(4190 \text{ J/kg} \cdot \text{K})(0.0°\text{C} - 40.0 \text{ C}°) = -3.352 \times 10^4 \text{ J}$

Heat that comes out of water produced from condensing the steam if it cools from 100.0°C (the temperature at which it is produced) to 0.0°C:
$Q = mc\,\Delta T = (0.012 \text{ kg})(4190 \text{ J/kg} \cdot \text{K})(0.0°\text{C} - 100.0°\text{C}) = -5028 \text{ J}$

The total heat that the rest of the system can give to the ice to melt it is $Q = -2.707 \times 10^4 \text{ J} - 3.352 \times 10^4 \text{ J} - 5028 \text{ J} = -6.562 \times 10^4 \text{ J}.$

This is less than the 1.169×10^5 J that it takes to melt all the ice so not all the ice melts and then the final temperature must be 0.0°C.

The mass of ice that melts is $m = Q/L_f = (6.562 \times 10^4 \text{ J})/(334 \times 10^3 \text{ J/kg}) = 0.196$ kg.

The mass of ice left is 0.350 kg − 0.196 kg = 0.154 kg.

The mass of liquid water at the end is the mass of liquid water at the start plus the mass of the steam (all condenses) plus the mass of the ice that melts: 0.200 kg + 0.012 kg + 0.196 kg = 0.408 kg.

No steam is left.

EVALUATE: If ice and liquid water are present in equilibrium, the temperature of the system is 0°C. If steam and liquid water are present in equilibrium, the temperature of the system is 100°C. If only liquid water is present at the end, the temperature can be in the range 0°C to 100°C.

17.111 IDENTIFY and **SET UP:** Use H written in terms of the thermal resistance R: $H = A\,\Delta T/R$, where $R = L/k$ and $R = R_1 + R_2 + \ldots$ (additive).

EXECUTE:

single pane

$R_s = R_{\text{glass}} + R_{\text{film}}$, where $R_{\text{film}} = 0.15 \text{ m}^2 \cdot \text{K/W}$ is the combined thermal resistance of the air films on the room and outdoor surfaces of the window.

$R_{\text{glass}} = L/k = (4.2 \times 10^{-3} \text{ m})/(0.80 \text{ W/m} \cdot \text{K}) = 0.00525 \text{ m}^2 \cdot \text{K/W}$

Thus $R_s = 0.00525 \text{ m}^2 \cdot \text{K/W} + 0.15 \text{ m}^2 \cdot \text{K/W} = 0.1553 \text{ m}^2 \cdot \text{K/W}.$

double pane

$R_d = 2R_{\text{glass}} + R_{\text{air}} + R_{\text{film}}$, where R_{air} is the thermal resistance of the air space between the panes.

$R_{\text{air}} = L/k = (7.0 \times 10^{-3} \text{ m})/(0.024 \text{ W/m} \cdot \text{K}) = 0.2917 \text{ m}^2 \cdot \text{K/W}$

Thus $R_{\text{d}} = 2(0.00525 \text{ m}^2 \cdot \text{K/W}) + 0.2917 \text{ m}^2 \cdot \text{K/W} + 0.15 \text{ m}^2 \cdot \text{K/W} = 0.4522 \text{ m}^2 \cdot \text{K/W}$

$H_{\text{s}} = A \Delta T/R_{\text{s}}$, $H_{\text{d}} = A \Delta T/R_{\text{d}}$, so $H_{\text{s}}/H_{\text{d}} = R_{\text{d}}/R_{\text{s}}$ (since A and ΔT are same for both)

$H_{\text{s}}/H_{\text{d}} = (0.4522 \text{ m}^2 \cdot \text{K/W})/(0.1553 \text{ m}^2 \cdot \text{K/W}) = 2.9$

EVALUATE: The heat loss is about a factor of 3 less for the double-pane window. The increase in R for a double-pane is due mostly to the thermal resistance of the air space between the panes.

17.113 a) EXECUTE: Heat must be conducted from the water to cool it to $0°\text{C}$ and to cause the phase transition. The entire volume of water is not at the phase transition temperature, just the upper surface that is in contact with the ice sheet.

b) IDENTIFY: The heat that must leave the water in order for it to freeze must be conducted through the layer of ice that has already been formed.

SET UP: Consider a section of ice that has area A. At time t let the thickness be h. Consider a short time interval t to $t + dt$. Let the thickness that freezes in this time be dh. The mass of the section that freezes in the time interval dt is $dm = \rho \, dV = \rho A \, dh$. The heat that must be conducted away from this mass of water to freeze it is

$dQ = dm \, L_{\text{f}} = (\rho A L_{\text{f}}) \, dh$.

$H = dQ/dt = kA(\Delta T/h)$, so the heat dQ conducted in time dt through the thickness h that is already there is

$$dQ = kA \left(\frac{T_{\text{H}} - T_{\text{C}}}{h} \right) dt.$$

Solve for dh in terms of dt and integrate to get an expression rlating h and t.

EXECUTE: Equate these expressions for dQ.

$$\rho A L_{\text{f}} \, dh = kA \left(\frac{T_{\text{H}} - T_{\text{C}}}{h} \right) dt$$

$$h \, dh = \left(\frac{k(T_{\text{H}} - T_{\text{C}})}{\rho L_{\text{f}}} \right) dt$$

Integrate from $t = 0$ to time t. At $t = 0$ the thickness h is zero.

$\int_0^h h \, dh = [k(T_{\text{H}} - T_{\text{C}})/\rho L_{\text{f}}] \int_0^t dt$

$\frac{1}{2}h^2 = \dfrac{k(T_{\text{H}} - T_{\text{C}})}{\rho L_{\text{f}}} t$ and $h = \sqrt{\dfrac{2k(T_{\text{H}} - T_{\text{C}})}{\rho L_{\text{f}}}} \sqrt{t}$

The thickness after time t is proportional to $\sqrt{t}$.

c) The expression in part (b) gives

$$t = \frac{h^2 \rho L_f}{2k(T_H - T_C)} = \frac{(0.25 \text{ m})^2 (920 \text{ kg/m}^3)(334 \times 10^3 \text{ J/kg})}{2(1.6 \text{ W/m} \cdot \text{K})(0^\circ\text{C} - (-10^\circ\text{C}))} = 6.0 \times 10^5 \text{ s}$$

$t = 170$ h.

d) Find t for $h = 40$ m. t is proportional to h^2, so

$t = (40 \text{ m}/0.25 \text{ m})^2 (6.00 \times 10^5 \text{ s}) = 1.5 \times 10^{10}$ s. This is about 500 years. With our current climate this will not happen.

EVALUATE: As the ice sheet gets thicker, the rate of heat conduction through it decreases. Part (d) shows that it takes a very long time for a moderately deep lake to totally freeze.

17.115 IDENTIFY and **SET UP:** Set the heat that enters the ice in time t due to the solar radiation (Eq.17.25) equal to the heat required to melt a certain volume of ice. Work with a 1.00 m² area.

EXECUTE:

The heat current into a 1.00 m² area of ice due to the absorbed solar radiation is $H = (0.70)(600 \text{ W/m}^2)(1.00 \text{ m}^2) = 420$ W.

The heat required to melt a $h = 2.50$ cm thick layer of ice that is initially at 0°C is $Q = mL_f = \rho h A L_f = (920 \text{ kg/m}^3)(0.0250 \text{ m})(1.00 \text{ m}^2)(334 \times 10^3 \text{ J/kg}) = 7.68 \times 10^6$ J.

$H = Q/t$, so the time t it takes the heat current from solar radiation to input this amount of heat into the ice is

$t = Q/H = 7.68 \times 10^6 \text{ J}/420 \text{ W} = 1.83 \times 10^4 \text{ s} = 305$ min.

EVALUATE: The answer is independent of the area we used; 1.00 m² was chosen just for convenience.

17.117 IDENTIFY and **SET UP:** Use Eq.(17.26) to find the net heat current into the can due to radiation. Use $Q = Ht$ to find the heat that goes into the liquid helium, set this equal to mL and solve for the mass m of helium that changes phase.

EXECUTE: Calculate the net rate of radiation of heat from the can. $H_{\text{net}} = Ae\sigma(T^4 - T_s^4)$.

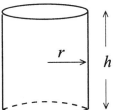

The surface area of the cylindrical can is $A = 2\pi r h + 2\pi r^2$.

$$A = 2\pi r(h + r) = 2\pi(0.045 \text{ m})(0.250 \text{ m} + 0.045 \text{ m}) = 0.08341 \text{ m}^2.$$

$$H_{\text{net}} = (0.08341 \text{ m}^2)(0.200)(5.67 \times 10^{-8} \text{ W/m}^2 \cdot \text{K}^4)(4.22 \text{ K})^4 - (77.3 \text{ K})^4)$$

$H_{\text{net}} = -0.0338 \text{ W}$ (the minus sign says that the net heat current is into the can).

The heat that is put into the can by radiation in one hour is
$$Q = -(H_{\text{net}})t = (0.0338 \text{ W})(3600 \text{ s}) = 121.7 \text{ J}.$$

This heat boils a mass m of helium according to the equation $Q = mL_f$, so

$$m = \frac{Q}{L_f} = \frac{121.7 \text{ J}}{2.09 \times 10^4 \text{ J/kg}} = 5.82 \times 10^{-3} \text{ kg} = 5.82 \text{ g}.$$

EVALUATE: In the expression for the net heat current into the can the temperature of the surroundings is raised to the fourth power. The rate at which the helium boils away increases by about a factor of $(293/77)^4 = 210$ if the walls surrounding the can are at room temperature rather than at the temperature of the liquid nitrogen.

CHAPTER 18
THERMAL PROPERTIES OF MATTER

Exercises 1, 7, 11, 15, 21, 25, 29, 37, 39, 41, 43, 47, 49
Problems 55, 57, 61, 63, 65, 69, 71, 73, 77, 81, 85

Exercises

18.1 **a) IDENTIFY:** We are asked about a single state of the system.

SET UP: Use Eq.(18.2) to calculate the number of moles and then apply the ideal-gas equation.

EXECUTE: $n = \dfrac{m_{tot}}{M} = \dfrac{0.225 \text{ kg}}{4.00 \times 10^{-3} \text{ kg/mol}} = 56.2 \text{ mol}$

b) $pV = nRT$ implies $p = nRT/V$

T must be in kelvins; $T = (18 + 273) \text{ K} = 291 \text{ K}$

$$p = \frac{(56.2 \text{ mol})(8.3145 \text{ J/mol} \cdot \text{K})(291 \text{ K})}{20.0 \times 10^{-3} \text{ m}^3} = 6.80 \times 10^6 \text{ Pa}$$

$p = (6.80 \times 10^6 \text{ Pa})(1.00 \text{ atm}/1.013 \times 10^5 \text{ Pa}) = 67.1 \text{ atm}$

EVALUATE: Example 18.1 shows that 1.0 mol of an ideal gas is about this volume at STP. Since there are 56.2 moles the pressure is about 60 times greater than 1 atm.

18.7 **IDENTIFY:** We are asked to compare two states. Use the ideal gas law to obtain T_2 in terms of T_1 and ratios of pressures and volumes of the gas in the two states.

SET UP: $pV = nRT$ n, R constant implies $pV/T = nR = $ constant and $p_1 V_1/T_1 = p_2 V_2/T_2$

EXECUTE: $T_1 = (27 + 273) \text{ K} = 300 \text{ K}$

$p_1 = 1.01 \times 10^5 \text{ Pa}$

$p_2 = 2.72 \times 10^6 \text{ Pa} + 1.01 \times 10^5 \text{ Pa} = 2.82 \times 10^6 \text{ Pa}$ (in the ideal gas equation the pressures must be absolute, not gauge, pressures)

$$T_2 = T_1 \left(\frac{p_2}{p_1}\right)\left(\frac{V_2}{V_1}\right) = 300 \text{ K} \left(\frac{2.82 \times 10^6 \text{ Pa}}{1.01 \times 10^5 \text{ Pa}}\right)\left(\frac{46.2 \text{ cm}^3}{499 \text{ cm}^3}\right) = 776 \text{ K}$$

$T_2 = (776 - 273)°\text{C} = 503°\text{C}$

EVALUATE: The units cancel in the V_2/V_1 volume ratio, so it was not necessary to convert the volumes in cm^3 to m^3. It was essential, however, to use T in kelvins.

18.11 IDENTIFY: We are aksed to compare two states. Use the ideal-gal law to obtain V_1 in terms of V_2 and the ratio of the temperatures in the two states.

SET UP: $pV = nRT$ n, R, p are constant so $V/T = nR/p$ = constant and $V_1/T_1 = V_2/T_2$

EXECUTE: $T_1 = (19 + 273)$ K $= 292$ K (T must be in kelvins)

$V_2 = V_1(T_2/T_1) = (0.600$ L$)(77.3$ K$/292$ K$) = 0.159$ L

EVALUATE: p is constant so the ideal-gas equation says that a decrease in T means a decrease in V.

18.15 IDENTIFY: We are asked to compare two states. First use $pV = nRT$ to calculate p_1. Then use it to obtain T_2 in terms of T_1 and the ratio of pressures in the two states.

a) SET UP: $pV = nRT$. Find the initial pressure p_1:

EXECUTE:
$$p_1 = \frac{nRT_1}{V} = \frac{(11.0 \text{ mol})(8.3145 \text{ J/mol} \cdot \text{K})((23.0 + 273.13) \text{ K})}{3.10 \times 10^{-3} \text{ m}^3} = 8.737 \times 10^6 \text{ Pa}$$

SET UP: $p_2 = 100$ atm$(1.013 \times 10^5$ Pa$/1$ atm$) = 1.013 \times 10^7$ Pa

$p/T = nR/V$ = constant, so $p_1/T_1 = p_2/T_2$

EXECUTE: $T_2 = T_1 \left(\dfrac{p_2}{p_1} \right) = (296.15 \text{ K}) \left(\dfrac{1.013 \times 10^7 \text{ Pa}}{8.737 \times 10^6 \text{ Pa}} \right) = 343.4 \text{ K} = 70.2°\text{C}$

b) EVALUATE: The coefficient of volume expansion for a gas is much larger than for a solid, so the expansion of the tank is negligible.

18.21 IDENTIFY: Use Eq.(18.5) and solve for p.

SET UP: $\rho = pM/RT$ and $p = RT\rho/M$

$T = (-56.5 + 273.15)$ K $= 216.6$ K

For air $M = 28.8 \times 10^{-3}$ kg/mol (Example 18.3)

EXECUTE: $p = \dfrac{(8.3145 \text{ J/mol} \cdot \text{K})(216.6 \text{ K})(0.364 \text{ kg/m}^3)}{28.8 \times 10^{-3} \text{ kg/mol}} = 2.28 \times 10^4$ Pa

EVALUATE: The pressure is about one-fifth the pressure at sea-level.

18.25 IDENTIFY: We are asked about a single state of the system.

SET UP: Use the ideal-gas law. Write n in terms of the number of moleucles N.

a) EXECUTE: $pV = nRT$, $n = N/N_A$ so $pV = (N/N_A)RT$

$$p = \left(\frac{N}{V} \right) \left(\frac{R}{N_A} \right) T$$

$$p = \left(\frac{80 \text{ molecules}}{1 \times 10^{-6} \text{ m}^3}\right)\left(\frac{8.3145 \text{ J/mol} \cdot \text{K}}{6.022 \times 10^{23} \text{ molecules/mol}}\right)(7500 \text{ K}) = 8.28 \times 10^{-12} \text{ Pa}$$

$p = 8.2 \times 10^{-17}$ atm. This is much lower than the laboratory pressure of 1×10^{-13} atm in Exercise 18.24.

b) EVALUATE: The Lagoon Nebula is a very rarefied low pressure gas. The gas would exert <u>very</u> little force on an object passing through it.

18.29 a) IDENTIFY and **SET UP:** Use the density and the mass of 5.00 mol to calculate the volume.

$\rho = m/V$ implies $V = m/\rho$, where $m = m_{\text{tot}}$, the mass of 5.00 mol of water.

EXECUTE: $m_{\text{tot}} = nM = (5.00 \text{ mol})(18.0 \times 10^{-3} \text{ kg/mol}) = 0.0900$ kg

Then $V = \dfrac{m}{\rho} = \dfrac{0.0900 \text{ kg}}{1000 \text{ kg/m}^3} = 9.00 \times 10^{-5} \text{ m}^3$

b) One mole contains $N_A = 6.022 \times 10^{23}$ molecules, so the volume occuppied by one molecule is

$$\frac{9.00 \times 10^{-5} \text{ m}^3/\text{mol}}{(5.00 \text{ mol})(6.022 \times 10^{23} \text{ molecules/mol})} = 2.989 \times 10^{-29} \text{ m}^3/\text{molecule}$$

$V = a^3$, where a is the length of each side of the cube occupied by a molecule. $a^3 = 2.989 \times 10^{-29} \text{ m}^3$, so $a = 3.1 \times 10^{-10}$ m.

c) EVALUATE: Atoms and molecules are on the order of 10^{-10} m in diameter, in agreement with the above estimates.

18.37 IDENTIFY and **SET UP:** Apply the analysis of Section 18.3.

EXECUTE:

a) $\frac{1}{2}m(v^2)_{\text{av}} = \frac{3}{2}kT = \frac{3}{2}(1.38 \times 10^{-23} \text{ J/molecule} \cdot \text{K})(300 \text{ K}) = 6.21 \times 10^{-21}$ J

b) We need the mass m of one atom:

$$m = \frac{M}{N_A} = \frac{32.0 \times 10^{-3} \text{ kg/mol}}{6.022 \times 10^{23} \text{ molecules/mol}} = 5.314 \times 10^{-26} \text{ kg/molecule}$$

Then $\frac{1}{2}m(v^2)_{\text{av}} = 6.21 \times 10^{-21}$ J (from part (a)) gives

$$(v^2)_{\text{av}} = \frac{2(6.21 \times 10^{-21} \text{ J})}{m} = \frac{2(6.21 \times 10^{-21} \text{ J})}{5.314 \times 10^{-26} \text{ kg}} = 2.34 \times 10^5 \text{ m}^2/\text{s}^2$$

c) $v_{\text{rms}} = \sqrt{(v^2)_{\text{rms}}} = \sqrt{2.34 \times 10^4 \text{ m}^2/\text{s}^2} = 484$ m/s

d) $p = mv_{\text{rms}} = (5.314 \times 10^{-26} \text{ kg})(484 \text{ m/s}) = 2.57 \times 10^{-23} \text{ kg} \cdot \text{m/s}$

e) Time betwen collisions with one wall is $t = \dfrac{0.20 \text{ m}}{v_{\text{rms}}} = \dfrac{0.20 \text{ m}}{484 \text{ m/s}} = 4.13 \times 10^{-4}$ s

In a collision $\vec{v}$ changes direction, so $\Delta p = 2mv_{\text{rms}} = 2(2.57 \times 10^{-23} \text{ kg} \cdot \text{m/s}) = 5.14 \times 10^{-23}$ kg $\cdot$ m/s

$F = \dfrac{dp}{dt}$ so $F_{\text{av}} = \dfrac{\Delta p}{\Delta t} = \dfrac{5.14 \times 10^{-23} \text{ kg} \cdot \text{m/s}}{4.13 \times 10^{-4} \text{ s}} = 1.24 \times 10^{-19}$ N

f) pressure $= F/A = 1.24 \times 10^{-19} \text{ N}/(0.10 \text{ m})^2 = 1.24 \times 10^{-17}$ Pa (due to one atom)

g) pressure $= 1$ atm $= 1.013 \times 10^5$ Pa

Number of atoms needed is $1.013 \times 10^5 \text{ Pa}/(1.24 \times 10^{-17} \text{ Pa/atom}) = 8.17 \times 10^{21}$ atoms

h) $pV = NkT$ (Eq.18.18), so

$$N = \frac{pV}{kT} = \frac{(1.013 \times 10^5 \text{ Pa})(0.10 \text{ m})^3}{(1.381 \times 10^{-23} \text{ J/molecule} \cdot \text{K})(300 \text{ K})} = 2.45 \times 10^{22} \text{ atoms}$$

i) From the factor of $\frac{1}{3}$ in $(v_x^2)_{\text{av}} = \frac{1}{3}(v^2)_{\text{av}}$.

EVALUATE: This Exercise shows that the pressure exerted by a gas arises from collisions of the moleucles of the gas with the walls.

18.39 **IDENTIFY** and **SET UP:** Use equal v_{rms} to relate T and M for the two gases.

$v_{\text{rms}} = \sqrt{3RT/M}$ (Eq.16-19), so

$v_{\text{rms}}^2/3R = T/M$, where T must be in kelvins.

Same v_{rms} so same T/M for the two gases and $T_{\text{N}_2}/M_{\text{N}_2} = T_{\text{H}_2}/M_{\text{H}_2}$.

EXECUTE:

$$T_{\text{N}_2} = T_{\text{H}_2}\left(\frac{M_{\text{N}_2}}{M_{\text{H}_2}}\right) = ((20 + 273) \text{ K})\left(\frac{28.014 \text{ g/mol}}{2.016 \text{ g/mol}}\right) = 4.071 \times 10^3 \text{ K}$$

$T_{\text{N}_2} = (4071 - 273)°\text{C} = 3800°\text{C}$

EVALUATE: A N_2 molecule has more mass so N_2 gas must be at a higher temperature to have the same v_{rms}.

18.41 **a)** **IDENTIFY** and **SET UP:** $\frac{1}{2}R$ contribution to C_V for each degree of freedom. The molar heat capacity C is related to the specific heat capacity c by $C = Mc$.

EXECUTE $C_V = 6(\frac{1}{2}R) = 3R = 3(8.3145 \text{ J/mol} \cdot \text{K}) = 24.9$ J/mol $\cdot$ K. The specific heat capacity is $c_V = C_V/M = (24.9 \text{ J/mol} \cdot \text{K})/(18.0 \times 10^{-3} \text{ kg/mol}) = 1380$ J/kg$\cdot$K.

b) For water vapor the specific heat capacity is $c = 2000$ J/kg $\cdot$ K. The molar heat capacity is $C = Mc = (18.0 \times 10^{-3} \text{ kg/mol})(2000 \text{ J/kg} \cdot \text{K}) = 36.0 \text{ J/mol} \cdot \text{K}$.

EVALUATE: The difference is $36.0 \text{ J/mol} \cdot \text{K} - 24.9 \text{ J/mol} \cdot \text{K} = 11.1 \text{ J/mol} \cdot \text{K}$, which is about $2.7(\frac{1}{2}R)$; the vibrational degrees of freedom make a significant contribution.

18.43 IDENTIFY: Use Eq.(18.24), applied to a finite temperature change.

SET UP: $C_V = 5R/2$ for a diatomic ideal gas and $C_V = 3R/2$ for a monatomic ideal gas.

EXECUTE:

a) $Q = nC_V \Delta T = n(\frac{5}{2}R) \Delta T$

$Q = (2.5 \text{ mol})(\frac{5}{2})(8.3145 \text{ J/mol} \cdot \text{K})(30.0 \text{ K}) = 1560 \text{ J}$

b $Q = nC_V \Delta T = n(\frac{3}{2}R) \Delta T$

$Q = (2.5 \text{ mol})(\frac{3}{2})(8.3145 \text{ J/mol} \cdot \text{K})(30.0 \text{ K}) = 935 \text{ J}$

EVALUATE: More heat is required for the diatomic gas; not all the heat that goes into the gas appears as translational kinetic energy, some goes into energy of the internal motion of the molecules (rotations).

18.47 IDENTIFY and **SET UP:** Eq.(18.33): $f(v) = \frac{8\pi}{m} \left(\frac{m}{2\pi kT} \right)^{3/2} \epsilon e^{-\epsilon/kT}$

At the maximum of $f(\epsilon)$, $\dfrac{df}{d\epsilon} = 0$.

EXECUTE: $\dfrac{df}{d\epsilon} = \dfrac{8\pi}{m} \left(\dfrac{m}{2\pi kT} \right)^{3/2} \dfrac{d}{d\epsilon} \left(\epsilon e^{-\epsilon/kT} \right) = 0$

This requires that $\dfrac{d}{d\epsilon} \left(\epsilon e^{-\epsilon/kT} \right) = 0$.

$e^{-\epsilon/kT} - (\epsilon/kT)e^{-\epsilon/kT} = 0$

$(1 - \epsilon/kT)e^{-\epsilon/kT} = 0$

This requires that $1 - \epsilon/kT = 0$ so $\epsilon = kT$, as was to be shown.

And then since $\epsilon = \frac{1}{2}mv^2$, this gives $\frac{1}{2}mv_{\text{mp}}^2 = kT$ and $v_{\text{mp}} = \sqrt{2kT/m}$, which is Eq.(18.4).

EVALUATE: $v_{\text{rms}} = \sqrt{\frac{3}{2}}v_{\text{mp}}$. The average of v^2 weights larger v.

18.49 IDENTIFY and **SET UP:** If the temperature at altitude y is below the freezing point only cirrus clouds can form. Use $T = T_0 - \alpha y$ to find the y that gives $T = 0.0°\text{C}$.

EXECUTE: $y = \dfrac{T_0 - T}{\alpha} = \dfrac{15.0°\text{C} - 0.0°\text{C}}{6.0\text{C}°/\text{km}} = 2.5 \text{ km}$

EVALUATE: The solid-liquid phase transition occurs at $0°\text{C}$ only for $p = 1.01 \times 10^5$ Pa. Use the results of Example 18.4 to estimate the pressure at an altitude of

2.5 km.

$$p_2 = p_1 e^{Mg(y_2-y_1)/RT}$$

$Mg(y_2-y_1)/RT = 1.10(2500 \text{ m}/8863 \text{ m}) = 0.310$ (using the calculation in Example 16-4)

Then $p_2 = (1.01 \times 10^5 \text{ Pa})e^{-0.31} = 0.74 \times 10^5 \text{ Pa}$.

This pressure is well above the triple point pressure for water. Figure 18.21 shows that the fusion curve has large slope and it takes a large change in pressure to change the phase transition temperature very much. Using $0.0°C$ introduces little error.

Problems

18.55 **IDENTIFY:** We are asked to compare two states. Use the ideal-gas law to obtain m_2 in terms of m_1 and the ratio of pressures in the two states. Apply Eq.(18.4) to the initial state to calculate m_1.

SET UP: $pV = nRT$ can be written $pV = (m/M)RT$

T, V, M, R are all constant, so $p/m = RT/MV = $ constant.

So $p_1/m_1 = p_2/m_2$, where m is the mass of the gas in the tank.

EXECUTE:

$p_1 = 1.30 \times 10^6 \text{ Pa} + 1.01 \times 10^5 \text{ Pa} = 1.40 \times 10^6 \text{ Pa}$

$p_2 = 2.50 \times 10^5 \text{ Pa} + 1.01 \times 10^5 \text{ Pa} = 3.51 \times 10^5 \text{ Pa}$

$m_1 = p_1 V M/RT$; $V = hA = h\pi r^2 = (1.00 \text{ m})\pi(0.060 \text{ m})^2 = 0.01131 \text{ m}^3$

$$m_1 = \frac{(1.40 \times 10^6 \text{ Pa})(0.01131 \text{ m}^3)(44.1 \times 10^{-3} \text{ kg/mol})}{(8.3145 \text{ J/mol} \cdot \text{K})((22.0 + 273.15) \text{ K})} = 0.2845 \text{ kg}$$

Then $m_2 = m_1 \left(\dfrac{p_2}{p_1}\right) = (0.2845 \text{ kg}) \left(\dfrac{3.51 \times 10^5 \text{ Pa}}{1.40 \times 10^6 \text{ Pa}}\right) = 0.0713 \text{ kg}$.

m_2 is the mass that remains in the tank. The mass that has been used is $m_1 - m_2 = 0.2848 \text{ kg} - 0.0713 \text{ kg} = 0.213 \text{ kg}$.

EVALUATE: Note that we have to use absolute pressures. The absolute pressure decreases by a factor of four and the mass of gas in the tank decreases by a factor of four.

18.57 **IDENTIFY:** Apply $p = p_0 + \rho g h$ to the mercury column. At a point in the column at the same level as the mercury surface outside the tube the pressure equals air pressure. Use this analysis to find p for the air at the top of the tube and then apply the ideal-gas law to find m_{tot} for the air.

SET UP:

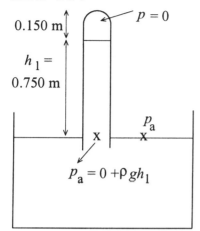

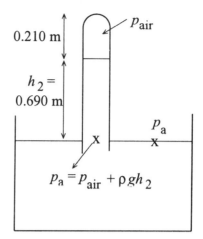

The pressure at points which are level with the surface of the mercury outside the tube is equal to atmospheric pressure, p_a. In the first sketch this gives $p_a = \rho g h_1$ and in the second sketch it gives $p_a = p_{air} + \rho g h_2$, where p_{air} is the pressure of the air in the space in the tube above the mercury.

Equating these two expressions for p_a gives $\rho g h_1 = p_{air} + \rho g h_2$.

EXECUTE: $p_{air} = \rho g(h_1 - h_2) =$

$(13.6 \times 10^3 \text{ kg/m}^3)(9.80 \text{ m/s}^2)(0.750 \text{ m} - 0.690 \text{ m}) = 7.997 \times 10^3 \text{ Pa}$

We know p, V, and T; calculate the mass m_{tot} of the air:

$$n = \frac{pV}{RT} = \frac{(7.997 \times 10^3 \text{ Pa})(0.210 \text{ m})(0.620 \times 10^{-4} \text{ m}^2)}{(8.3145 \text{ J/mol} \cdot \text{K})(293.15 \text{ K})} = 4.272 \times 10^{-5} \text{ mol}$$

$m_{tot} = nM = (4.272 \times 10^{-5} \text{ mol})(28.8 \times 10^{-3} \text{ kg/mol}) = 1.23 \times 10^{-6} \text{ kg}$

EVALUATE: The volume occupied by the gas is very small and the number of moles of gas is very small.

18.61 **a) IDENTIFY:** Consider the gas in one cylinder. Calculate the volume to which this volume of gas expands when the pressure is decreased from

$(1.20 \times 10^6 \text{ Pa} + 1.01 \times 10^5 \text{ Pa}) = 1.30 \times 10^6 \text{ Pa}$ to $1.01 \times 10^5 \text{ Pa}$.

Apply the ideal-gas law to the two states of the system to obtain an expression for V_2 in terms of V_1 and the ratio of the pressures in the two states.

SET UP: $pV = nRT$

n, R, T constant implies $pV = nRT = $ constant, so $p_1 V_1 = p_2 V_2$.

EXECUTE: $V_2 = V_1(p_1/p_2) = (1.90 \text{ m}^3)\left(\dfrac{1.30 \times 10^6 \text{ Pa}}{1.01 \times 10^5 \text{ Pa}}\right) = 24.46 \text{ m}^3$

The number of cylinders required to fill a 750 m^3 balloon is 750 $\text{m}^3/24.46 \text{ m}^3 = 30.7$ cylinders.

EVALUATE: The ratio of the volume of the balloon to the volume of a cylinder is about 400. Fewer cylinders than this are required because of the large factor by which the gas is compressed in the cylinders.

b) IDENTIFY: The upward force on the balloon is given by Archimedes' principle (Chapter 14):

B = weight of air displaced by balloon = $\rho_{air} V g$.

Apply Newton's 2nd law to the balloon and solve for the weight of the load that can be supported. Use the ideal-gas equation to find the mass of the gas in the balloon.

SET UP: Free-body diagram for the balloon

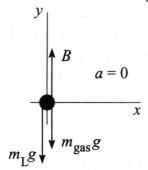

m_{gas} is the mass of the gas that is inside the balloon; m_L is the mass of the load that is supported by the balloon

EXECUTE:
$$\sum F_y = ma_y$$
$$B - m_L g - m_{gas} g = 0$$

$$\rho_{air} V g - m_L g - m_{gas} g = 0$$
$$m_L = \rho_{air} V - m_{gas}$$

Calculate m_{gas}, the mass of hydrogen that occupies 750 m^3 at 15°C and $p = 1.01 \times 10^5$ Pa.

$pV = nRT = (m_{gas}/M)RT$ gives

$$m_{gas} = pVM/RT = \frac{(1.01 \times 10^5 \text{ Pa})(750 \text{ m}^3)(2.02 \times 10^{-3} \text{ kg/mol})}{(8.3145 \text{ J/mol} \cdot \text{K})(288 \text{ K})} = 63.9 \text{ kg}$$

Then $m_L = (1.23 \text{ kg/m}^3)(750 \text{ m}^3) - 63.9 \text{ kg} = 859 \text{ kg}$, and the weight that can be supported is $w_L = m_L g = (859 \text{ kg})(9.80 \text{ m/s}^2) = 8420 \text{ N}$.

c) $m_L = \rho_{air} V - m_{gas}$

$m_{gas} = pVM/RT = (63.9 \text{ kg})((4.00 \text{ g/mol})/(2.02 \text{ g/mol})) = 126.5 \text{ kg}$ (using the results of part (b)).

Then $m_L = (1.23 \text{ kg/m}^3)(750 \text{ m}^3) - 126.5 \text{ kg} = 796 \text{ kg}$.

$w_L = m_L g = (796 \text{ kg})(9.80 \text{ m/s}^2) = 7800 \text{ N}$.

EVALUATE: A greater weight can be supported when hydrogen is used because its density is less.

18.63 IDENTIFY: Apply Bernoulli's equation to relate the efflux speed of water out the hose to the height of water in the tank and the pressure of the air above the water in the tank. Use the ideal-gas equation to relate the volume of the air in the tank to the pressure of the air.

a) SET UP:

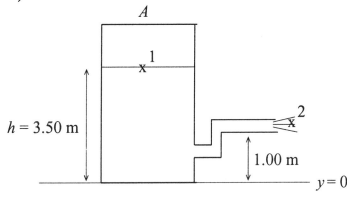

$$p_1 = 4.20 \times 10^5 \text{ Pa}$$
$$p_2 = p_{\text{air}} = 1.00 \times 10^5 \text{ Pa}$$
large tank implies $v_1 \approx 0$

EXECUTE:

$$p_1 + \rho g y_1 + \tfrac{1}{2}\rho v_1^2 = p_2 + \rho g y_2 + \tfrac{1}{2}\rho v_2^2$$
$$\tfrac{1}{2}\rho v_2^2 = p_1 - p_2 + \rho g(y_1 - y_2)$$
$$v_2 = \sqrt{(2/\rho)(p_1 - p_2) + 2g(y_1 - y_2)}$$
$$v_2 = 26.2 \text{ m/s}$$

b) $\underline{h = 3.00 \text{ m}}$

The volume of the air in the tank increases so its pressure decreases.

$pV = nRT = $ constant, so $pV = p_0 V_0$ (p_0 is the pressure for $h_0 = 3.50$ m and p is the pressure for $h = 3.00$ m)

$$p(4.00 \text{ m} - h)A = p_0(4.00 \text{ m} - h_0)A$$

$$p = p_0 \left(\frac{4.00 \text{ m} - h_0}{4.00 \text{ m} - h} \right) = (4.20 \times 10^5 \text{ Pa}) \left(\frac{4.00 \text{ m} - 3.50 \text{ m}}{4.00 \text{ m} - 3.00 \text{ m}} \right) = 2.10 \times 10^5 \text{ Pa}$$

Repeat the calculation of part (a), but now $p_1 = 2.10 \times 10^5$ Pa and $y_1 = 3.00$ m.

$$v_2 = \sqrt{(2/\rho)(p_1 - p_2) + 2g(y_1 - y_2)}$$
$$v_2 = 16.1 \text{ m/s}$$

$\underline{h = 2.00 \text{ m}}$

$$p = p_0 \left(\frac{4.00 \text{ m} - h_0}{4.00 \text{ m} - h} \right) = (4.20 \times 10^5 \text{ Pa}) \left(\frac{4.00 \text{ m} - 3.50 \text{ m}}{4.00 \text{ m} - 2.00 \text{ m}} \right) = 1.05 \times 10^5 \text{ Pa}$$

$$v_2 = \sqrt{(2/\rho)(p_1 - p_2) + 2g(y_1 - y_2)}$$
$$v_2 = 5.44 \text{ m/s}$$

c) $v_2 = 0$ means $(2/\rho)(p_1 - p_2) + 2g(y_1 - y_2) = 0$

$$p_1 - p_2 = -\rho g(y_1 - y_2)$$
$$y_1 - y_2 = h - 1.00 \text{ m}$$

$$p = p_0 \left(\frac{0.50 \text{ m}}{4.00 \text{ m} - h} \right) = (4.20 \times 10^5 \text{ Pa}) \left(\frac{0.50 \text{ m}}{4.00 \text{ m} - h} \right)$$

$(4.20 \times 10^5 \text{ Pa}) \left(\dfrac{0.50 \text{ m}}{4.00 \text{ m} - h} \right) - 1.00 \times 10^5 \text{ Pa} = (9.80 \text{ m/s}^2)(1000 \text{ kg/m}^3)(1.00 \text{ m} - h)$

$(210/(4.00 - h)) - 100 = 9.80 - 9.80h$, with h in meters.

$210 = (4.00 - h)(109.8 - 9.80h)$

$9.80h^2 - 149h + 229.2 = 0$ and $h^2 - 15.20h + 23.39 = 0$

quadratic formula: $h = \frac{1}{2}(15.20 \pm \sqrt{(15.20)^2 - 4(23.39)}) = (7.60 \pm 5.86) \text{ m}$

h must be less than 4.00 m, so the only acceptable value is

$h = 7.60 \text{ m} - 5.86 \text{ m} = 1.74 \text{ m}$

EVALUATE: The flow stops when $p + \rho g(y_1 - y_2)$ equals air pressure. For $h = 1.74$ m, $p = 9.3 \times 10^4$ Pa and $\rho g(y_1 - y_2) = 0.7 \times 10^4$ Pa, so $p + \rho g(y_1 - y_2) = 1.0 \times 10^5$ Pa, which is air pressure.

18.65 **IDENTIFY** and **SET UP:** Apply Eq.(18.2) to find n and then use Avogadro's number to find the number of molecules.
EXECUTE: Calculate the number of water molecules N.

Number of moles: $n = \dfrac{m_{\text{tot}}}{M} = \dfrac{50 \text{ kg}}{18.0 \times 10^{-3} \text{ kg/mol}} = 2.778 \times 10^3 \text{ mol}$

$N = nN_A = (2.778 \times 10^3 \text{ mol})(6.022 \times 10^{23} \text{ molecules/mol}) = 1.7 \times 10^{27}$ molecules

Each water molecule has three atoms, so the number of atoms is

$3(1.7 \times 10^{27}) = 5.1 \times 10^{27}$ atoms

EVALUATE: We could also use the masses in Example 18.5 to find the mass m of one H_2O molecule: $m = 2.99 \times 10^{-26}$ kg. Then $N = m_{\text{tot}}/m = 1.7 \times 10^{27}$ molecules, which checks.

18.69 **IDENTIFY** and **SET UP:** At equilibrium $F(r) = 0$.
The work done to increase the separation from r_2 to ∞ is $U(\infty) - U(r_2)$.

a) EXECUTE: $U(r) = U_0[(R_0/r)^{12} - 2(R_0/r)^6]$
Eq.(13.26): $F(r) = 12(U_0/R_0)[(R_0/r)^{13} - (R_0/r)^7]$

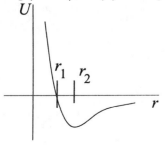
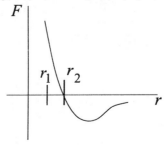

b) equilibrium requires $F = 0$; occurs at point r_2. r_2 is where U is a minimum (stable equilibirum).

c) $U = 0$ implies $[(R_0/r)^{12} - 2(R_0/r)^6] = 0$

$(r_1/R_0)^6 = 1/2$ and $r_1 = R_0/(2)^{1/6}$

$F = 0$ implies $[(R_0/r)^{13} - (R_0/r)^7] = 0$

$(r_2/R_0)^6 = 1$ and $r_2 = R_0$

Then $r_1/r_2 = (R_0/2^{1/6})/R_0 = 2^{-1/6}$

d) $W_{other} = \Delta U$

At $r \to \infty$, $U = 0$, so $W = -U(R_0) = -U_0[(R_0/R_0)^{12} - 2(R_0/R_0)^6] = +U_0$

EVALUATE: The answer to part (d), U_0, is the depth of the potential well shown in the graph of $U(r)$.

18.71 **IDENTIFY** and **SET UP:** Apply Eq.(18.19) for v_{rms}. The equation preceeding Eq.(18.12) relates v_{rms} and $(v_x)_{rms}$.

EXECUTE:

a) $v_{rms} = \sqrt{3RT/M}$

$$v_{rms} = \sqrt{\frac{3(8.3145\ \text{J/mol}\cdot\text{K})(300\ \text{K})}{28.0 \times 10^{-3}\ \text{kg/mol}}} = 517\ \text{m/s}$$

b) $(v_x^2)_{av} = \frac{1}{3}(v^2)_{av}$ so $\sqrt{(v_x^2)_{av}} = (1/\sqrt{3})\sqrt{(v^2)_{av}} = (1/\sqrt{3})v_{rms} = (1/\sqrt{3})(517\ \text{m/s}) = 298\ \text{m/s}$

EVALUATE: The speed of sound is approximately equal to $(v_x)_{rms}$ since it is the motion along the direction of propagation of the wave that transmits the wave.

18.73 **a) IDENTIFY** and **SET UP:** Apply conservation of energy
$K_1 + U_1 + W_{other} = K_2 + U_2$, where $U = -Gmm_p/r$. Let point 1 be at the surface of the planet, where the projectile is launched, and let point 2 be far from the earth. Just barely escapes says $v_2 = 0$.

EXECUTE: Only gravity does work says $W_{other} = 0$.

$U_1 = -Gmm_p/R_p$; $r_2 \to \infty$ so $U_2 = 0$; $v_2 = 0$ so $K_2 = 0$.

The conservation of energy equation becomes

$K_1 - Gmm_p/R_p = 0$ and $K_1 = Gmm_p/R_p$.

But $g = Gm_p/R_p^2$ so $Gm_p/R_p = R_p g$ and $K_1 = mgR_p$, as was to be shown.

EVALUATE: The greater gR_p is the more initial kinetic energy is required for escape.

b) IDENTIFY and **SET UP:** Set K_1 from part (a) equal to the average kinetic energy of a molecule as given by Eq.(18.16). $\frac{1}{2}m(v^2)_{av} = mgR_p$ (from part (a)). But also, $\frac{1}{2}m(v^2)_{av} = \frac{3}{2}kT$, so $mgR_p = \frac{3}{2}kT$

EXECUTE: $T = \dfrac{2mgR_p}{3k}$

<u>nitrogen</u>

$m_{N_2} = (28.0 \times 10^{-3} \text{ kg/mol})/(6.022 \times 10^{23} \text{ molecules/mol}) = 4.65 \times 10^{-26} \text{ kg/molecule}$

$T = \dfrac{2mgR_p}{3k} = \dfrac{2(4.65 \times 10^{-26} \text{ kg/molecule})(9.80 \text{ m/s}^2)(6.38 \times 10^6 \text{ m})}{3(1.381 \times 10^{-23} \text{ J/molecule} \cdot \text{K})} =$

$1.40 \times 10^5 \text{ K}$

<u>hydrogen</u>

$m_{H_2} = (2.02 \times 10^{-3} \text{ kg/mol})/(6.022 \times 10^{23} \text{ molecules/mol}) = 3.354 \times 10^{-27} \text{ kg/molecule}$

$T = \dfrac{2mgR_p}{3k} = \dfrac{2(3.354 \times 10^{-27} \text{ kg/molecule})(9.80 \text{ m/s}^2)(6.38 \times 10^6 \text{ m})}{3(1.381 \times 10^{-23} \text{ J/molecule} \cdot \text{K})} =$

$1.01 \times 10^4 \text{ K}$

c) $T = \dfrac{2mgR_p}{3k}$

<u>nitrogen</u>

$T = \dfrac{2(4.65 \times 10^{-26} \text{ kg/molecule})(1.63 \text{ m/s}^2)(1.74 \times 10^6 \text{ m})}{3(1.381 \times 10^{-23} \text{ J/molecule} \cdot \text{K})} = 6370 \text{ K}$

<u>hydrogen</u>

$T = \dfrac{2(3.354 \times 10^{-27} \text{ kg/molecule})(1.63 \text{ m/s}^2)(1.74 \times 10^6 \text{ m})}{3(1.381 \times 10^{-23} \text{ J/molecule} \cdot \text{K})} = 459 \text{ K}$

d) EVALUATE: The "escape temperatures" are much less for the moon than for the earth. For the moon a larger fraction of the molecules at a given temperature will have speeds in the Maxwell-Boltzmann distribution larger than the escape speed. After a long time most of the molecules will have escaped from the moon.

18.77 IDENTIFY: Relate the average energy per atom to kT.

a) SET UP: The atoms have two degrees of freedom and hence each atom has an average kinetic energy $2(\frac{1}{2}kT) = kT$. Each atom also has potential energy due to the Hooke's law force that binds it at the surface to its position parallel to the surface. The average potential energy for this two-dimensional harmonic oscillator equals the average kinetic energy, just as for a one-dimensional or three-dimensional oscillator. The average total energy per atom is $kT + kT = 2kT$.

EXECUTE: For n moles of atoms the average total energy is $2nRT$ and
$C_V = 2R = 2(8.3145 \text{ J/mol} \cdot \text{K}) = 16.6 \text{ J/mol} \cdot \text{K}$.

b) EVALUATE: At low temperatures the average energy per atom is less than $2kT$ because most atoms remain in their lowest energy state. Thus the molar heat capacity will be less than the value found in part (a).

18.81 IDENTIFY and **SET UP:** Evaluate the integral in Eq.(18.31) as specified in the problem.

EXECUTE: $\int_0^\infty v^2 f(v) \, dv = 4\pi (m/2\pi kT)^{3/2} \int_0^\infty v^4 e^{-mv^2/2kT} \, dv$

The integral formula with $n = 2$ gives $\int_0^\infty v^4 e^{-av^2} \, dv = (3/8a^2)\sqrt{\pi/a}$

Apply with $a = m/2kT$,
$\int_0^\infty v^2 f(v) \, dv = 4\pi (m/2\pi kT)^{3/2}(3/8)(2kT/m)^2 \sqrt{2\pi kT/m} =$
$(3/2)(2kT/m) = 3kT/m$

EVALUATE: Equation (18.16) says $\frac{1}{2}m(v^2)_{\text{av}} = 3kT/2$, so $(v^2)_{\text{av}} = 3kT/m$, in agreement with our calculation.

18.85 IDENTIFY: The measurement gives the dew point. Relative humidity is defined in Problem 18.84.

SET UP:

relative humidity $= \dfrac{\text{partial pressure of water vapor at temperature } T}{\text{vapor pressure of water at temperature } T}$

EXECUTE: The experiment shows that the dew point is 16.0°C, so the partial pressure of water vapor at 30.0°C is equal to the vapor pressure at 16.0°C, which is 1.81×10^3 Pa.

Thus the relative humidity $= \dfrac{1.81 \times 10^3 \text{ Pa}}{4.25 \times 10^3 \text{ Pa}} = 0.426 = 42.6\%$.

EVALUATE: The lower the dew point is compared to the air temperature, the smaller the relative humidity.

CHAPTER 19
THE FIRST LAW OF THERMODYNAMICS

Exercises 1, 7, 11, 13, 17, 19, 25, 27, 31, 33, 35, 37
Problems 41, 43, 47, 49, 51, 55, 59, 61, 63, 67

Exercises

19.1 **a) IDENTIFY** and **SET UP:** The pressure is constant and the volume increases.

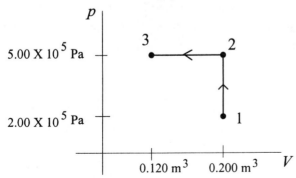

b) $W = \int_{V_1}^{V_2} p \, dV$

Since p is constant, $W = p \int_{V_1}^{V_2} dV = p(V_2 - V_1)$

The problem gives T rather than p and V, so use the ideal gas law to rewrite the expression for W.

EXECUTE: $pV = nRT$ so $p_1 V_1 = nRT_1$, $p_2 V_2 = nRT_2$; subracting the two equations gives

$p(V_2 - V_1) = nR(T_2 - T_1)$

Thus $W = nR(T_2 - T_1)$ is an alternative expression for the work in a constant pressure process for an ideal gas.

Then $W = nR(T_2 - T_1) = (2.00 \text{ mol})(8.3145 \text{ J/mol} \cdot \text{K})(107°\text{C} - 27°\text{C}) = +1330$ J.

EVALUATE: The gas expands when heated and does positive work.

19.7 **a) IDENTIFY** and **SET UP:** pV-diagram

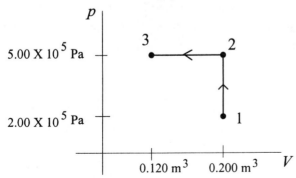

b) Calculatwe W for each process, using the expression for W that applies to the speicific type of process.

EXECUTE:

$1 \rightarrow 2$, $\Delta V = 0$, so $W = 0$

$2 \rightarrow 3$

p is constant, so $W = p\Delta V = (5.00 \times 10^5 \text{ Pa})(0.120 \text{ m}^3 - 0.200 \text{ m}^3) = -4.00 \times 10^4 \text{ J}$ (W is negative since the volume decreases in the process.)

$W_{\text{tot}} = W_{1 \rightarrow 2} + W_{2 \rightarrow 3} = -4.00 \times 10^4 \text{ J}$

EVALUATE: The volume decreases so the total work done is negative.

19.11 IDENTIFY: The type of process is not specified. We can use $\Delta U = Q - W$ because this applies to all processes. Calculate ΔU and then from it calculate ΔT.

SET UP: Q is positive since heat goes into the gas; $Q = +1200 \text{ J}$

W positive since gas expands; $W = +2100 \text{ J}$

EXECUTE: $\Delta U = 1200 \text{ J} - 2100 \text{ J} = -900 \text{ J}$

We can also use $\Delta U = n(\frac{3}{2}R)\Delta T$ since this is true for any process for an ideal gas.

$$\Delta T = \frac{2\Delta U}{3nR} = \frac{2(-900 \text{ J})}{3(5.00 \text{ mol})(8.3145 \text{ J/mol} \cdot \text{K})} = -14.4\text{C}°$$

$T_2 = T_1 + \Delta T = 127°\text{C} - 14.4\text{C}° = 113°\text{C}$

EVALUATE: More energy leaves the gas in the expansion work than enters as heat. The internal energy therefore decreases, and for an ideal gas this means the temperature decreases. We didn't have to convert ΔT to kelvins since ΔT is the same on the Kelvin and Celsius scales.

19.13 IDENTIFY and **SET UP:** Calculate W using the equation for a constant pressure process. Then use $\Delta U = Q - W$ to calculate Q.

a) EXECUTE: $W = \int_{V_1}^{V_2} p\,dV = p(V_2 - V_1)$ for this constant pressure process.

$W = (2.3 \times 10^5 \text{ Pa})(1.20 \text{ m}^3 - 1.70 \text{ m}^3) = -1.15 \times 10^5 \text{ J}$ (The volume decreases in the process, so W is negative.)

b) $\Delta U = Q - W$

$Q = \Delta U + W = -1.40 \times 10^5 \text{ J} + (-1.15 \times 10^5 \text{ J}) = -2.55 \times 10^5 \text{ J}$

Q negative means heat flows out of the gas.

c) EVALUATE: $W = \int_{V_1}^{V_2} p\,dV = p(V_2 - V_1)$ (constant pressure) and $\Delta U = Q - W$ apply to <u>any</u> system, not just to an ideal gas. We did not use the ideal gas equation, either directly or indirectly, in any of the calculations, so the results are the same whether the gas is ideal or not.

19.17 IDENTIFY and **SET UP:** Apply the first law, $\Delta U = Q - W$. For a cycle $\Delta U = 0$. W is the area under the pressure versus V curve and is positive when V increases, negative when V decreases.

a) EXECUTE: For one cycle, $\Delta U = 0$.

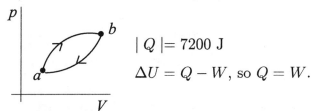

$|Q| = 7200$ J

$\Delta U = Q - W$, so $Q = W$.

The magnitude of the positive work done from a to b is larger than the magnitude of the negative work done from b to a, so the net work done in one cycle is positive. Then from $Q = W$, Q must be positive.

$Q = +7200$ J; the system absorbs heat.

b) $W = Q = +7200$ J

c)

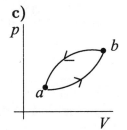

Now the net work done is negative.

Since $Q = W$, Q is negative; system liberates heat.

$|W|$ is the same as in part (b) and $Q = W$, so $Q = -7200$ J. Q is negative; the system liberates heat.

EVALUATE: For any cycle, $Q = W$. For a cycle, W is positive if the system goes around the cycle in the clockwise direction and negative if it goes around in the counterclockwise direction.

19.19 IDENTIFY and **SET UP:** Deduce information about Q and W from the problem statement and then apply the first law, $\Delta U = Q - W$, to infer whether Q is positive or negative.

EXECUTE:

a) For the water $\Delta T > 0$, so by $Q = mc\,\Delta T$ heat has been added to the water. Thus heat energy comes from the burning fuel-oxygen mixture, and Q for the system (fuel and oxygen) is negative.

b) Constant volume implies $W = 0$.

c) The 1st law (Eq.19.4) says $\Delta U = Q - W$.

$Q < 0$, $W = 0$ so by the 1st law $\Delta U < 0$. The internal energy of the fuel-oxygen mixture decreased.

EVALUATE: In this process internal energy from the fuel-oxygen mixture was transferred to the water, raising its temperature.

19.25 **IDENTIFY** and **SET UP:** Use information about the pressure and volume in the ideal gas law to determine the sign of ΔT, and from that the sign of Q.

EXECUTE: For constant p, $Q = nC_p\Delta T$

Since the gas is ideal, $pV = nRT$ and for constant p, $p\Delta V = nR\,\Delta T$.

$$Q = nC_p\left(\frac{p\,\Delta V}{nR}\right) = \left(\frac{C_p}{R}\right)p\,\Delta V$$

Since the gas expands, $\Delta V > 0$ and therefore $Q > 0$. $Q > 0$ means heat goes into gas.

EVALUATE: Heat flows into the gas, W is positive and the internal energy increases. It must be that $Q > W$.

19.27 **IDENTIFY:** Calculate W and ΔU and then use the first law to calculate Q.

a) SET UP: $W = \int_{V_1}^{V_2} p\,dV$

$pV = nRT$ so $p = nRT/V$

$W = \int_{V_1}^{V_2}(nRT/V)\,dV = nRT\int_{V_1}^{V_2} dV/V = nRT\ln(V_2/V_1)$ (work done during an isothermal process).

EXECUTE: $W = (0.150 \text{ mol})(8.3145 \text{ J/mol}\cdot\text{K})(350 \text{ K})\ln(0.25V_1/V_1) = (436.5 \text{ J})\ln(0.25) = -605 \text{ J}$.

EVALUATE: W for the gas is negative, since the volume decreases.

b) EXECUTE: $\Delta U = nC_V\,\Delta T$ for any ideal gas process.

$\Delta T = 0$ (isothermal) so $\Delta U = 0$.

EVALUATE: $\Delta U = 0$ for any ideal gas process in which T doesn't change.

c) EXECUTE: $\Delta U = Q - W$

$\Delta U = 0$ so $Q = W = -605 \text{ J}$. (Q is negative; the gas liberates 605 J of heat to the surroundings.)

EVALUATE: $Q = nC_V\,\Delta T$ is only for a constant volume process so doesn't apply here.

$Q = nC_p\,\Delta T$ is only for a constant pressure process so doesn't apply here.

19.31 **a) IDENTIFY** and **SET UP:** $Q = nC_p\,\Delta T$, since it is a constant pressure process. Need to calculate C_p from $\gamma = C_p/C_V$ and $C_p = C_V + R$.

EXECUTE: Combining these equations gives $\gamma = (C_V + R)/C_V = 1 + R/C_V$

$C_V = R/(\gamma - 1) = (8.3145 \text{ J/mol}\cdot\text{K})/(1.220 - 1) = 37.79 \text{ J/mol}\cdot\text{K}$

$C_p = C_V + R = 37.79 \text{ J/mol}\cdot\text{K} + 8.3145 \text{ J/mol}\cdot\text{K} = 46.10 \text{ J/mol}\cdot\text{K}$

Then $Q = nC_p\,\Delta T = 2.40(46.10 \text{ J/mol}\cdot\text{K})(25.0°\text{C} - 20.0°\text{C}) = +553 \text{ J}$

b) IDENTIFY and **SET UP:** For a constant pressure process, $W = p\,\Delta V$. We can use the ideal gas law to write this as $W = nR\Delta T$.

EXECUTE: $W = (2.40 \text{ mol})(8.3145 \text{ J/mol} \cdot \text{K})(25.0°\text{C} - 20.0°\text{C}) = +99.7 \text{ J}$.

Then $\Delta U = Q - W = 553 \text{ J} - 99.7 \text{ J} = 453 \text{ J}$.

EVALUATE: When the temperature of an ideal gas is increased at constant pressure, $pV = nRT$ says that V increases. 533 J of heat energy flows into the gas. 100 J of this energy input is used to do work as the gas expands and the remaining 453 J remians in the gas as increased internal energy.

19.33 **IDENTIFY** and **SET UP: a)** In the process the pressure increases and the volume decreases.

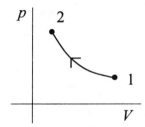

b) For an adiabatic process for an ideal gas

$T_1 V_1^{\gamma-1} = T_2 V_2^{\gamma-1}$, $p_1 V_1^{\gamma} = p_2 V_2^{\gamma}$,   and $pV = nRT$

EXECUTE:

From the first equation, $T_2 = T_1(V_1/V_2)^{\gamma-1} = (293 \text{ K})(V_1/0.0900V_1)^{1.4-1}$

$T_2 = (293 \text{ K})(11.11)^{0.4} = 768 \text{ K} = 495°\text{C}$

(Note: In the equation $T_1 V_1^{\gamma-1} = T_2 V_2^{\gamma-1}$ the temperature **must** be in kelvins.)

$p_1 V_1^{\gamma} = p_2 V_2^{\gamma}$ implies $p_2 = p_1(V_1/V_2)^{\gamma} = (1.00 \text{ atm})(V_1/0.0900V_1)^{1.4}$

$p_2 = (1.00 \text{ atm})(11.11)^{1.4} = 29.1 \text{ atm}$

EVALUATE: Alternatively, we can use $pV = nRT$ to calculate p_2:

n, R constant implies $pV/T = nR = $ constant so $p_1 V_1/T_1 = p_2 V_2/T_2$

$p_2 = p_1(V_1/V_2)(T_2/T_1) = (1.00 \text{ atm})(V_1/0.0900V_1)(768 \text{ K}/293 \text{ K}) = 29.1 \text{ atm}$, which checks.

19.35 **IDENTIFY** and **SET UP:** For an ideal gas $\Delta U = nC_V \Delta T$. The sign of ΔU is the same as the sign of ΔT. Combine Eq.(19.22) and the ideal gas law to obtain an equation relating T and p, and use it to determine the sign of ΔT.

EXECUTE: $T_1 V_1^{\gamma-1} = T_2 V_2^{\gamma-1}$ and $V = nRT/p$ so,

$T_1^{\gamma} p_1^{1-\gamma} = T_2^{\gamma} p_2^{1-\gamma}$ and $T_2^{\gamma} = T_1^{\gamma}(p_2/p_1)^{\gamma-1}$

$p_2 < p_1$ and $\gamma - 1$ is positive so $T_2 < T_1$. ΔT is negative so ΔU is negative; the energy of the gas decreases.

EVALUATE: Eq.(19.24) shows that the volume increases for this process, so it is an adiabatic expansion. In an adiabatic expansion the temperature decreases.

19.37 a) IDENTIFY and **SET UP:** In the expansion the pressure decreases and the volume increases.

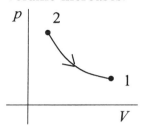

b) Adiabatic means $Q = 0$.

Then $\Delta U = Q - W$ gives $W = -\Delta U = -nC_V \,\Delta T = nC_V(T_1 - T_2)$ (Eq.19.25).

$C_V = 12.47$ J/mol $\cdot$ K (Table 19.1)

EXECUTE: $W = (0.450 \text{ mol})(12.47 \text{ J/mol} \cdot \text{K})(50.0°\text{C} - 10.0°\text{C}) = +224$ J

W positive for $\Delta V > 0$ (expansion)

c) $\Delta U = -W = -224$ J.

EVALUATE: There is no heat energy input. The energy for doing the expansion work comes from the internal energy of the gas, which therefore decreases. For an ideal gas, when T decreases, U decreases.

Problems

19.41 IDENTIFY: Calculate W and ΔU and use the first law to find ΔU

SET UP: $|W|$ is the area under the path from A to B in the pV-graph. The volume decreases, so $W < 0$.

EXECUTE: $W = -\frac{1}{2}(500 \times 10^3 \text{ Pa} + 150 \times 10^3 \text{ Pa})(0.60 \text{ m}^3) = -1.95 \times 10^5$ J

SET UP: $\Delta U = nC_V \,\Delta T$. Use the ideal gas law to write ΔT in terms of p and V in the two states.

EXECUTE: $T_1 = \dfrac{p_1 V_1}{nR}$, $T_2 = \dfrac{p_2 V_2}{nR}$ so $\Delta T = T_2 - T_1 = \dfrac{p_2 V_2 - p_1 V_1}{nR}$

$\Delta U = (C_V / R)(p_2 V_2 - p_1 V_1)$

$\Delta U = (20.85/8.315)[(500 \times 10^3 \text{ Pa})(0.20 \text{ m}^3) - (150 \times 10^3 \text{ Pa})(0.80 \text{ m}^3)] = -5.015 \times 10^4$ J

Then $\Delta U = Q - W$ gives

$Q = \Delta U + W = -5.015 \times 10^4 \text{ J} - 1.95 \times 10^5 \text{ J} = -2.45 \times 10^5$ J

Q is negative, so heat flows out of the gas.

EVALUATE: W, ΔU, and Q are all negative. The heat energy that flows out of the gas comes partly from the energy added by work done on the gas and partly from a decrease in internal energy of the gas.

19.43 IDENTIFY: Use $\Delta U = Q - W$ and the fact that ΔU is path indepedent.

$W > 0$ when the volume increases, $W < 0$ when the volume decreases, and $W = 0$ when the volume is constant. $Q > 0$ if heat flows into the system.

SET UP:

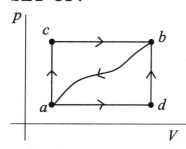

$Q_{acb} = +90.0$ J (positive since heat flows in)

$W_{acb} = +60.0$ J (positive since $\Delta V > 0$)

EXECUTE:

a) $\Delta U = Q - W$

ΔU is path independent; Q and W depend on the path.

$\Delta U = U_b - U_a$

This can be calculated for any path from a to b, in particular for path acb:

$\Delta U_{a \to b} = Q_{acb} - W_{acb} = 90.0$ J $- 60.0$ J $= 30.0$ J.

Now apply $\Delta U = Q - W$ to path adb; $\Delta U = 30.0$ J for this path also.

$W_{adb} = +15.0$ J (positive since $\Delta V > 0$)

$\Delta U_{a \to b} = Q_{adb} - W_{adb}$ so $Q_{adb} = \Delta U_{a \to b} + W_{adb} = 30.0$ J $+ 15.0$ J $= +45.0$ J

b) Apply $\Delta U = Q - W$ to path ba:

$\Delta U_{b \to a} = Q_{ba} - W_{ba}$

$W_{ba} = -35.0$ J (negative since $\Delta V < 0$)

$\Delta U_{b \to a} = U_a - U_b = -(U_b - U_a) = -\Delta U_{a \to b} = -30.0$ J

Then $Q_{ba} = \Delta U_{b \to a} + W_{ba} = -30.0$ J $- 35.0$ J $= -65.0$ J.

($Q_{ba} < 0$; the system liberates heat.)

c) $U_a = 0$, $U_d = 8.0$ J

$\Delta U_{a \to b} = U_b - U_a = +30.0$ J, so $U_b = +30.0$ J.

process $a \to d$

$\Delta U_{a \to d} = Q_{ad} - W_{ad}$

$\Delta U_{a \to d} = U_d - U_a = +8.0$ J

$W_{adb} = +15.0$ J and $W_{adb} = W_{ad} + W_{db}$. But the work W_{db} for the process $d \to b$ is zero since $\Delta V = 0$ for that process. Therefore $W_{ad} = W_{adb} = +15.0$ J.

Then $Q_{ad} = \Delta U_{a \to d} + W_{ad} = +8.0$ J $+ 15.0$ J $= +23.0$ J (positive implies heat absorbed).

process $d \rightarrow b$

$\Delta U_{d \rightarrow b} = Q_{db} - W_{db}$

$W_{db} = 0$, as already noted.

$\Delta U_{d \rightarrow b} = U_b - U_d = 30.0 \text{ J} - 8.0 \text{ J} = +22.0 \text{ J}.$

Then $Q_{db} = \Delta U_{d \rightarrow b} + W_{db} = +22.0 \text{ J}$ (positive; heat absorbed).

EVALUATE: The signs of our calculated Q_{ad} and Q_{db} agree with the problem statement that heat is absorbed in these processes.

19.47 **IDENTIFY:** Use the 1st law to relate Q_{tot} to W_{tot} for the cycle.
Calculate W_{ab} and W_{bc} and use what we know about W_{tot} to deduce W_{ca}

a) SET UP: We aren't told whether the pressure increases or decreases in process bc. The cycle could be

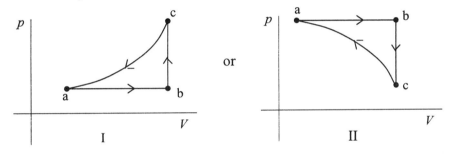

In cycle I, the total work is negative and in cycle II the total work is positive. For a cycle, $\Delta U = 0$, so $Q_{\text{tot}} = W_{\text{tot}}$

The net heat flow for the cycle is out of the gas, so heat $Q_{\text{tot}} < 0$ and $W_{\text{tot}} < 0$. Sketch I is correct.

b) EXECUTE: $W_{\text{tot}} = Q_{\text{tot}} = -800 \text{ J}$

$W_{\text{tot}} = W_{ab} + W_{bc} + W_{ca}$

$W_{bc} = 0$ since $\Delta V = 0$.

$W_{ab} = p \, \Delta V$ since p is constant. But since it is an ideal gas, $p \, \Delta V = nR \, \Delta T$

$W_{ab} = nR(T_b - T_a) = 1660 \text{ J}$

$W_{ca} = W_{\text{tot}} - W_{ab} = -800 \text{ J} - 1660 \text{ J} = -2460 \text{ J}$

EVALUATE: In process ca the volume decreases and the work W is negative.

19.49 **IDENTIFY:** Use the 1st law to relate Q_{tot} to W_{tot}.
Calculate W for each process.

a) SET UP:

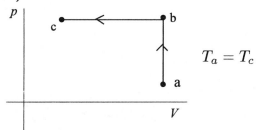

$T_a = T_c$

b) EXECUTE: $\Delta U_{\text{tot}} = 0$ since $\Delta T = 0$

$Q_{\text{tot}} = W_{\text{tot}}$

$W_{ab} = 0$, since $\Delta V = 0$

$W_{bc} = p\,\Delta V = (6.00 \times 10^4 \text{ Pa})(0.300 \text{ m}^3 - 0.700 \text{ m}^3) = -2.40 \times 10^4 \text{ J}$

$W_{\text{tot}} = -2.40 \times 10^4 \text{ J}$ so $Q_{\text{tot}} = -2.40 \times 10^4 \text{ J}$

EVALUATE: $W_{\text{tot}} < 0$ since the volume decreases. Work done on the system adds energy to the system and this same amount of energy flows out of the system as heat energy.

19.51 **IDENTIFY** and **SET UP:** Use the first law to calculate W and then use $W = p\,\Delta V$ for the constant pressure process to calculate ΔV.

EXECUTE: $\Delta U = Q - W$

$Q = -2.15 \times 10^5 \text{ J}$ (negative since heat energy goes out of the system)

$\Delta U = 0$ so $W = Q = -2.15 \times 10^5 \text{ J}$

Constant pressure, so $W = \int_{V_1}^{V_2} p\,dV = p(V_2 - V_1) = p\,\Delta V$.

Then $\Delta V = \dfrac{W}{p} = \dfrac{-2.15 \times 10^5 \text{ J}}{9.50 \times 10^5 \text{ J}} = -0.226 \text{ m}^3$.

EVALUATE: Positive work is done on the system by its surroundings; this inputs to the system the energy that then leaves the system as heat. Both Eq.(19.4) and (19.2) apply to all processes for any system, not just to an ideal gas.

19.55 **IDENTIFY** and **SET UP:** The heat produced from the reaction is $Q_{\text{reaction}} = mL_{\text{reaction}}$, where L_{reaction} is the heat of reaction of the chemicals.

$Q_{\text{reaction}} = W + \Delta U_{\text{spray}}$

EXECUTE: For a mass m of spray, $W = \frac{1}{2}mv^2 = \frac{1}{2}m(19 \text{ m/s})^2 = (180.5 \text{ J/kg})m$ and

$\Delta U_{\text{spray}} = Q_{\text{spray}} = mc\,\Delta T = m(4190 \text{ J/kg·K})(100°\text{C}-20°\text{C}) = (335, 200 \text{ J/kg})m$.

Then $Q_{\text{reaction}} = (180 \text{ J/kg} + 335, 200 \text{ J/kg})m = (335, 380 \text{ J/kg})m$ and

$Q_{\text{reaction}} = mL_{\text{reaction}}$ implies $mL_{\text{reaction}} = (335, 380 \text{ J/kg})m$.

The mass m divides out and $L_{\text{reaction}} = 3.4 \times 10^5$ J/kg

EVALUATE: The amount of energy converted to work is negligible for the two significant figures to which the answer should be expressed. Almost all of the energy produced in the reaction goes into heating the compound.

19.59 IDENTIFY Assume that the gas is ideal and that the process is adiabatic. Apply Eqs.(19.22) and (19.24) to relate pressure and volume and temperature and volume. The distance the piston moves is related to the volume of the gas. Use Eq.(19.25) to calculate W.

a) SET UP: $\gamma = C_p/C_V = (C_V + R)/C_V = 1 + R/C_V = 1.40$

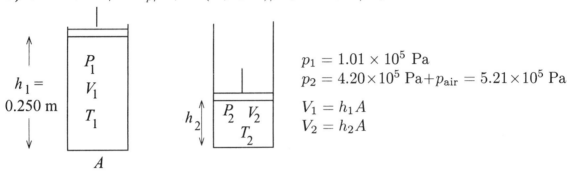

$$p_1 = 1.01 \times 10^5 \text{ Pa}$$
$$p_2 = 4.20 \times 10^5 \text{ Pa} + p_{\text{air}} = 5.21 \times 10^5 \text{ Pa}$$
$$V_1 = h_1 A$$
$$V_2 = h_2 A$$

EXECUTE: adiabatic process: $p_1 V_1^\gamma = p_2 V_2^\gamma$

$$p_1 h_1^\gamma A^\gamma = p_2 h_2^\gamma A^\gamma$$

$$h_2 = h_1 \left(\frac{p_1}{p_2}\right)^{1/\gamma} = (0.250 \text{ m}) \left(\frac{1.01 \times 10^5 \text{ Pa}}{5.21 \times 10^5 \text{ Pa}}\right)^{1/1.40} = 0.0774 \text{ m}$$

The piston has moved a distance $h_1 - h_2 = 0.250 \text{ m} - 0.0774 \text{ m} = 0.173 \text{ m}$.

b) $T_1 V_1^{\gamma-1} = T_2 V_2^{\gamma-1}$

$$T_1 h_1^{\gamma-1} A^{\gamma-1} = T_2 h_2^{\gamma-1} A^{\gamma-1}$$

$$T_2 = T_1 \left(\frac{h_1}{h_2}\right)^{\gamma-1} = 300.1 \text{ K} \left(\frac{0.250 \text{ m}}{0.0774 \text{ m}}\right)^{0.40} = 479.7 \text{ K} = 207°\text{C}$$

c) $W = nC_V (T_1 - T_2)$ (Eq.19.25)

$$W = (20.0 \text{ mol})(20.8 \text{ J/mol} \cdot \text{K})(300.1 \text{ K} - 479.7 \text{ K}) = -7.47 \times 10^4 \text{ J}$$

EVALUATE: In an adiabatic compression of an ideal gas the temperature increases. In any compression the work W is negative.

19.61 IDENTIFY: In each case calculate either ΔU or Q for the specific type of process and then apply the first law.

a) SET UP: <u>isothermal</u> $(\Delta T = 0)$ $\Delta U = Q - W$; $W = +300$ J

For any process of an ideal gas, $\Delta U = nC_V \Delta T$.

EXECUTE: Therefore, for an ideal gas, if $\Delta T = 0$ then $\Delta U = 0$ and $Q = W = +300$ J.

b) SET UP: <u>adiabatic</u> $(Q = 0)$

$\Delta U = Q - W; \; W = +300$ J

EXECUTE: $Q = 0$ says $\Delta U = -W = -300$ J

c) SET UP: <u>isobaric</u> $\Delta p = 0$

Use W to calculate ΔT and then calculate Q.

EXECUTE: $W = p \Delta V = nR \Delta T; \quad \Delta T = W/nR$

$Q = nC_p \Delta T$ and for a monatomic ideal gas $C_p = \frac{5}{2}R$

Thus $Q = n\frac{5}{2}R\Delta T = (5Rn/2)(W/nR) = 5W/2 = +750$ J.

$\Delta U = nC_V \Delta T$ for any ideal gas process and $C_V = C_p - R = \frac{3}{2}R$.

Thus $\Delta U = 3W/2 = +450$ J

EVALUATE: 300 J of energy leaves the gas when it performs expansion work. In the isothermal process this energy is replaced by heat flow into the gas and the internal energy remains the same. In the adiabatic process the energy used in doing the work decreases the internal energy. In the isobaric process 750 J of heat energy enters the gas, 300 J leaves as the work done and 450 J remains in the gas as increased internal energy.

19.63 IDENTIFY and SET UP: Use the ideal gas law, the first law and expressions for Q and W for specific types of processes.

EXECUTE:

a) initial expansion (state 1 → state 2)

$p_1 = 2.40 \times 10^5$ Pa, $T_1 = 355$ K, $p_2 = 2.40 \times 10^5$ Pa, $V_2 = 2V_1$

$pV = nRT; \; T/V = p/nR = $ constant, so $T_1/V_1 = T_2/V_2$ and $T_2 = T_1(V_2/V_1) = 355$ K$(2V_1/V_1) = 710$ K

$\Delta p = 0$ so $W = p\Delta V = nR\Delta T = (0.250 \text{ mol})(8.3145 \text{ J/mol} \cdot \text{K})(710 \text{ K} - 355 \text{ K}) = +738$ J

$Q = nC_p \Delta T = (0.250 \text{ mol})(29.17 \text{ J/mol} \cdot \text{K})(710 \text{ K} - 355 \text{ K}) = +2590$ J

$\Delta U = Q - W = 2590 \text{ J} - 738 \text{ J} = 1850$ J

b) At the beginning of the final cooling process (cooling at constant volume), $T = 710$ K. The gas returns to its original volume and pressure, so also to its original temperature of 355 K.

$\Delta V = 0$ so $W = 0$

$Q = nC_V \Delta T = (0.250 \text{ mol})(20.85 \text{ J/mol} \cdot \text{K})(355 \text{ K} - 710 \text{ K}) = -1850$ J

$\Delta U = Q - W = -1850$ J.

c) For any ideal gas process $\Delta U = nC_V\,\Delta T$. For an isothermal process $\Delta T = 0$, so $\Delta U = 0$.

EVALUATE: The three processes return the gas to its initial state, so $\Delta U_{\text{total}} = 0$; our results agree with this.

19.67 **IDENTIFY** and **SET UP:** Calculate Q for each process. $\Delta U_{\text{total}} = 0$ since the net change in T is zero and the system is an ideal gas. Then apply the first law to calculate W_{total}. Calculate the final volume by finding the volume change in each process.

a) In the adiabatic expansion the pressure decreases.

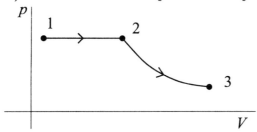

EXECUTE:

b) <u>process $1 \rightarrow 2$</u> $(\Delta p = 0)$

$pV = nRT$ and p constant implies $V_1/T_1 = V_2/T_2$

$T_2 = T_1(V_2/V_1) = (300\text{ K})(2V_1/V_1) = 600$ K

<u>process $2 \rightarrow 3$</u> $(Q = 0)$

$T_2 = 600$ K

Temperature returns to its initial value says that $T_3 = T_1 = 300$ K

$Q_{12} = nC_p\,\Delta T$, since $\Delta p = 0$ for this process.

$Q_{12} = (0.350\text{ mol})(34.60\text{ J/mol} \cdot \text{K})(600\text{ K} - 300\text{ K}) = 3630$ J

$Q_{23} = 0$ since process $2 \rightarrow 3$ is adiabatic.

Then $Q = Q_{12} + Q_{23} = 3630$ J.

c) $\Delta U_{12} = nC_V\,\Delta T = nC_V(T_2 - T_1)$

$\Delta U_{23} = nC_V\,\Delta T = nC_V(T_3 - T_2)$

$\Delta U_{\text{tot}} = \Delta U_{12} + \Delta U_{23} = nC_V(T_2 - T_1 + T_3 - T_2) = nC_V(T_3 - T_1) = 0$, since $T_1 = T_3$; $\Delta U_{\text{tot}} = 0$ since the ideal gas ends up with the same temperature as at the start.

d) $\Delta U = Q - W$

$W = Q - \Delta U$ and $\Delta U = 0$ give $W = Q = 3630$ J

e) In process $1 \to 2$ the volume doubles, so $V_2 = 0.0140$ m^3.

Process $2 \to 3$ is adiabatic and we know T_2 and T_3. To find what happens to the volume use $T_2 V_2^{\gamma-1} = T_3 V_3^{\gamma-1}$.

$V_3^{\gamma-1} = V_2^{\gamma-1}(T_2/T_3)$ so $V_3 = V_2(T_2/T_3)^{1/(\gamma-1)}$

For H$_2$S $\gamma = 1.33$ (from Table 19.1), so $V_3 = (0.0140 \text{ m}^3)(600 \text{ K}/300 \text{ K})^{1/0.33} = 0.114$ m^3.

(Note that $V_3 > V_2$ as it should, since process $2 \to 3$ is an expansion.)

EVALUATE: The volume increases and the total work is positive. The net heat flow into the gas equals the total work done.

CHAPTER 20
THE SECOND LAW OF THERMODYNAMICS

Exercises 3, 7, 9, 11, 13, 15, 25, 27, 29, 33, 35
Problems 39, 41, 45, 47, 53, 55, 57

Exercises

20.3 **IDENTIFY** and **SET UP:** The problem deals with a heat engine.
$W = +3700$ W and $Q_H = +16,100$ J. Use Eq.(20.4) to calculate the efficiency e
and Eq.(20.2) to calculate $|Q_C|$. Power $= W/t$.

EXECUTE:

a) $e = \dfrac{\text{work output}}{\text{heat energy input}} = \dfrac{W}{Q_H} = \dfrac{3700 \text{ J}}{16,100 \text{ J}} = 0.23 = 23\%.$

b) $W = Q = |Q_H| - |Q_C|$

Heat discarded is $|Q_C| = |Q_H| - W = 16,100 \text{ J} - 3700 \text{ J} = 12,400 \text{ J}.$

c) Q_H is supplied by burning fuel; $Q_H = mL_c$ where L_c is the heat of combustion.

$m = \dfrac{Q_H}{L_c} = \dfrac{16,100 \text{ J}}{4.60 \times 10^4 \text{ J/kg}} = 0.350 \text{ g}.$

d) $W = 3700$ J per cycle

In $t = 1.00$ s the engine goes through 60.0 cycles.

$P = W/t = 60.0(3700 \text{ J})/1.00 \text{ s} = 222 \text{ kW}$

$P = (2.22 \times 10^5 \text{ W})(1 \text{ hp}/746 \text{ W}) = 298 \text{ hp}$

EVALUATE: $Q_C = -12,400$ J. In one cycle $Q_{tot} = Q_C + Q_H = 3700$ J. This
equals W_{tot} for one cycle.

20.7 **IDENTIFY** and **SET UP:** The compression stroke is adiabatic. The
compression ratio r tells us the factor by which the volume changes. Assume an
ideal gas and use Eq.(19.22) to relate T and V and solve for T_b. We can then use
the ideal gas law to calculate p_b.

EXECUTE:

a) Process $a \rightarrow b$ is adiabatic so $T_a V_a^{\gamma-1} = T_b V_b^{\gamma-1}$

$V_b = V, \quad V_a = rV$ (as given in Fig.18-3)

$T_a = 22.0°\text{C} = 295$ K

$T_b = T_a(V_a/V_b)^{\gamma-1} = (295 \text{ K})(rV/V)^{1.40-1} = (295 \text{ K})(9.50)^{0.4} = 726 \text{ K} = 453°\text{C}$

b) $pV = nRT$, $pV/T =$ constant, so $p_a V_a/T_a = p_b V_b/T_b$

$p_b = p_a(V_a/V_b)(T_b/T_a) = 8.50 \times 10^4 \text{ Pa}(rV/V)(726 \text{ K}/295 \text{ K}) = 1.99 \times 10^6 \text{ Pa}$

EVALUATE: In an adiabatic compression the temperature increases and the pressure increases.

20.9 **IDENTIFY** and **SET UP:** For the refrigerator $K = 2.10$ and $Q_C = +3.4 \times 10^4$ J. Use Eq.(20.9) to calculate $|W|$ and then Eq.(20.2) to calculate Q_H.

a) EXECUTE: Performance coefficient $K = Q_C/|W|$ (Eq.20.9)

$|W| = Q_C/K = 3.40 \times 10^4 \text{ J}/2.10 = 1.62 \times 10^4$ J

b) SET UP:

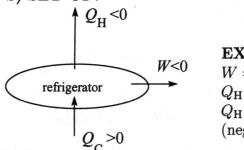

EXECUTE:

$W = Q_C + Q_H$

$Q_H = W - Q_C$

$Q_H = -1.62 \times 10^4 \text{ J} - 3.40 \times 10^4 \text{ J} = -5.02 \times 10^4$ J

(negative because heat goes out of the system)

EVALUATE: $|Q_H| = |W| + |Q_C|$. The heat $|Q_H|$ delivered to the high temperature reservoir is greater than the heat taken in from the low temperature reservoir.

20.11 **IDENTIFY** and **SET UP:** Apply Eq.(20.2) to the cycle and calculate $|W|$ and then $P = |W|/t$. Section 20.4 shows that EER $= (3.413)K$.

a)

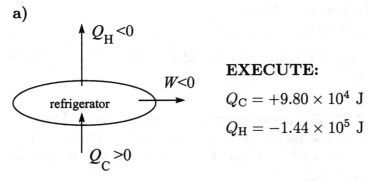

EXECUTE:

$Q_C = +9.80 \times 10^4$ J

$Q_H = -1.44 \times 10^5$ J

$W = Q_C + Q_H = +9.80 \times 10^4 \text{ J} - 1.44 \times 10^5 \text{ J} = -4.60 \times 10^4$ J

$P = W/t = -4.60 \times 10^4 \text{ J}/60.0 \text{ s} = -767$ W

b) EER $= (3.413)K$

$K = |Q_C|/|W| = 9.80 \times 10^4 \text{ J}/4.60 \times 10^4 \text{ J} = 2.13$

EER $= (3.413)(2.13) = 7.27$

EVALUATE: W negative means power is consumed, not produced, by the device. $|Q_H| = |W| + |Q_C|$.

20.13 IDENTIFY: Use Eq.(20.2) to calculate $|W|$. Since it is a Carnot device we can use Eq.(20.13) to relate the heat flows out of the reservoirs. The reservoir temperatures can be used in Eq.(20.14) to calculate e.

a) SET UP:

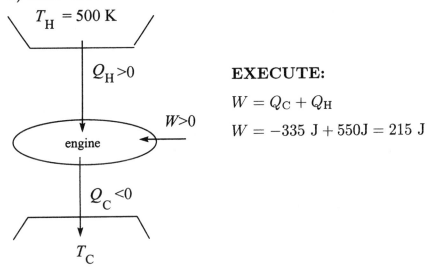

EXECUTE:

$W = Q_C + Q_H$

$W = -335\ \text{J} + 550\text{J} = 215\ \text{J}$

b) For a Carnot cycle, $\dfrac{|Q_C|}{|Q_H|} = \dfrac{T_C}{T_H}$ (Eq.18-13)

$T_C = T_H \dfrac{|Q_C|}{|Q_H|} = 620\ \text{K}\left(\dfrac{335\ \text{J}}{550\ \text{J}}\right) = 378\ \text{K}$

c) $e(\text{Carnot}) = 1 - T_C/T_H = 1 - 378\ \text{K}/620\ \text{K} = 0.390 = 39.0\%$

EVALUATE: We could use the underlying definition of e (Eq.20.4):
$e = W/Q_H = (215\ \text{J})/(550\ \text{J}) = 39\%$, which checks.

20.15 IDENTIFY and **SET UP:** The device is a Carnot refrigerator. We can use Eqs.(20.2) and (20.13).

a)

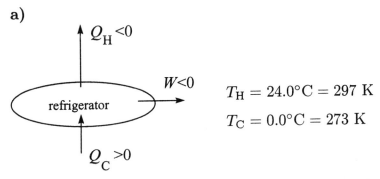

$T_H = 24.0°\text{C} = 297\ \text{K}$

$T_C = 0.0°\text{C} = 273\ \text{K}$

The amount of heat taken out of the water to make the liquid $\rightarrow$ solid phase change is $Q = -mL_f = -(85.0\ \text{kg})(334 \times 10^3\ \text{J/kg}) = -2.84 \times 10^7\ \text{J}$. This amount of heat must go into the working substance of the refrigerator, so $Q_C = +2.84 \times 10^7\ \text{J}$.

For Carnot cycle $|Q_C|/|Q_H| = T_C/T_H$

EXECUTE:

$|Q_H| = |Q_C|(T_H/T_C) = 2.84 \times 10^7 \text{ J}(297 \text{ K}/273 \text{ K}) = 3.09 \times 10^7 \text{ J}$

b) $W = Q_C + Q_H = +2.84 \times 10^7 \text{ J} - 3.09 \times 10^7 \text{ J} = -2.5 \times 10^6 \text{ J}$

EVALUATE: W is negative because this much energy must be supplied to the refrigerator rather than obtained from it. Note that in Eq.(20.13) we must use Kelvin temperatures.

20.25 a) IDENTIFY and **SET UP:** Heat flows out of the 80.0°C water into the ocean water and the 80.0°C water cools to 20.0°C (the ocean warms very, very slightly). Heat flow for an isolated system is always in this direction, from warmer objects into cooler objects, so this process is irreversible.

b) Use $Q = mc\,\Delta T$ to find the heat flow out of the water; this heat flows into the ocean. Use $\Delta S = mc\ln(T_2/T_1)$ (Example 20.6) to calculate ΔS for the water when it undergoes the temperature change. The ocean is at constant temperature, so for it $\Delta S = Q/T$.

EXECUTE: 0.100 kg of water goes from 80.0°C to 20.0°C and the heat flow is $Q = mc\,\Delta T = (0.100 \text{ kg})(4190 \text{ J/kg} \cdot \text{K})(-60.0\text{C}°) = -2.514 \times 10^4 \text{ J}$

This Q comes out of the 0.100 kg of water and goes into the ocean.

For the 0.100 kg of water,

$\Delta S = mc\ln(T_2/T_1) = (0.100 \text{ kg})(4190 \text{ J/kg} \cdot \text{K})\ln(293.15/353.15) = -78.02 \text{ J/K}$

For the ocean the heat flow is $Q = +2.154 \times 10^4 \text{ J}$ and occurs at constant T:

$$\Delta S = \frac{Q}{T} = \frac{2.514 \times 10^4 \text{ J}}{293.15 \text{ K}} = +85.76 \text{ J/K}$$

$\Delta S_\text{net} = \Delta S_\text{water} + \Delta S_\text{ocean} = -78.02 \text{ J/K} + 85.76 \text{ J/K} = +7.7 \text{ J/K}$

EVALUATE: The ocean and water have the same $|Q|$ but the water has a higher temperature than the ocean as heat flows from it, so by $dS = dQ/T$ the positive ΔS for the ocean is greater than the negative ΔS for the water. For the total isolated system $\Delta S > 0$ and the process is irreversible.

20.27 IDENTIFY and **SET UP:** First use $Q_\text{system} = 0$ to find the final temperature of the mixed water.

EXECUTE:

The 1.00 kg of water warms from 20.0°C to T:

$Q = mc\,\Delta T = (1.00 \text{ kg})(4190 \text{ J/kg} \cdot \text{K})(T - 20.0°\text{C})$

The 2.00 kg of water cools from 80.0°C to T:

$Q = mc\,\Delta T = (2.00 \text{ kg})(4190 \text{ J/kg} \cdot \text{K})(T - 80.0°\text{C})$

$Q_\text{system} = 0$ says

$(1.00 \text{ kg})(4190 \text{ J/kg} \cdot \text{K})(T - 20.0°\text{C}) + (2.00 \text{ kg})(4190 \text{ J/kg} \cdot \text{K})(T - 80.0°\text{C}) = 0$

$(T - 20.0°C) + 2(T - 80.0°C) = 0$

$3T = 180°C$ and $T = 60.0°C$

IDENTIFY and **SET UP:** Now we can calculate the entropy change of each mass of water. The process for each is a temperature change so use the method of Example 20.6.

EXECUTE:

The 1.00 kg of water: $\Delta S = \int_1^2 dQ/T$

$dQ = mc\,dT$ so $\Delta S = mc \int_1^2 dT/T = mc \ln(T_2/T_1)$ (Example 20.6).

$\Delta S = (1.00 \text{ kg})(4190 \text{ J/kg} \cdot \text{K}) \ln \left(\dfrac{(60.0 + 273.15) \text{ K}}{(20.0 + 273.15) \text{ K}} \right) = 535.9 \text{ J/K}$

The 2.00 kg of water:

$\Delta S = mc \ln(T_2/T_1) = (2.00 \text{ kg})(4190 \text{ J/kg} \cdot \text{K}) \ln \left(\dfrac{(60.0 + 273.15) \text{ K}}{(80.0 + 273.15) \text{ K}} \right) =$

-488.6 J/K

The entropy change of the system is $\Delta S_{\text{tot}} = 535.9 \text{ J/K} - 488.6 \text{ J/K} = 47 \text{ J/K}$.

EVALUATE: We assumed the two masses of water to be an isolated system. $\Delta S_{\text{tot}} > 0$ so the mixing process is irreversible.

20.29 **IDENTIFY** and **SET UP:** The initial and final states are at the same temperature, at the normal boiling point of 4.216 K. Calculate the entropy change for the irreversible process by considering a reversible isothermal process that connects the same two states, since ΔS is path independent and depneds only on the initial and final states. For the reversible isothermal process we can use Eq.(20.18). The heat flow for the helium is $Q = -mL_{\text{v}}$, negative since in condensation heat flows out of the helium. The heat of vaporization L_{v} is given in Table 17.4 and is $L_v = 20.9 \times 10^3$ J/kg.

EXECUTE:

$Q = -mL_v = -(0.130 \text{ kg})(20.9 \times 10^3 \text{ J/kg}) = -2717 \text{ J}$

$\Delta S = Q/T = -2717 \text{ J}/4.216 \text{ K} = -644 \text{ J/K}$.

EVALUATE: The system we considered is the 0.130 kg of helium; ΔS is the entropy change of the helium. This is not an isolated system since heat must flow out of it into some other material. Our result that $\Delta S < 0$ doesn't violate the 2nd law since it is not an isolated system. The material that receives the heat that flows out of the helium would have a positive entropy change and the total entropy change would be positive.

20.33 **IDENTIFY** and **SET UP:** Reversible implies $\Delta S = \int_1^2 dQ/T$. Isothermal means T is constant, and then $\Delta S = Q/T$. Calculate W, note that $\Delta U = 0$ for an ideal gas when $\Delta T = 0$, and use the 1st law to calculate Q.

EXECUTE: Calculate Q.

$\Delta U = Q - W$

$\Delta U = nC_V \Delta T$ (any ideal gas process); $\Delta U = 0$ since $\Delta T = 0$

$W = nRT \int_{V_1}^{V_2} dV/V = nRT \ln(V_2/V_1)$

$W = (2.00 \text{ mol})(8.3145 \text{ J/mol} \cdot \text{K})(298 \text{ K}) \ln(0.0420 \text{ m}^3/0.0280 \text{ m}^3) = 2.01 \times 10^3 \text{ J}$

Then $Q = \Delta U + W = 0 + 2.01 \times 10^3 \text{ J} = 2.01 \times 10^3 \text{ J}$.

$\Delta S = Q/T = 2.01 \times 10^3 \text{ J}/298 \text{ K} = +6.74 \text{ J/K}$.

EVALUATE: Heat flows into the gas and the same amount of energy leaves the system in the expansion work. $Q > 0$ so $\Delta S > 0$. This is not an isolated system, so $\Delta S > 0$ for a reversible process doesn't violate the 2nd law.

20.35 a) IDENTIFY and **SET UP:** The velocity distribution of Eq.(16-32) depends only on T, so in an isothermal process it does not change.

b) EXECUTE: Calculate the change in the number of available microscopic states and apply Eq.(20.23). Following the reasoning of Example 20.11, the number of possible positions available to each molecule is altered by a factor of 3 (becomes larger). Hence the number of microscopic states the gas occupies at volume $3V$ is $w_2 = (3)^N w_1$, where N is the number of molecules and w_1 is the number of possible microscopic states at the start of the process, where the volume is V. Then, by Eq.(20.23),

$\Delta S = k \ln(w_2/w_1) = k \ln(3)^N = Nk \ln(3) = nN_A k \ln(3) = nR \ln(3)$

$\Delta S = (2.00 \text{ mol})(8.3145 \text{ J/mol} \cdot \text{K}) \ln(3) = +18.3 \text{ J/K}$

c) IDENTIFY and **SET UP:** For an isothermal reversible process $\Delta S = Q/T$.

EXECUTE: Calculate W and then use the first law to calculate Q.

$\Delta T = 0$ implies $\Delta U = 0$, since system is an ideal gas.

Then by $\Delta U = Q - W$, $Q = W$.

For an isothermal process, $W = \int_{V_1}^{V_2} p \, dV = \int_{V_1}^{V_2} (nRT/V) \, dV = nRT \ln(V_2/V_1)$

Thus $Q = nRT \ln(V_2/V_1)$ and $\Delta S = Q/T = nR \ln(V_2/V_1)$

$\Delta S = (2.00 \text{ mol})(8.3145 \text{ J/mol} \cdot \text{K}) \ln(3V_1/V_1) = +18.3 \text{ J/K}$

EVALUATE: This is the same result as obtained in part (b).

Problems

20.39 IDENTIFY: Use the ideal gas law to calculate p and V for each state.

Use the first law and specific expressions for Q, W, and ΔU for each process. Use

Eq.(20.4) to calculate e. Q_H is the net heat flow into the gas.

SET UP: $\gamma = 1.40$

$C_V = R/(\gamma - 1) = 20.70 \text{ J/mol} \cdot \text{K}; \quad C_p = C_V + R = 29.10 \text{ J/mol} \cdot \text{K}$

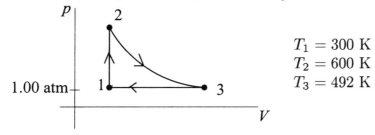

$T_1 = 300 \text{ K}$
$T_2 = 600 \text{ K}$
$T_3 = 492 \text{ K}$

EXECUTE:

a) point 1

$p_1 = 1.00 \text{ atm} = 1.013 \times 10^5 \text{ Pa}$ (given); $\quad pV = nRT$;

$V_1 = \dfrac{nRT_1}{p_1} = \dfrac{(0.350 \text{ mol})(8.3145 \text{ J/mol} \cdot \text{K})(300 \text{ K})}{1.013 \times 10^5 \text{ Pa}} = 8.62 \times 10^{-3} \text{ m}^3$

point 2

process $1 \to 2$ at constant volume so $V_2 = V_1 = 8.62 \times 10^{-3} \text{ m}^3$

$pV = nRT$ and n, R, V constant implies $p_1/T_1 = p_2/T_2$

$p_2 = p_1(T_2/T_1) = (1.00 \text{ atm})(600 \text{ K}/300 \text{ K}) = 2.00 \text{ atm} = 2.03 \times 10^5 \text{ Pa}$

point 3

Consider the process $3 \to 1$, since it is simpler than $2 \to 3$.

Process $3 \to 1$ is at constant pressure so $p_3 = p_1 = 1.00 \text{ atm} = 1.013 \times 10^5 \text{ Pa}$

$pV = nRT$ and n, R, p constant implies $V_1/T_1 = V_3/T_3$

$V_3 = V_1(T_3/T_1) = (8.62 \times 10^{-3} \text{ m}^3)(492 \text{ K}/300 \text{ K}) = 14.1 \times 10^{-3} \text{ m}^3$

b) process $1 \to 2$

constant volume ($\Delta V = 0$)

$Q - nC_V \Delta T = (0.350 \text{ mol})(20.79 \text{ J/mol} \cdot \text{K})(600 \text{ K} - 300 \text{ K}) = 2180 \text{ J}$

$\Delta V = 0$ and $W = 0$. Then $\Delta U = Q - W = 2180 \text{ J}$

process $2 \to 3$

Adiabatic means $Q = 0$.

$\Delta U = nC_V \Delta T$ (any process), so

$\Delta U = (0.350 \text{ mol})(20.70 \text{ J/mol} \cdot \text{K})(492 \text{ K} - 600 \text{ K}) = -780 \text{ J}$

Then $\Delta U = Q - W$ gives $W = Q - \Delta U = +780 \text{ J}$. (It is correct for W to be positive since ΔV is positive.)

process $3 \to 1$

For constant pressure

$W = p\,\Delta V = (1.013 \times 10^5\ \text{Pa})(8.62 \times 10^{-3}\ \text{m}^3 - 14.1 \times 10^{-3}\ \text{m}^3) = -560\ \text{J}$

or $W = nR\,\Delta T = (0.350\ \text{mol})(8.3145\ \text{J/mol} \cdot \text{K})(300\ \text{K} - 492\ \text{K}) = -560\ \text{J}$, which
checks. (It is correct for W to be negative, since ΔV is negative for this process.)

$Q = nC_p\,\Delta T = (0.350\ \text{mol})(29.10\ \text{J/mol} \cdot \text{K})(300\ \text{K} - 455\ \text{K}) = -1580\ \text{J}$

$\Delta U = Q - W = -1960\ \text{J} - (-560\ \text{K}) = -1400\ \text{J}$

or $\Delta U = nC_V\,\Delta T = (0.350\ \text{mol})(20.79\ \text{J/mol} \cdot \text{K})(300\ \text{K} - 492\ \text{K}) = -1400\ \text{J}$,
which checks

c) $W_{\text{net}} = W_{1\to2} + W_{2\to3} + W_{3\to1} = 0 + 780\ \text{J} - 560\ \text{J} = +220\ \text{J}$

d) $Q_{\text{net}} = Q_{1\to2} + Q_{2\to3} + Q_{3\to1} = 2180\ \text{J} + 0 - 1960\ \text{J} = +220\ \text{J}$

e) $e = \dfrac{\text{work output}}{\text{heat energy input}} = \dfrac{W}{Q_\text{H}} = \dfrac{220\ \text{J}}{2180\ \text{J}} = 0.102 = 10.2\%.$

$e(\text{Carnot}) = 1 - T_\text{C}/T_\text{H} = 1 - 300\ \text{K}/600\ \text{K} = 0.500.$

EVALUATE: For a cycle $\Delta U = 0$, so by $\Delta U = Q - W$ it must be that $Q_{\text{net}} = W_{\text{net}}$
for a cycle. We can also check that $\Delta U_{\text{net}} = 0$:

$\Delta U_{\text{net}} = \Delta U_{1\to2} + \Delta U_{2\to3} + \Delta U_{3\to1} = 2180\ \text{J} - 1050\ \text{J} - 1130\ \text{J} = 0$

$e < e(\text{Carnot})$, as it must.

20.41 **a) IDENTIFY** and **SET UP:** Combine Eqs.(20.13) and (20.2) to
eliminate Q_C and obtain an expression for Q_H in terms of W, T_C, and T_H.

$W = 1.00\ \text{J},\ T_\text{C} = 268.15\ \text{K},\ T_\text{H} = 290.15\ \text{K}$

For the heat pump $Q_\text{C} > 0$ and $Q_\text{H} < 0$

EXECUTE: $W = Q_\text{C} + Q_\text{H}$; combining this with $\dfrac{Q_\text{C}}{Q_\text{H}} = -\dfrac{T_\text{C}}{T_\text{H}}$ gives

$Q_\text{H} = \dfrac{W}{1 - T_\text{C}/T_\text{H}} = \dfrac{1.00\ \text{J}}{1 - (268.15/290.15)} = 13.2\ \text{J}$

b) Electrical energy is converted directly into heat, so an electrical energy input of
13.2 J would be required.

c) **EVALUATE:** From part (a), $Q_\text{H} = \dfrac{W}{1 - T_\text{C}/T_\text{H}}$. Q_H decreases as T_C decreases.

The heat pump is less efficient as the temperature difference through which the
heat has to be "pumped" increases. In an engine, heat flows from T_H to T_C and
work is extracted. The engine is more efficient the larger the temperature difference
through which the heat flows.

20.45 **IDENTIFY:** Use Eq.(20.4) to calculate e.

SET UP:

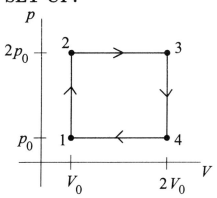

$C_V = 5R/2$

for an ideal gas $C_p = C_V + R = 7R/2$

SET UP: Calculate Q and W for each process.

process $1 \to 2$

$\Delta V = 0$ implies $W = 0$

$\Delta V = 0$ implies $Q = nC_V \Delta T = nC_V(T_2 - T_1)$

But $pV = nRT$ and V constant says $p_1 V = nRT_1$ and $p_2 V = nRT_2$.

Thus $(p_2 - p_1)V = nR(T_2 - T_1)$; $V \Delta p = nR \Delta T$ (true when V is constant).

Then $Q = nC_V \Delta T = nC_V(V \Delta p/nR) = (C_V/R)V \Delta p = (C_V/R)V_0(2p_0 - p_0) = (C_V/R)p_0 V_0$. $Q > 0$; heat is absorbed by the gas.)

process $2 \to 3$

$\Delta p = 0$ so $W = p \Delta V = p(V_3 - V_2) = 2p_0(2V_0 - V_0) = 2p_0 V_0$ (W is positive since V increases.)

$\Delta p = 0$ implies $Q = nC_p \Delta T = nC_p(T_2 - T_1)$

But $pV = nRT$ and p constant says $pV_1 = nRT_1$ and $pV_2 = nRT_2$.

Thus $p(V_2 - V_1) = nR(T_2 - T_1)$; $p \Delta V = nR \Delta T$ (true when p is constant).

Then $Q = nC_p \Delta T = nC_p(p \Delta V/nR) = (C_p/R)p \Delta V = (C_p/R)2p_0(2V_0 - V_0) = (C_p/R)2p_0 V_0$. ($Q > 0$; heat is absorbed by the gas.)

process $3 \to 4$

$\Delta V = 0$ implies $W = 0$

$\Delta V = 0$ so

$Q = nC_V \Delta T = nC_V(V \Delta p/nR) = (C_V/R)(2V_0)(p_0 - 2p_0) = -2(C_V/R)p_0 V_0$ ($Q < 0$ so heat is rejected by the gas.)

process $4 \to 1$

$\Delta p = 0$ so $W = p\Delta V = p(V_1 - V_4) = p_0(V_0 - 2V_0) = -p_0 V_0$ (W is negative since V decreases)

$\Delta p = 0$ so $Q = nC_p \Delta T = nC_p(p \Delta V/nR) = (C_p/R)p \Delta V = (C_p/R)p_0(V_0 - 2V_0) = -(C_p/R)p_0 V_0$ ($Q < 0$ so heat is rejected by the gas.)

total work performed by the gas during the cycle:

$$W_{\text{tot}} = W_{1\to2} + W_{2\to3} + W_{3\to4} + W_{4\to1} = 0 + 2p_0V_0 + 0 - p_0V_0 = p_0V_0$$

(Note that W_{tot} equals the area enclosed by the cycle in the pV-diagram.)

total heat absorbed by the gas during the cycle (Q_H):

Heat is absorbed in processes $1 \to 2$ and $2 \to 3$.

$$Q_H = Q_{1\to2} + Q_{2\to3} = \frac{C_V}{R} p_0V_0 + 2\frac{C_p}{R} p_0V_0 = \left(\frac{C_V + 2C_p}{R}\right) p_0V_0$$

But $C_p = C_V + R$ so $Q_H = \dfrac{C_V + 2(C_V + R)}{R} p_0V_0 = \left(\dfrac{3C_V + 2R}{R}\right) p_0V_0.$

total heat rejected by the gas during the cycle (Q_C):

Heat is rejected in processes $3 \to 4$ and $4 \to 1$.

$$Q_C = Q_{3\to4} + Q_{4\to1} = -2\frac{C_V}{R} p_0V_0 - \frac{C_p}{R} p_0V_0 = -\left(\frac{2C_V + C_p}{R}\right) p_0V_0$$

But $C_p = C_V + R$ so $Q_C = -\dfrac{2C_V + (C_V + R)}{R} p_0V_0 = -\left(\dfrac{3C_V + R}{R}\right) p_0V_0.$

efficiency

$$e = \frac{W}{Q_H} = \frac{p_0V_0}{([3C_V + 2R]/R)(p_0V_0)} = \frac{R}{3C_V + 2R} = \frac{R}{3(5R/2) + 2R} = \frac{2}{19}.$$

$e = 0.105 = 10.5\%$

EVALUATE: As a check on the calculations note that

$$Q_C + Q_H = -\left(\frac{3C_V + R}{R}\right) p_0V_0 + \left(\frac{3C_V + 2R}{R}\right) p_0V_0 = p_0V_0 = W, \text{ as it should.}$$

20.47   **IDENTIFY** and **SET UP:** For the constant pressure processes ab and cd calculate W and use the first law to calculate Q. Calculate Q_{tot} and use that $W_{\text{tot}} = Q_{\text{tot}}$ for a cycle. The coefficient of performance is given by Eq.(20.9); Q_C is the net heat that goes into the system.

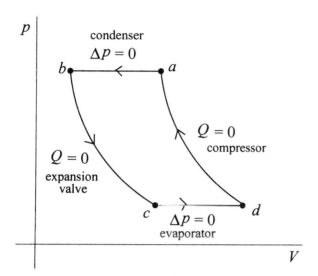

EXECUTE:

a) process $c \to d$

$\Delta U = U_d - U_c = 1657 \times 10^3 \text{ J} - 1005 \times 10^3 \text{ J} = 6.52 \times 10^5 \text{ J}$

$W = \int_{V_c}^{V_d} p\, dV = p \Delta V$ (since is a constant pressure process)

$W = (363 \times 10^3 \text{ Pa})(0.4513 \text{ m}^3 - 0.2202 \text{ m}^3) = +8.39 \times 10^4 \text{ J}$ (positive since process is an expansion)

$\Delta U = Q - W$ so $Q = \Delta U + W = 6.52 \times 10^5 \text{ J} + 8.39 \times 10^4 \text{ J} = 7.36 \times 10^5 \text{ J}$.

(Q positive so heat goes into the coolant)

b) process $a \to b$

$\Delta U = U_b - U_a = 1171 \times 10^3 \text{ J} - 1969 \times 10^3 \text{ J} = -7.98 \times 10^5 \text{ J}$

$W = p \Delta V = (2305 \times 10^3 \text{ Pa})(0.00946 \text{ m}^3 - 0.0682 \text{ m}^3) = -1.35 \times 10^5 \text{ J}$

(negative since $\Delta V < 0$ for the process)

$Q = \Delta U + W = -7.98 \times 10^5 \text{ J} - 1.35 \times 10^5 \text{ J} = -9.33 \times 10^5 \text{ J}$

(negative so heat comes out of coolant).

c) The coolant cannot be treated as an ideal gas, so we can't calculate W for the adiabatic processes. But $\Delta U = 0$ (for cycle) so $W_{\text{net}} = Q_{\text{net}}$.

$Q = 0$ for the two adiabatic processes, so

$Q_{\text{net}} = Q_{cd} + Q_{ab} = 7.36 \times 10^5 \text{ J} - 9.33 \times 10^5 \text{ J} = -1.97 \times 10^5 \text{ J}$

Thus $W_{\text{net}} = -1.97 \times 10^5 \text{ J}$ (negative since work is done on the coolant, the working substance).

d) $K = Q_C / |W| = (+7.36 \times 10^5 \text{ J}) / (+1.97 \times 10^5 \text{ J}) = 3.74$.

EVALUATE: $W_{\text{net}} < 0$ when the cycle is taken in the counterclockwise direction, as is the case here.

20.53 a) IDENTIFY and **SET UP:** Calculate e from Eq.(20.6), Q_C from Eq.(20.4) and then W from Eq.(20.2).

EXECUTE: $e = 1 - 1/(r^{\gamma-1}) = 1 - 1/(10.6^{0.4}) = 0.6111$

$e = (Q_H + Q_C)/Q_H$ and we are given $Q_H = 200$ J; calculate Q_C.

$Q_C = (e - 1)Q_H = (0.6111 - 1)(200 \text{ J}) = -78$ J (negative since corresponds to heat leaving)

Then $W = Q_C + Q_H = -78 \text{ J} + 200 \text{ J} = 122$ J. (Positive, in agreement with Fig.20.5.)

EVALUATE: Q_H, $W > 0$, and $Q_C < 0$ for an engine cycle.

b) IDENTIFY and **SET UP:** The stroke times the bore equals the change in volume. The initial volume is the final volume V times the compression ratio r. Combining these two expressions gives an equation for V. For each cylinder of area $A = \pi(d/2)^2$ the piston moves 0.0864 m and the volume changes from rV to V.

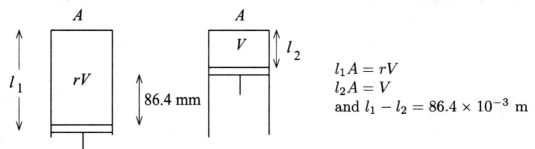

$l_1 A = rV$
$l_2 A = V$
and $l_1 - l_2 = 86.4 \times 10^{-3}$ m

EXECUTE:

$l_1 A - l_2 A = rV - V$ and $(l_1 - l_2)A = (r - 1)V$

$$V = \frac{(l_1 - l_2)A}{r - 1} = \frac{(86.4 \times 10^{-3} \text{ m})\pi(41.25 \times 10^{-3} \text{ m})^2}{10.6 - 1} = 4.811 \times 10^{-5} \text{ m}^3$$

At point a the volume is $rV = 10.6(4.811 \times 10^{-5} \text{ m}^3) = 5.10 \times 10^{-4}$ m³.

c) IDENTIFY and **SET UP:** The processes in the Otto cycle are either constant volume or adiabatic. Use the Q_H that is given to calculate ΔT for process bc. Use Eq.(19.22) and $pV = nRT$ to relate p, V and T for the adiabatic processes ab and cd.

EXECUTE:

point a: $T_a = 300$ K, $p_a = 8.50 \times 10^4$ Pa, and $V_a = 5.10 \times 10^{-4}$ m³

point b:

$V_b = V_a/r = 4.81 \times 10^{-5}$ m³

process $a \rightarrow b$ is adiabatic, so $T_a V_a^{\gamma-1} = T_b V_b^{\gamma-1}$

$T_a(rV)^{\gamma-1} = T_b V^{\gamma-1}$

$T_b = T_a r^{\gamma-1} = 300 \text{ K}(10.6)^{0.4} = 771$ K

$pV = nRT$ so $pV/T = nR = $ constant, so $p_aV_a/T_a = p_bV_b/T_b$

$p_b = p_a(V_a/V_b)(T_b/T_a) = (8.50 \times 10^4 \text{ Pa})(rV/V)(771 \text{ K}/300 \text{ K}) = 2.32 \times 10^6$ Pa

point c

process $b \rightarrow c$ is at constant volume, so $V_c = V_b = 4.81 \times 10^{-5} \text{ m}^3$

$Q_\text{H} = nC_V \, \Delta T = nC_V(T_c - T_b)$. The problem specifies $Q_\text{H} = 200$ J; use to calculate T_c. First use the p, V, T values at point a to calculate the number of moles n.

$$n = \frac{pV}{RT} = \frac{(8.50 \times 10^4 \text{ Pa})(5.10 \times 10^{-4} \text{ m}^3)}{(8.3145 \text{ J/mol} \cdot \text{K})(300 \text{ K})} = 0.01738 \text{ mol}$$

Then $T_c - T_b = \dfrac{Q_\text{H}}{nC_V} = \dfrac{200 \text{ J}}{(0.01738 \text{ mol})(20.5 \text{ J/mol} \cdot \text{K})} = 561.3$ K, and

$T_c = T_b + 561.3 \text{ K} = 771 \text{ K} + 561 \text{ K} = 1332$ K

$p/T = nR/V = $ constant so $p_b/T_b = p_c/T_c$

$p_c = p_b(T_c/T_b) = (2.32 \times 10^6 \text{ Pa})(1332 \text{ K}/771 \text{ K}) = 4.01 \times 10^6$ Pa

point d:

$V_d = V_a = 5.10 \times 10^{-4} \text{ m}^3$

process $c \rightarrow d$ is adiabatic, so $T_dV_d^{\gamma-1} = T_cV_c^{\gamma-1}$

$T_d(rV)^{\gamma-1} = T_cV^{\gamma-1}$

$T_d = T_c/r^{\gamma-1} = 1332 \text{ K}/10.6^{0.4} = 518$ K

$p_cV_c/T_c = p_dV_d/T_d$

$p_d = p_c(V_c/V_d)(T_d/T_c) = (4.01 \times 10^6 \text{ Pa})(V/rV)(518 \text{ K}/1332 \text{ K}) = 1.47 \times 10^5$ Pa

EVALUATE: Can look at process $d \rightarrow a$ as a check.

$Q_\text{C} = nC_V(T_a - T_d) = (0.01738 \text{ mol})(20.5 \text{ J/mol} \cdot \text{K})(300 \text{ K} - 518 \text{ K}) = -78$ J, which agrees with part (a)

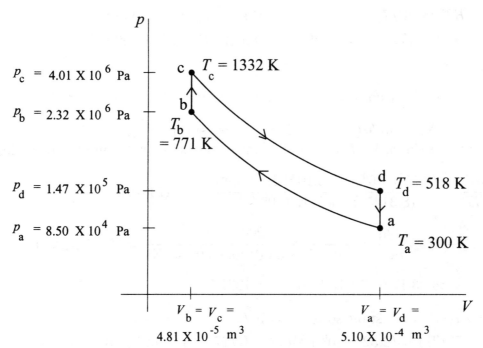

$P_c = 4.01 \times 10^6$ Pa

$P_b = 2.32 \times 10^6$ Pa

$P_d = 1.47 \times 10^5$ Pa

$P_a = 8.50 \times 10^4$ Pa

$T_c = 1332$ K

$T_b = 771$ K

$T_d = 518$ K

$T_a = 300$ K

$V_b = V_c = 4.81 \times 10^{-5}$ m^3

$V_a = V_d = 5.10 \times 10^{-4}$ m^3

d) IDENTIFY and **SET UP:** The Carnot efficiency is given by Eq.(20.14). T_H is the highest temperature reached in the cycle and T_C is the lowest.

EXECUTE: From part (a) the efficiency of this Otto cycle is $e = 0.611 = 61.1\%$.

The efficiency of a Carnot cycle operating between 1332 K and 300 K is

e(Carnot) $= 1 - T_C/T_H = 1 - 300$ K$/1332$ K $= 0.775 = 77.5\%$, which is larger.

EVALUATE: The 2nd law requires that $e \geq e$(Carnot), and our result obeys this law.

20.55 IDENTIFY and **SET UP:** Use eq.(20.13) for an infinitesimal heat flow dQ_H from the hot reservoir and use that expression with Eq.(20.19) to relate ΔS_H, the entropy change of the hot reservoir, to $|Q_C|$

a) EXECUTE: Consider an infinitesimal heat flow dQ_H that occurs when the temperature of the hot reservoir is T':

$dQ_C = -(T_C/T')dQ_H$

$\int dQ_C = -T_C \int \dfrac{dQ_H}{T'}$

$|Q_C| = |T_C| \int \dfrac{dQ_H}{|T'|} = T_C|\Delta S_H|$

b) The 1.00 kg of water (the high-temperature reservoir) goes from 373 K to 273 K.

$Q_H = mc\,\Delta T = (1.00 \text{ kg})(4190 \text{ J/kg} \cdot \text{K})(100 \text{ K}) = 4.19 \times 10^5$ J

$\Delta S_H = mc\ln(T_2/T_1) = (1.00 \text{ kg})(4190 \text{ J/kg} \cdot \text{K})\ln(273/373) = -1308$ J/K

The result of part (a) gives $|Q_C| = (273 \text{ K})(1308 \text{ J/K}) = 3.57 \times 10^5 \text{ J}$

Q_C comes out of the engine, so $Q_C = -3.57 \times 10^5 \text{ J}$

Then $W = Q_C + Q_H = -3.57 \times 10^5 \text{ J} + 4.19 \times 10^5 \text{ J} = 6.2 \times 10^4 \text{ J}$.

c) 2.00 kg of water goes from 323 K to 273 K

$Q_H = mc\,\Delta T = (2.00 \text{ kg})(4190 \text{ J/kg} \cdot \text{K})(50 \text{ K}) = 4.19 \times 10^5 \text{ J}$

$\Delta S_h = mc\ln(T_2/T_1) = (2.00 \text{ kg})(4190 \text{ J/kg} \cdot \text{K})\ln(273/323) = -1.41 \times 10^3 \text{ J/K}$

$Q_C = -T_C|\Delta S_h| = -3.85 \times 10^5 \text{ J}$

$W = Q_C + Q_H = 3.4 \times 10^4 \text{ J}$

d) EVALUATE: More work can be extracted from 1.00 kg of water at 373 K than from 2.00 kg of water at 323 K even though the energy that comes out of the water as it cools to 273 K is the same in both cases. The energy in the 323 K water is less available for conversion into mechanical work.

20.57 IDENTIFY and **SET UP:** First use the methods of Chapter 17 to calculate the final temperature T of the system.

EXECUTE:

<u>0.600 kg of water</u> (cools from 45.0°C to T)

$Q = mc\,\Delta T =$

$(0.600 \text{ kg})(4190 \text{ J/kg} \cdot \text{K})(T - 45.0°\text{C}) = (2514 \text{ J/K})T - 1.1313 \times 10^5 \text{ J}$

<u>0.0500 kg of ice</u> (warms to 0°C, melts, and water warms from 0°C to T)

$Q = mc_{\text{ice}}(0°\text{C} - (-15.0°\text{C})) + mL_f + mc_{\text{water}}(T - 0°\text{C})$

$Q = 0.0500 \text{ kg}[(2100 \text{ J/kg} \cdot \text{K})(15.0°\text{C}) + 334 \times 10^3 \text{ J/kg} + (4190 \text{ J/kg} \cdot \text{K})(T - 0°\text{C})]$

$Q = 1575 \text{ J} + 1.67 \times 10^4 \text{ J} + (209.5 \text{ J/K})T = 1.828 \times 10^4 \text{ J} + (209.5 \text{ J/K})T$

$Q_{\text{system}} = 0$ gives $(2514 \text{ J/K})T - 1.1313 \times 10^5 \text{ J} + 1.828 \times 10^4 \text{ J} + (209.5 \text{ J/K})T = 0$

$(2.724 \times 10^3 \text{ J/K})T = 9.485 \times 10^4 \text{ J}$

$T = (9.485 \times 10^4 \text{ J})/(2.724 \times 10^3 \text{ J/K}) = 34.82°\text{C} = 308 \text{ K}$

EVALUATE: The final temperature must lie between $-15.0°\text{C}$ and $45.0°\text{C}$. A final temperature of 34.8°C is consistent with only liquid water being present at equilibrium.

IDENTIFY and **SET UP:** Now we can calculate the entropy changes. Use $\Delta S = Q/T$ for phase changes and the method of Example 20.6 to calculate ΔS for temperature changes.

EXECUTE:

<u>ice:</u>

The process takes ice at $-15°\text{C}$ and produces water at 34.8°C. Calculate ΔS for a reversible process between these two states, in which heat is added very slowly. ΔS is path independent, so ΔS for a reversible process is the same as ΔS for the actual (irreversible) process as long as the initial and final states are the same.

$\Delta S = \int_1^2 dQ/T$, where T must be in kelvins

For a temperature change $dQ = mc\, dT$ so $\Delta S = \int_{T_1}^{T_2} (mc/T)\, dT = mc\ln(T_2/T_1)$.

For a phase change, since it occurs at constant T,

$\Delta S = \int_1^2 dQ/T = Q/T = \pm mL/T$.

Therefore $\Delta S_{ice} = mc_{ice}\ln(273\text{ K}/258\text{ K}) + mL_f/273K + mc_{water}\ln(308\text{ K}/273\text{ K})$

$\Delta S_{ice} = (0.0500\text{ kg})[(2100\text{ J/kg}\cdot\text{K})\ln(273\text{ K}/258\text{ K}) + (334\times10^3\text{ J/kg})/273\text{ K} + (4190\text{ J/kg}\cdot\text{K})\ln(308\text{ K}/273\text{ K})]$

$\Delta S_{ice} = 5.93\text{ J/K} + 61.17\text{ J/K} + 25.27\text{ J/K} = 92.4\text{ J/K}$

water

$\Delta S_{water} = mc\ln(T_2/T_1) = (0.600\text{ kg})(4190\text{ J/kg}\cdot\text{K})\ln(308\text{ K}/318\text{ K}) = -80.3\text{ J/K}$

For the system, $\Delta S = \Delta S_{ice} + \Delta S_{water} = 92.4\text{ J/K} - 80.3\text{ J/K} = +12\text{ J/K}$

EVALUATE: Our calculation gives $\Delta S > 0$, as it must for an irreversible process of an isolated system.